Engineering and Sustainable Development

Renewable Energy: Research, Development and Policies

nova science publishers
www.novapublishers.com

Engineering and Sustainable Development

Intelligent Decision Support System for IoT-Enabling Technologies: Opportunities, Challenges and Applications
Subrata Sahana, PhD (Editor), Anil Kumar Sagar, PhD (Editor),
Sanjoy Das, PhD (Editor) and Vishal Jain, PhD (Editor)
2024. ISBN: 979-8-89113-249-8 (Hardcover)
2021. ISBN: 979-8-89113-333-4 (eBook)

Modeling for Sustainable Development: A Multidisciplinary Approach
Rajendra Kumar, PhD (Editor), R. C. Singh, PhD (Editor),
Rohit Khokher, PhD (Editor) and Vishal Jain, PhD (Editor)
2022. ISBN: 979-8-89113-015-9 (Softcover)
2021. ISBN: 979-8-89113-056-2 (eBook)

More information about this series can be found at
https://novapublishers.com/product-category/series/engineering-and-sustainable-development/

Renewable Energy: Research, Development and Policies

Photovoltaic Systems: Advances in Research and Applications
Sudip Mandal, PhD (Editor)
Pijush Dutta, PhD (Editor)
2023. ISBN: 979-8-89113-102-6 (eBook)

The Future of Solar Power
Professor Dr. Hussain H. Al-Kayiem (Editor)
2023. ISBN: 979-8-88697-709-7 (eBook)

More information about this series can be found at
https://novapublishers.com/product-category/series/renewable-energy-research-development-and-policies/

Pawan Kumar Pathak
and Anil Kumar Yadav
Editors

The Future of Hybrid-Electric Vehicles

nova
science publishers
www.novapublishers.com

DOI: https://doi.org/10.52305/XOKX2676

NOTICE TO THE READER

Library of Congress Cataloging-in-Publication Data

ISBN: 979-8-89113-515-4 (hardcover)
ISBN: 979-8-89113-574-1 (e-book)

Published by Nova Science Publishers, Inc. † New York

Contents

Preface vii

Chapter 1 **Power Electronic Converters for Electric Vehicles** **1**
Trilochan Penthia

Chapter 2 **Battery Management System of EVs** **105**
Trilochan Penthia

Chapter 3 **Probabilistic Load Demand Modeling of HEVs in the Distribution System** **181**
Jaber Pouladi, Taher Abedinzadeh and Ali Daghigh

Chapter 4 **Electric Vehicle: Types, Charging Topologies, Energy Storage Management System and Applicability Micro-Grids – A Review** **227**
Ashish Kumar Sinha, Prashant Kumar and Prince Kumar

Chapter 5 **Electric Vehicle Applications in Frequency Regulation of Standalone Microgrid Systems** **245**
Maloth Ramesh, Anil Kumar Yadav, Sneha V. Sah and Sampath Kumar Bodapatla

Chapter 6 **A Comprehensive Review of Battery Management System for Electric Vehicles** **269**
Mufaddal Huzefa Shakir and Akant Kumar Singh

Chapter 7 **Development of Effective Charging Station for EVs Using Multiport Converter and Photovoltaic Cell Integration** **289**
Parimalasundar Ezhilvannan, Suresh Krishnan, Kumar Reddy Cheepati and Busireddy Hemanth Kumar

Index **319**

Preface

The future of transportation is rapidly changing, and hybrid electric vehicles (HEVs) are at the forefront of that change. As technological advances continue to make HEVs more efficient and sustainable, they are becoming increasingly popular as a viable alternative to traditional vehicles. HEVs are excellent solutions for making travel more sustainable. In order to achieve an emission-free future by 2050, HEVs are the critical factor. The contents of this book are presented in seven lucid chapters. The book deals with the fundamentals of HEVs, requirements of power electronic converters and challenges associated with their implementation in electric vehicles (EVs), battery management systems of HEVs, load demand modelling of HEVs, charging topologies, applications of EVs in microgrids, the role of HEVs in the load frequency control of standalone microgrid, energy storage management systems of HEVs, and the development of influential charging stations for HEVs with the advent of new topologies of power electronic converters.

Chapter 1 - Power electronics converters play a critical role in the operation and control of electric vehicles (EVs). They convert the electrical energy stored in the battery to the required voltage and current levels for the different components of the vehicle. These converters also control the power flow between the battery, motor, and other components of the vehicle to optimize the performance and efficiency of the system. In this way, power electronic converters are essential for the reliable and efficient operation of electric vehicles. This chapter provides an overview of power electronic converters used in EVs. It begins with an introduction to the importance of power electronic converters in electric vehicles, followed by a classification of converters and an overview of the chapter. The chapter then delves into the three main types of power electronic converters used in electric vehicles: DC-to-DC converters, AC-to-DC converters, and DC-to-AC converters. Each type of converter is explained, along with its advantages and disadvantages. The chapter also covers the requirements and protection of converters in electric vehicles, such as voltage and current ratings, efficiency, thermal management, electromagnetic interference (EMI) & electromagnetic compatibility (EMC), reliability, and cost. Additionally, the design considerations for power electronic converters are discussed, including component selection, topology selection, control and modulation strategies, cooling and thermal management techniques, EMI/EMC mitigation techniques, reliability and durability considerations, and cost optimization. The chapter then explores the various applications of power electronic converters in electric vehicles, such as battery charging, motor drive, regenerative braking, and on-board power supply. Future trends and challenges in power electronic converters for electric vehicles are also discussed, including emerging technologies and materials, advanced control and modulation techniques, integration with renewable energy sources, standardization and interoperability, and cybersecurity and safety. Finally, the chapter concludes with a summary of the key points covered in the chapter, future research directions, and recommendations. The chapter also includes a list of references for further reading.

Chapter 2 - Energy storage systems are becoming increasingly important for electric vehicle (EV) charging infrastructure due to the significant benefits they can provide. By storing energy during off-peak hours, energy storage systems can reduce the strain on the electricity grid during peak hours when electricity demand is highest, helping to maintain grid stability. They can also help to reduce the cost of EV charging by allowing EV owners to charge their vehicles during periods when electricity prices are lower. Additionally, energy storage systems can play a key role in supporting the integration of renewable energy sources into the grid, as they can store excess energy generated by solar or wind power during times of low demand and supply it during periods of high demand. This can help to increase the utilization of renewable energy and reduce greenhouse gas emissions from the electricity sector. Overall, energy storage systems are critical for enabling the widespread adoption of EVs and supporting the transition to a more sustainable energy system. This book chapter provides an overview of energy storage systems (ESSs) and their importance for EV chargers. It discusses different types of ESSs, including battery-based, capacitor-based, flywheel-based, and hydrogen fuel cell-based ESSs, and outlines design considerations such as power and energy capacity, charging and discharging rates, scalability, efficiency, safety, reliability, and cost. The chapter also explores the integration of ESSs with EV chargers, including DC fast charging systems, AC level 2 chargers, and bi-directional charging systems, and examines the benefits of ESSs for EV chargers, such as peak shaving, load levelling, renewable energy integration, resiliency and backup power, grid stabilization, and frequency regulation. Several case studies are presented, including Tesla's supercharger, Duke Energy's energy storage project in North Carolina, and Pacific Gas & Electric's smart grid program in California. The chapter concludes with a discussion of the challenges and opportunities associated with ESSs for EV chargers, including regulatory and policy barriers, technical and engineering challenges, and market and economic opportunities, and suggests future research directions. Overall, the chapter highlights the significant implications of ESSs for the future of EV charging and the transition to a more sustainable energy system.

Chapter 3 - During the past few years, the Hybrid Electric Vehicle (HEV) has gained momentum as a capable solution in overcoming energy and environmental issues. Designing suitable residential distribution systems using a wide array of HEVs requires prediction of the load in existing networks. This chapter offers a comprehensive and exact model in regard to the actual statistics for HEVs aggregation. For this reason, Cumulative Distribution Functions (CDF) is obtained from analysing existing . Also, calculating the initial State of Charge (SOC) in batteries of HEVs are discussed in different models. The chapter explores the influence of Heating, Ventilation, and Air Conditioning (HVAC) power usage on the charging demand of HEVs. It introduces a novel model for estimating the SOC that takes into account vehicle fuel usage and various driving factors, including speed, road slope, and more. Subsequently, based on the point estimate method, the probabilistic load flow simulates the stochastic aspects of HEV charging and examines HEV impact on the residential distribution system. In contrast to existing models that provide fixed outputs for the impacts of HEVs, this chapter examines the probability of HEV effects. Additionally, the proposed models are employed to assess the performance and accuracy by applying them to the IEEE-37 node test feeder.

Chapter 4 - With the looming scenario of an alarming rate of conventional energy level depletion, snowballing crude oil prices and ever-increasing environmental constraints, such as global warming and ozone layer depletion, associated with the same, energy efficiency has become a priority numero-uno. A substantial proportion of overall crude oil consumption is dedicated to road vehicle transportation systems. Therefore, battery-powered electric vehicles are becoming ever more popular all over the world in the ongoing scenario.

In recent years, electric vehicles, whose requirements have grown rapidly in accordance with green energy demand, are being recognised as a viable alternative to conventional vehicle engines, i.e., internal combustion engines (ICE).

Electric vehicles in various configurations and capacities like Pure/ Battery Electric Vehicles (PEV/ BEV), Plug-in Hybrid Electric Vehicles (PHEV), and Fuel-Cell Electric Vehicles (FCEV) are being researched and developed to meet current market demand. Moreover, emphasis is being given to battery technology for electric vehicles to obtain a faster rate of charging and increase the discharging time, i.e., increasing the range of travel in a single charging cycle of an EV battery.

Even while electric vehicles have a broad spectrum of applications and requirements, the ongoing research spectrums are also accompanied by some major roadblocks, disadvantages and glitches. These include time-constraint of charging, the longevity of battery lifetime and standardisation. Significant research in recent years has been dedicated towards accelerating the charging time of EV batteries through various mechanisms. Herein, the charging mechanism is typically seen to adopt two major techniques, namely constant current (CC) technique and constant voltage (CV) technique. Furthermore, the charging topologies include unidirectional and bi-directional, wherein vehicle-to-grid and grid-to-vehicle schemes are the present areas of concentrated research. Moreover, in accordance with the available literature, the bi-directional power flow model is clearly seen to have an advantage over the traditional unidirectional power flow topology w.r.t. the charging techniques generally employed for electric vehicles. In this regard, an effective battery management system also becomes indispensable and a comprehendible area of vital research.

Research shows that bi-directional topology with the incorporation of microgrids into the scheme forms a highly exciting and comprehendible area of future research and may potentially overcome the existing drawbacks of the schemes and methodologies that are currently being used in an attempt to accelerate the battery charging time and improve the overall working performance of the electric vehicles.

Chapter 5 - Owing to the seasonal behavior of the solar-wind-based energy system utilized in standalone microgrids, wide frequency variations occur when the frequency regulation capacity is insufficient to regulate the imbalance of power generation and demand. This issue may become more severe when the system is operating in standalone mechanism. In the meantime, Electric vehicles (EVs) are gaining more and more popularity, and they are currently being employed as scattered energy storage units rather than compact batteries in the systems. The EVs are used in this study to aid microgrids in quickly arresting fluctuations in the system frequency after power demands. As a result, the limited frequency regulation capacity can be made up for by using vehicle-to-grid (V2G) power control, which also enhances the frequency constancy of microgrids, particularly in the standalone operation mode. However, Large-scale solar-wind systems and sophisticated modern microgrids are subject to a broad-spectrum of uncertainty. To handle these uncertainties, this work proposes a metaheuristic based such as artificial gorilla troop (GTO) based sliding mode control (SMC) approach for frequency control of a standalone microgrid. The results of simulation illustrate the perfection and effectiveness of the suggested controller.

Chapter 6 - The development and use of electric vehicles can significantly minimize difficulties with energy deficit and environmental pollution. To achieve the varied battery performance requirements, a good battery management system (BMS) is essential in maintaining the safe and dependable battery operations, which has been extensively utilized in variety of power-intensive applications, for instance such as electric and hybrid electric vehicles. Battery status estimation, cell balance, temperature control, and fault diagnosis are among the BMS's primary duties. This article tries to provide a brief summary of a number of crucial BMS technologies. Before discussing the essential technologies implemented in BMS, a survey of the most common battery types employed in EVs is carried out. Thereafter, a thorough survey of battery condition assessments for the inside temperature, state of charge (SOC), and state of health (SOH) is conducted. Also, a literature survey has been carried out on various works done to improve BMS of electric vehicles.

Chapter 7 - The adoption of electric vehicles (EVs) is increasing rapidly as a response to the urgent need to reduce greenhouse gas emissions and dependence on fossil fuels. However, the deployment of a reliable and cost-effective EV charging infrastructure remains a challenge. To address this challenge, this paper proposes the development of an effective charging station using multiport converter and photovoltaic cell integration. A promising approach in the transportation sector is an efficient charging station with multiport converter and photovoltaic cell integration. A multiport converter, photovoltaic cells, a battery energy storage system, and an electric vehicle charging port constitute the proposed charging station. The power flow between the photovoltaic cells, the battery energy storage system, and the electric vehicle charging port is controlled by the multiport converter. The excess energy generated by the photovoltaic cells is stored in the battery energy storage system. The proposed charging station is designed and analysed using MATLAB/Simulink. The simulation results show that the proposed charging station is capable of providing an electric vehicle a reliable and consistent power supply. In contrast to conventional charging stations, the proposed charging station is also capable of offering a faster charging rate.

Chapter 1

Power Electronic Converters for Electric Vehicles

Trilochan Penthia*
Department of Electrical Engineering, National Institute of Technology Patna, Bihar, India

Abstract

Power electronics converters play a critical role in the operation and control of electric vehicles (EVs). They convert the electrical energy stored in the battery to the required voltage and current levels for the different components of the vehicle. These converters also control the power flow between the battery, motor, and other components of the vehicle to optimize the performance and efficiency of the system. In this way, power electronic converters are essential for the reliable and efficient operation of electric vehicles. This chapter provides an overview of power electronic converters used in EVs. It begins with an introduction to the importance of power electronic converters in electric vehicles, followed by a classification of converters and an overview of the chapter. The chapter then delves into the three main types of power electronic converters used in electric vehicles: DC-to-DC converters, AC-to-DC converters, and DC-to-AC converters. Each type of converter is explained, along with its advantages and disadvantages. The chapter also covers the requirements and protection of converters in electric vehicles, such as voltage and current ratings, efficiency, thermal management, electromagnetic interference (EMI) & electromagnetic compatibility (EMC), reliability, and cost. Additionally, the design considerations for power electronic converters are discussed, including component

* Corresponding Author's Email: trilochankoraput@gmail.com.

In: The Future of Hybrid-Electric Vehicles
Editors: Pawan Kumar Pathak and Anil Kumar Yadav
ISBN: 979-8-89113-515-4

selection, topology selection, control and modulation strategies, cooling and thermal management techniques, EMI/EMC mitigation techniques, reliability and durability considerations, and cost optimization. The chapter then explores the various applications of power electronic converters in electric vehicles, such as battery charging, motor drive, regenerative braking, and on-board power supply. Future trends and challenges in power electronic converters for electric vehicles are also discussed, including emerging technologies and materials, advanced control and modulation techniques, integration with renewable energy sources, standardization and interoperability, and cybersecurity and safety. Finally, the chapter concludes with a summary of the key points covered in the chapter, future research directions, and recommendations. The chapter also includes a list of references for further reading.

Keywords: electric vehicles, EV charging, DC-to-DC converter, AC-to-DC converter, DC-to-AC converter

1.1. Introduction

Electric vehicles (EVs) are becoming increasingly popular as a more environmentally-friendly and sustainable mode of transportation. Power electronic converters are an essential component of EVs, responsible for converting the battery's direct current (DC) to the alternating current (AC) required by the electric motor. As the adoption of EVs continues to grow worldwide, the demand for efficient and reliable power electronic converters is also increasing. EVs rely on power electronic converters to control and convert electrical power between different subsystems such as the battery, motor drive, and on-board power supply. These converters play a critical role in the performance and efficiency of the EV system (Maroti, Padmanaban, Bhaskar, Ramachandaramurthy, & Blaabjerg, 2022).

According to (International Energy Agency, 2021) report, the global electric car stock exceeded 10 million in 2020, a 43% increase from 2019. Furthermore, the International Energy Agency projects that electric car sales could reach 145 million by 2030, representing a 7% share of the global car market. This growth in EV adoption is expected to drive the demand for power electronic converters. In addition to passenger vehicles, the electrification of other modes of transportation such as buses and trucks is also on the rise. For instance, the market for electric buses is expected to grow at a compound annual growth rate (CAGR) of 26.01% between 2022 and 2027, according to a report by (Research and Markets, 2022). Similarly, the market for electric trucks is projected to grow at a CAGR of 14.3% between 2020 and 2027, according to the same report.

The increasing demand for power electronic converters is also being driven by government policies and regulations aimed at reducing greenhouse gas emissions and improving air quality. For instance, many countries have set targets to phase out internal combustion engine vehicles in the coming years, which is expected to further boost the adoption of EVs and the demand for power electronic converters. Furthermore, the global market for power electronic converters for electric vehicles is expected to grow significantly in the coming years, driven by the increasing adoption of EVs and the electrification of other modes of transportation, as well as government policies and regulations aimed at reducing emissions.

1.1.1. Importance of Power Electronic Converters in Electric Vehicle

Power electronic converters are essential components of an EV system, as they control and regulate the flow of electrical power between different subsystems. They provide a high degree of efficiency and precision control, which is necessary for the performance and reliability of an EV. Without power electronic converters, EVs would not be able to operate efficiently or effectively (Sakka, Mierlo, & Gualous, 2011). Converters are responsible for converting the DC power from the battery to AC power that can be used by the electric motor. This conversion process is essential for the operation of the motor, which drives the wheels of the vehicle. In this sense, power electronic converters play a vital role in the overall performance and efficiency of an EV. A block diagram of various converters for EV application is shown in Figure 1.1.

Here are some of the reasons why power electronic converters are so important in EVs (İnci, Büyük, Demir, & İlbey, 2021):

- *Efficiency:* Power electronic converters can operate at high efficiency levels, reducing the amount of energy lost during the conversion process. This means that more of the energy stored in the battery can be used to power the electric motor, improving the overall efficiency of the vehicle.

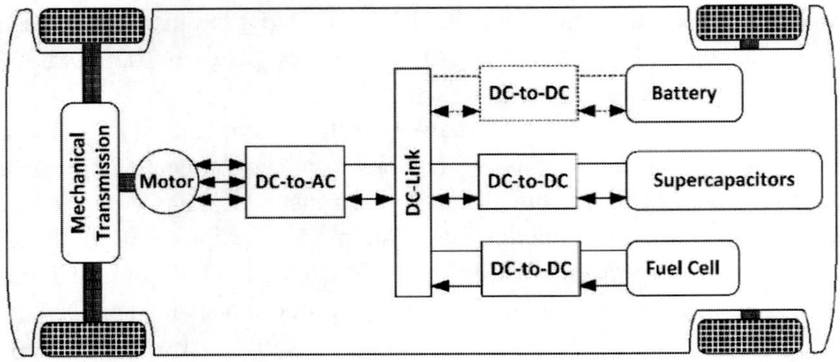

Figure 1.1. Different power electronic converters used for EVs.

- *Control:* Power electronic converters enable precise control over the power delivered to the electric motor, allowing for smooth and efficient operation. This control can be used to optimize the performance of the motor and improve the driving experience.
- *Regenerative braking:* Many EVs use regenerative braking systems, which use the electric motor to slow down the vehicle and convert the kinetic energy of the moving vehicle back into electrical energy to be stored in the battery. Power electronic converters are necessary for this process, as they allow the motor to act as a generator and convert the kinetic energy into electrical energy.
- *Safety:* Power electronic converters are essential for maintaining the safety of the vehicle and its occupants. They can monitor and control the power delivered to the motor, preventing overloading and overheating. In addition, they can also provide protection against short circuits and other electrical faults.
- *Compatibility:* Power electronic converters enable compatibility between different types of batteries and motors, allowing EV manufacturers to use a range of battery chemistries and motor types in their vehicles. This flexibility can help to reduce costs and improve the overall performance of the vehicle.

Overall, power electronic converters play a vital role in the operation, efficiency, and safety of electric vehicles. They enable precise control over the power delivered to the electric motor, allowing for efficient and smooth operation. They also enable regenerative braking and provide essential safety features to protect the vehicle and its occupants. As the demand for EVs continues to grow, the importance of power electronic converters in these vehicles will only continue to increase.

1.1.2. Classification of Power Converters

EV power electronics systems require a range of converters to manage power flow and voltage regulation between various subsystems. These converters can be classified based on various factors for EV applications. Some common types of power electronic converters used in EVs include DC-to-DC converters, AC-to-DC converters, and DC-to-AC converters as illustrted in Figure 1.2. Each type of converter has its own advantages and disadvantages, depending on the specific application and requirements (Safayatullah, Elrais, Ghosh, Rezaii, & Batarseh, 2022). DC-to-DC converters are an important component of electric vehicle power electronics systems. These are used to step up or step down the voltage of the battery to match the requirements of the motor, as well as to regulate the voltage of the battery during charging. These converters can be classified based on various factors for EV applications.

DC-to-DC converters can be classified based on topology, input and output voltage range, power rating, efficiency, cooling method, and integration with other components. Some of the common classification factors include:

- *Topology:* The topology of a DC-to-DC converter refers to its configuration and arrangement of power electronics components. Common topologies for EV applications include buck, boost, and buck-boost converters.
- *Input voltage range:* The input voltage range of a DC-to-DC converter refers to the range of DC voltages that can be accepted by the converter. In EV applications, the input voltage range is typically from 200 to 800 volts DC, depending on the battery voltage.
- *Output voltage range:* The output voltage range of a DC-to-DC converter refers to the range of DC voltages that can be produced by the converter. In EV applications, the output voltage range is typically from 12 to 48 volts DC, depending on the voltage requirements of the various vehicle systems.

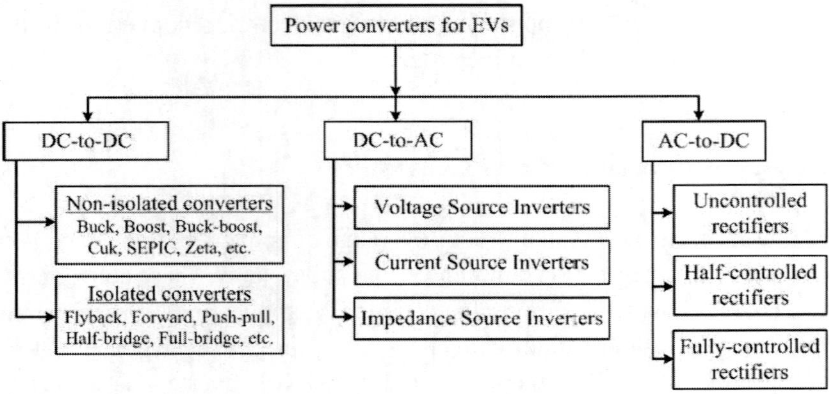

Figure 1.2. Classification of power electronic converters for EV applications.

- *Power rating:* The power rating of a DC-to-DC converter refers to its maximum output power rating, which is determined by the power requirements of the various vehicle systems. In EV applications, DC-to-DC converters typically have power ratings ranging from a few hundred watts to several kilowatts.
- *Efficiency:* The efficiency of a DC-to-DC converter refers to the ratio of output power to input power, and is an important factor in EV applications due to the need for maximum range and efficiency. In EV applications, DC-to-DC converters typically have efficiencies ranging from 90% to 98%.
- *Cooling method:* The cooling method of a DC-to-DC converter refers to the way in which it dissipates heat generated during operation. Common cooling methods for EV DC-to-DC converters include air-cooled and liquid-cooled designs.
- *Integration:* DC-to-DC converters can also be classified based on their integration with other components in the EV system. Integrated DC-to-DC converters are combined with other components, such as the motor controller, while standalone DC-to-DC converters are separate components that are connected to the vehicle systems.

DC-to-AC converters, also known as inverters, can be classified based on various factors for EV applications. Some of the common classification factors include:

- *Topology:* The topology of an inverter refers to its configuration and arrangement of power electronics components. Common inverter topologies for EV applications include voltage source inverters, current source inverters, and impedance source inverters.

- *Control method:* The control method of an inverter refers to the way in which it regulates the output voltage and frequency. Common control methods for EV inverters include pulse width modulation (PWM) control and sinusoidal pulse width modulation (SPWM) control.
- *Voltage rating:* The voltage rating of an inverter refers to its maximum input voltage rating, which is determined by the voltage rating of the EV battery pack. Inverters for EV applications typically have voltage ratings ranging from 200 to 800 volts DC.
- *Power rating:* The power rating of an inverter refers to its maximum output power rating, which is determined by the power requirements of the EV motor. Inverters for EV applications typically have power ratings ranging from 5 to 500 kilowatts (kW).
- *Efficiency:* The efficiency of an inverter refers to the ratio of output power to input power, and is an important factor in EV applications due to the need for maximum range and efficiency. Inverters for EV applications typically have efficiencies ranging from 90% to 98%.
- *Cooling method:* The cooling method of an inverter refers to the way in which it dissipates heat generated during operation. Common cooling methods for EV inverters include air-cooled and liquid-cooled designs.
- *Integration:* Inverters can also be classified based on their integration with other components in the EV system. Integrated inverters are combined with other components, such as the motor controller, while standalone inverters are separate components that are connected to the motor controller.

AC-to-DC converters, also known as rectifiers, are used in electric vehicle charging systems to convert AC power from the grid to DC power for charging the vehicle battery. These converters can be classified based on various factors for EV applications. Some of the common classification factors include:

- *Topology:* The topology of an AC-to-DC converter refers to its configuration and arrangement of power electronics components. Common topologies for EV applications include single-phase and three-phase rectifiers.

- *Input voltage range:* The input voltage range of an AC-to-DC converter refers to the range of AC voltages that can be accepted by the converter. In EV charging systems, the input voltage range is typically from 110 to 480 volts AC, depending on the charging station and grid voltage.
- *Output voltage range:* The output voltage range of an AC-to-DC converter refers to the range of DC voltages that can be produced by the converter for charging the vehicle battery. In EV charging systems, the output voltage range is typically from 200 to 1000 volts DC, depending on the battery voltage and charging requirements.
- *Power rating:* The power rating of an AC-to-DC converter refers to its maximum output power rating, which is determined by the charging requirements of the EV. In EV charging systems, AC-to-DC converters typically have power ratings ranging from a few kilowatts to several hundred kilowatts.
- *Efficiency:* The efficiency of an AC-to-DC converter refers to the ratio of output power to input power, and is an important factor in EV charging systems due to the need for maximum charging speed and efficiency. In EV charging systems, AC-to-DC converters typically have efficiencies ranging from 90% to 98%.
- *Cooling method:* The cooling method of an AC-to-DC converter refers to the way in which it dissipates heat generated during operation. Common cooling methods for EV AC-to-DC converters include air-cooled and liquid-cooled designs.
- *Integration:* AC-to-DC converters can also be classified based on their integration with other components in the EV charging system. Integrated AC-to-DC converters are combined with other components, such as the charging station and battery management system, while standalone AC-to-DC converters are separate components that are connected to the charging station.

The choice of converter for a particular EV application will depend on the specific requirements of the system, including voltage and power requirements, efficiency, cooling needs, and integration needs. The efficiency of the converters is particularly important in EV applications, as it can affect the driving range and overall performance of the vehicle. Additionally, the integration of these converters with other components in the EV power electronics system is critical for the efficient and reliable operation of the vehicle.

Following steps can be applied to model power electronic converters for EV applications:

Step 1: Define the Converter Topology

- Select the appropriate power electronic converter topology suitable for EV application, such as a DC-DC converter (e.g., buck, boost, buck-boost, full-bridge, etc.) or an AC-DC converter.

Step 2: Gather Parameters and Specifications

- Collect the necessary parameters and specifications for the selected converter topology, such as input voltage range, output voltage/current requirements, switching frequency, efficiency targets, etc.

Step 3: Mathematical Equations

- Formulate the mathematical equations governing the operation of the power electronic converter based on its topology, including voltage/current relationships, power losses, and switching dynamics.

Step 4: Select and Implement Control Strategy

- Choose a suitable control strategy for the converter to regulate the output voltage/current and achieve the desired performance. Common control techniques include Pulse Width Modulation (PWM) control, hysteresis control, etc.

Step 5: Modeling of Semiconductor Devices

- Develop models for the power semiconductor devices (e.g., MOSFETs, IGBTs) used in the converter. These models should accurately represent their switching behavior and losses.

Step 6: Implement Control Algorithm

- Integrate the selected control strategy into the converter model to achieve the desired voltage/current regulation and overall converter behavior.

Step 7: Model Validation

- Validate the developed converter model by comparing simulation results with experimental data or reference models to ensure accuracy and reliability.

Step 8: Efficiency Analysis

- Perform efficiency analysis to evaluate the converter's performance under different operating conditions, and identify potential areas for improvement.

Step 9: Thermal Analysis

- Conduct thermal analysis to assess the heat dissipation and temperature distribution within the converter components, ensuring safe and reliable operation.

Step 10: Optimization (Optional)

- If needed, perform optimization techniques to fine-tune the converter design and control strategy to meet specific performance targets.

Step 11: Model Integration

- Integrate the developed power electronic converter model into the larger EV system simulation to evaluate its overall impact on the vehicle's performance and efficiency.

Step 12: Simulation and Testing

- Simulate the entire EV system, including the power electronic converter, under various driving scenarios and conditions to evaluate its performance and efficiency.

1.1.3. Overview of the Chapter

This chapter provides an overview of power electronic converters used in EVs. It covers the different types of converters, their advantages and disadvantages, and the requirements and challenges associated with their use in EVs. The chapter also discusses design considerations for power electronic converters, including component selection, topology selection, control and modulation strategies, cooling and thermal management techniques, electromagnetic interference/electromagnetic compatibility (EMI/EMC) mitigation techniques, reliability and durability considerations, and cost optimization. The chapter then explores the various applications of power electronic converters in EVs, such as battery charging, motor drive, regenerative braking, and on-board power supply. Finally, the chapter concludes with future trends and challenges in power electronic converters for EVs. In addition, the objectives of the chapter are as follows:

1. To highlight the importance of power electronic converters in electric vehicles and explain their role in the overall EV system.
2. To provide a classification of power electronic converters used in EVs based on their input and output voltage levels, as well as their topology.
3. To discuss the advantages and disadvantages of different types of power electronic converters used in EVs.
4. To examine the requirements and protection of power electronic converters used in EVs, such as voltage and current ratings, efficiency, thermal management, EMI/EMC, reliability, and cost.
5. To explore the design considerations for power electronic converters used in EVs, including component selection, topology selection, control and modulation strategies, cooling and thermal management techniques, EMI/EMC mitigation techniques, reliability and durability considerations, and cost optimization.

6. To investigate the various applications of power electronic converters in EVs, such as battery charging, motor drive, regenerative braking, and on-board power supply.
7. To identify future trends and challenges in power electronic converters for EVs, including emerging technologies and materials, advanced control and modulation techniques, integration with renewable energy sources, standardization and interoperability, and cybersecurity and safety.
8. To provide a comprehensive overview of power electronic converters used in EVs, including their benefits, challenges, and future prospects, for researchers, engineers, and practitioners in the field of EVs.

1.2. DC-to-DC Converters for Electric Vehicles

DC-to-DC converters play a crucial role in electric vehicle power electronics by efficiently converting the high-voltage battery DC voltage to the required lower voltage for the vehicle's electrical subsystems. This section will discuss the various types and topologies of DC-to-DC converters used in electric vehicles, including bidirectional DC-to-DC converters and the control techniques used to ensure efficient and reliable operation. DC-to-DC converters are used to regulate and control the voltage and current levels from the battery pack to the motor drive, onboard power supply, and other auxiliary loads. The efficiency and reliability of DC-to-DC converters play a key role in the overall performance and range of an EV. They are available in various topologies, including buck, boost, and buck-boost, and can operate in different modes such as continuous conduction mode (CCM) and discontinuous conduction mode (DCM). The selection of the appropriate DC-DC converter topology and mode depends on the specific application and requirements, such as the input and output voltage levels, power rating, efficiency, size and weight constraints, and cost.

1.2.1. Types of DC-to-DC Converters

There are various types of DC-to-DC converters used in EVs, each with their own advantages and limitations depending on the specific application requirements. these converters work by converting the high voltage from the battery to a lower or higher voltage required by different EV components, with a high level of efficiency and precision. Broadly, DC-to-DC converters are of two types: non-isolated and isolated converters as depicted in Figure 1.2. A comparison table between non-isolated and isolated DC-to-DC converters is shown in Table 1.1. DC-to-DC converters play a critical role in enabling efficient power management and optimizing the performance of EVs. Some of the popular DC-to-DC converters are:

a) Buck converters

Buck converters are simple, efficient, and widely used in EVs. They step down the battery voltage (V_{in}) to a lower voltage using a switch (SW) and an inductor (L). Buck converters are ideal for applications with a fixed output voltage, such as powering the vehicle's 12V battery. The circuit diagram of a typical buck converter is shown in Figure 1.3.

Buck converters are a type of DC-to-DC converter commonly used in electric vehicles to step down the high voltage from the battery pack to a lower voltage required by various components, such as the electric motor drive and onboard electronics. A buck converter works by switching the battery voltage on and off at a high frequency, using a switch and an inductor, to create a lower voltage output. The output voltage is controlled by varying the duty cycle of the switch, which determines the percentage of time it is on versus off. One of the main advantages of buck converters is their high efficiency, which is important for electric vehicles to conserve energy and maximize range. Buck converters also offer a compact and lightweight solution for voltage regulation in EVs. Overall, buck converters are a key component in the power electronics system of electric vehicles, enabling efficient voltage regulation and helping to optimize the performance and range of electric vehicles.

b) Boost converters

Boost converters step up the battery voltage to a higher voltage, making them useful for applications that require a higher voltage than the battery voltage. Boost converters are often used in electric vehicle charging systems. A boost converter works by switching the low voltage on and off at a high frequency using a switch and an inductor, to create a higher voltage output. The output voltage is controlled by varying the duty cycle of the switch, which determines the percentage of time it is on versus off. Circuit diagram of a boost converter is shown in Figure 1.4. One of the main advantages of boost converters is their ability to efficiently step up the voltage with minimal power loss, which is important for EVs to conserve energy and maximize range. Boost converters also offer a compact and lightweight solution for voltage regulation in EVs. Boost converters are a critical component in the power electronics system of electric vehicles, enabling efficient voltage regulation and helping to optimize the performance and range of electric vehicles.

Table 1.1. A comparison between non-isolated and isolated DC-to-DC converters

Comparison factor	Non-isolated DC-to-DC converters	Isolated DC-to-DC converters
Galvanic isolation	No galvanic isolation between input and output	Provides galvanic isolation between input and output
Voltage regulation	Generally easier to achieve voltage regulation	Voltage regulation can be more complex due to isolation
Size and weight	Compact and lightweight design	Typically, larger and heavier due to additional components
Efficiency	Higher efficiency due to fewer components and lower losses	Efficiency can be lower due to additional isolation devices
Noise and EMI	More susceptible to noise and electromagnetic interference	Improved noise and EMI performance due to isolation
Safety	No isolation increases risk of electric shock	Provides safety by isolating high voltage from the output
Voltage transients	More sensitive to voltage transients and disturbances	Better ability to handle voltage transients and disturbances
Cost	Generally lower cost due to simpler design and components	Higher cost due to additional components and complexity
Cooling requirements	Typically, lower cooling requirements	May require additional cooling due to higher power levels
Input-Output compatibility	Limited flexibility in terms of input-output voltage matching	Provides flexibility to match input and output voltages
Load regulation	Can be affected by load variations and changes	Improved load regulation due to isolation
System integration	Easier integration into the overall EV system	Requires careful integration due to isolation requirements
Power density	Higher power density due to compact design	Lower power density due to larger size and additional parts
Voltage level conversion	Suitable for same-level voltage conversion	Suitable for voltage step-up or step-down conversions
Isolation voltage requirements	Not applicable	Meets safety standards for isolation voltage requirements

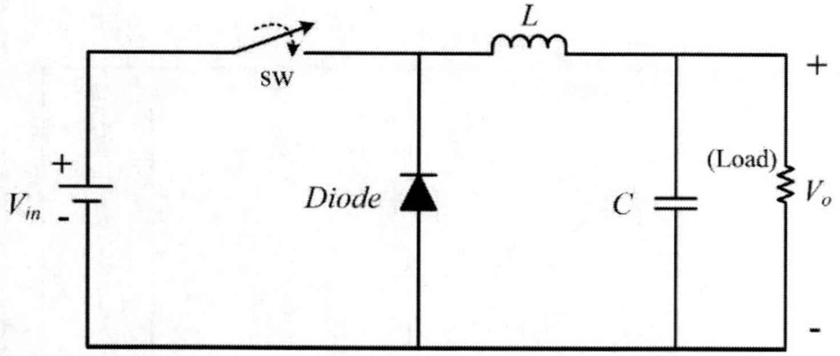

Figure 1.3. Buck converter.

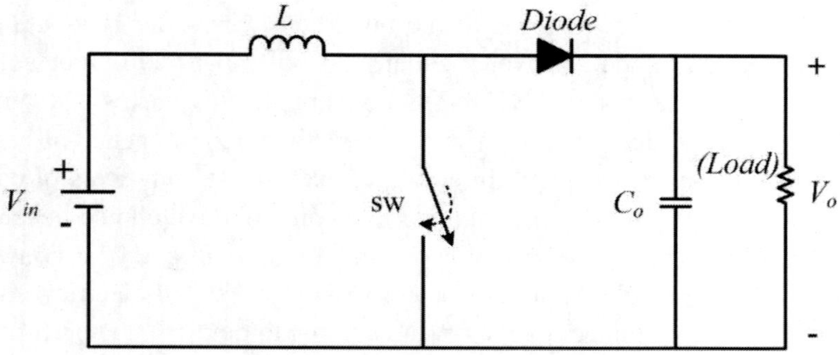

Figure 1.4. Boost converter.

c) Buck-boost converters

Buck-boost converters can step up or step down the battery voltage, making them ideal for applications with variable output voltage. Buck-boost converters are commonly used in electric vehicle power electronics, such as motor drives and inverters. A buck-boost converter works by using a combination of a buck converter (which steps down the voltage) and a boost converter (which steps up the voltage) to regulate the output voltage as depicted in Figure 1.5. By combining these two modes, the buck-boost converter can provide either a higher or lower output voltage than the input voltage, depending on the application requirements.

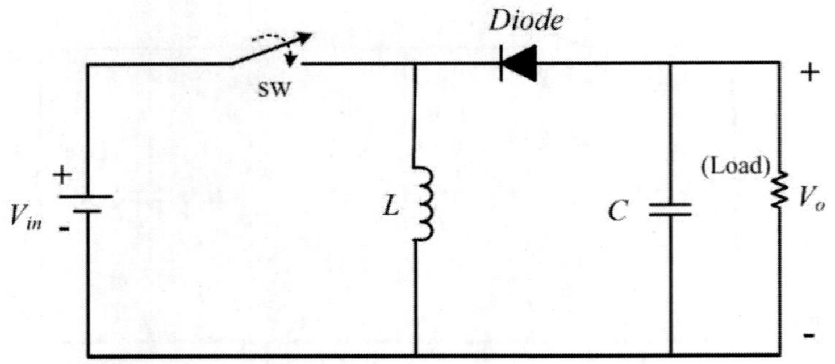

Figure 1.5. Buck-boost converter.

One of the key advantages of the buck-boost converter is its ability to provide voltage regulation over a wider range of input and output voltages compared to other types of DC-to-DC converters. This makes it a versatile solution for EV applications where the voltage requirements of various components may vary widely. In addition, buck-boost converters offer high efficiency and low power loss, making them an attractive choice for electric vehicles where energy efficiency is a critical factor. Buck-boost converters play an important role in the power electronics system of electric vehicles, enabling efficient voltage regulation and helping to optimize the performance and range of EVs.

d) SEPIC converters

A Single Ended Primary Inductance Converter (SEPIC) converter is a type of DC-DC converter that can be used in a variety of applications. SEPIC converters are similar to buck-boost converters in that they can step up or step down the voltage, but they offer several advantages over traditional buck-boost converters. For example, SEPIC converters provide a continuous output current, which helps to reduce the output ripple current and improve efficiency.

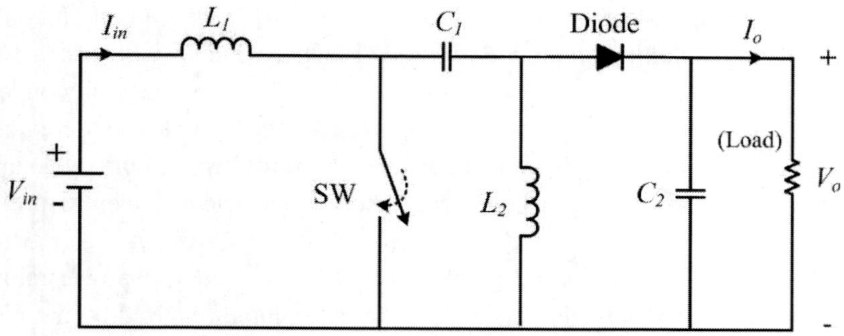

Figure 1.6. SEPIC converter.

SEPIC converters are versatile components that can be used in a variety of EV applications, including battery charging, power conversion, voltage regulation, and energy recovery. They can regulate the charging current and voltage during battery charging, convert high voltage to lower voltage for various electrical systems, and regulate voltage in the EV's electrical systems to prevent damage to components. SEPIC converters can also be used to recover energy from the regenerative braking system, allowing for the efficient storage and usage of recovered energy. SEPIC converters use a capacitor and an inductor in series with the input voltage to provide energy storage and to smooth out the input current, as depicted in Figure 1.6. The output voltage is regulated by controlling the duty cycle of the switch. SEPIC converters are capable of providing high efficiency, low output ripple, and precise voltage regulation over a wide range of input and output voltages. Overall, SEPIC converters play a crucial role in improving the efficiency, safety, and reliability of EV's electrical systems.

e) Cuk converters

A Cuk converter is a type of DC-to-DC converter that is commonly used in electric vehicles to step up or step down the voltage of the high-voltage battery to match the voltage requirements of various vehicle subsystems. The Cuk converter is named after its inventor, Slobodan Čuk, and is a type of non-isolated switching regulator. Cuk converters are similar to SEPIC converters in that they can handle input voltage variations, but they have some unique advantages as well.

The Cuk converter offers several advantages over other DC-to-DC converter topologies, including a simple circuit design, high efficiency, and low output ripple. The Cuk converter also allows for a wide input and output voltage range, making it suitable for use in electric vehicles where the battery voltage can vary widely depending on the state of charge. The Cuk converter uses a combination of an inductor, a capacitor, and a switch as depicted in Figure 1.7. The output voltage is regulated by controlling the duty cycle of the switch. Cuk converters are commonly used in electric vehicles to provide power to subsystems such as lighting, infotainment systems, and battery charging. They are also used in hybrid and electric vehicle powertrains to convert the high-voltage battery voltage to the voltage required by the motor controller.

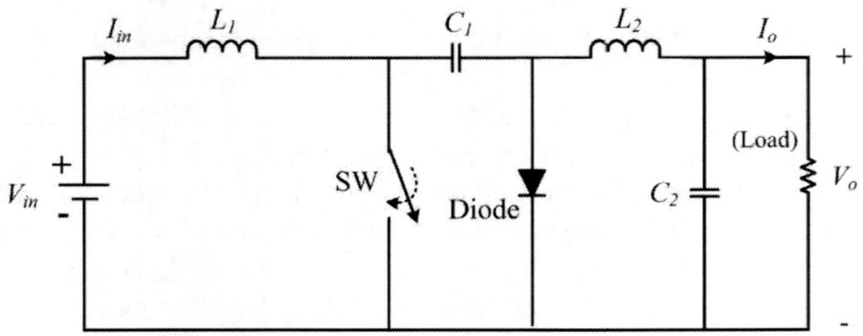

Figure 1.7. Cuk converter.

f) Zeta converters

A Zeta converter is a type of DC-to-DC converter that is used in electric vehicles to step up or step down the voltage of the high-voltage battery to match the voltage requirements of various vehicle subsystems. The Zeta converter is a modified version of the conventional buck-boost converter and offers several advantages over other DC-to-DC converter topologies. The Zeta converter allows for a higher efficiency compared to conventional buck-boost converters, as it can provide galvanic isolation between the input and output. This isolation helps to reduce the effect of noise and other disturbances on the output voltage. Additionally, the Zeta converter provides a low input ripple current, which helps to improve the efficiency of the converter.

The Zeta converter uses a capacitor and an inductor in series with the input voltage to provide energy storage and to smooth out the input current as shown in Figure 1.8. The output voltage is regulated by controlling the duty cycle of the switch. The Zeta converter can be designed to provide either a step-up or a step-down voltage conversion.

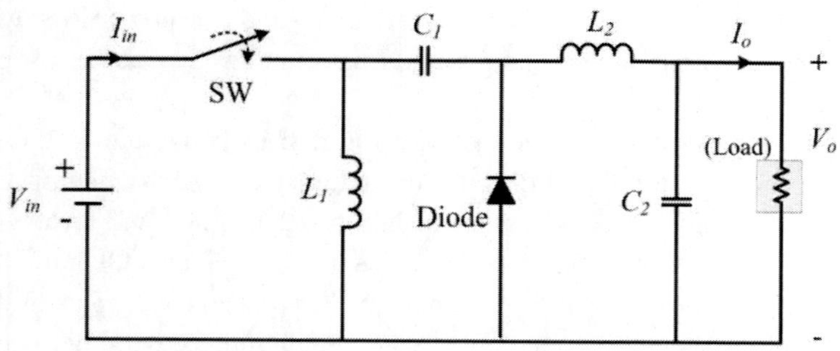

Figure 1.8. Zeta converter.

g) Flyback converters

Flyback converters are isolated DC-to-DC converter converters that use a transformer to step down or step up the voltage. They are commonly used in EVs to convert the high voltage from the battery into a lower voltage required by certain components such as lighting systems, audio systems, and dashboard displays. Circuit diagram of a flyback converter is shown in Figure 1.9.

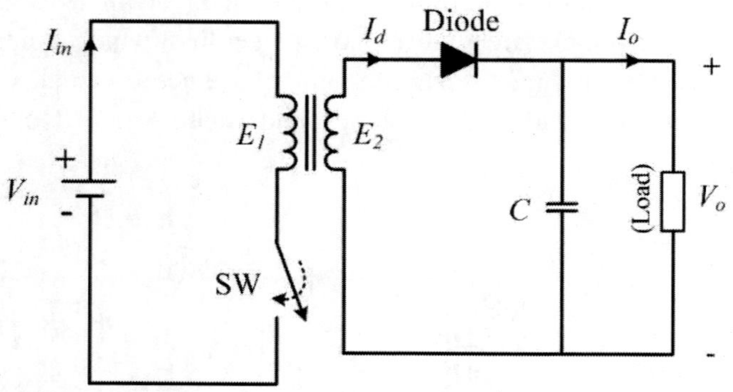

Figure 1.9. Flyback converter.

The flyback converter works by storing energy in a transformer during the "on" phase, and then releasing it during the "off" phase to produce the desired output voltage. The output voltage is controlled by varying the number of times the energy is stored and released per second, which is controlled by the frequency of the switching circuit. One of the main advantages of flyback converters is their ability to handle both step-up and step-down voltage conversion, which makes them suitable for a wide range of applications in electric vehicles. They also offer high efficiency and low power loss, which is important for conserving energy and maximizing the range of electric vehicles.

However, one of the challenges of using flyback converters in EVs is that they can generate significant EMI due to the high frequency switching. Careful design and shielding are necessary to minimize the EMI and ensure proper operation of other electronic components in the vehicle. Overall, flyback converters are an important component in the power electronics system of electric vehicles, enabling efficient voltage regulation and helping to optimize the performance and range of EVs.

h) Forward converters

Forward converters are isolated converters that use a transformer to step down the voltage. They are used in high-power applications such as electric vehicle charging systems. The circuit diagram of a forward converter is shown in Figure 1.10. A forward converter works by storing energy in a transformer during the "on" phase, and then releasing it during the "off" phase to produce the desired output voltage. The output voltage is controlled by varying the duty cycle of the switching circuit. One of the main advantages of forward converters is their high efficiency and low power loss, which is important for conserving energy and maximizing the range of electric vehicles. They also offer good regulation and output accuracy, and can be designed to handle high current loads.

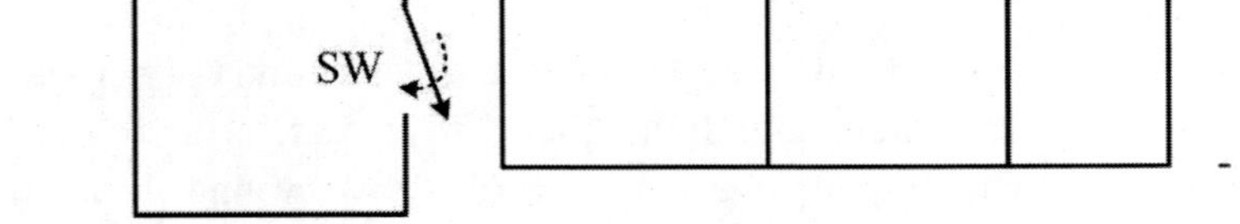

Figure 1.10. A typical forward converter.

However, one of the challenges of using forward converters in EVs is their relatively large size and weight compared to other types of DC-to-DC converters. This can make them less suitable for applications where space and weight are critical factors. Forward converters are an important component in the power electronics system of electric vehicles, enabling efficient voltage regulation and helping to optimize the performance and range of EVs. Similarly, Push-pull converters are a type of DC-DC converter used in EV applications to convert high voltage DC power from the EV battery to a lower voltage DC power for the vehicle's electrical system. They work by transforming the input voltage to a high voltage, which is then switched by a pair of transistors to create an AC waveform that is transformed to a lower voltage output using a transformer and rectified to provide the desired DC output. Push-pull converters are relatively simple, inexpensive, and efficient at handling high power levels, and are relatively immune to EMI. However, they can generate significant switching noise and require careful design to operate efficiently and reliably, particularly at high power levels.

1.2.2. Bidirectional DC-to-DC Converters

In addition to unidirectional converters that step down the high-voltage battery voltage to lower voltage, bidirectional DC-to-DC converters are used in EVs to allow power flow in both directions. Bidirectional DC-to-DC converters are critical components for EV applications because they allow for efficient power transfer between the high-voltage battery pack and the low-voltage electrical system, which includes the motor, auxiliary systems, and the vehicle's onboard electronics (Wang, et al., 2022). In an EV, the high-voltage battery pack typically operates at a voltage that is much higher than the voltage required by the low-voltage electrical system. A DC-to-DC converter is a power electronic circuit that converts a DC voltage at one level to a DC voltage at another level. Bidirectional DC-to-DC converters allow for the conversion of DC power in both directions, enabling power to flow from the battery to the electrical system and vice versa. This bi-directionality is essential in EVs, where the battery must be charged when the vehicle is regenerating energy during braking or coasting. A bidirectional DC-to-DC converter provides a way to step down the voltage from the battery pack to the voltage required by the low-voltage electrical system. Additionally, the converter can also step up the voltage from the low-voltage electrical system to charge the high-voltage battery pack.

Bidirectional DC-to-DC converters are particularly important for EVs because they allow for regenerative braking, which is a key feature of many

EVs. During regenerative braking, the motor acts as a generator, converting kinetic energy back into electrical energy that can be used to recharge the battery. The bidirectional converter allows this electrical energy to be efficiently transferred back to the battery pack. They are also used in other applications, such as solar power systems, where they enable the transfer of power between DC sources and loads.

Bidirectional DC-to-DC converters are classified based on the basic and advanced level of converter topology used as illustrated in Figure 1.11. Boost, Cuk, SEPIC, Zeta converter are basic non-isolated bidirectional DC-to-DC converters whereas, interleaved and other cascaded interleaved converters come under advanced non-isolated bidirectional DC-to-DC converters (Babazadeh, Sabahi, Babaei, & Kai, 2022). They can also be classified on basis of number of conversion stages they employ. Two-stage converters are the most commonly used type in EVs. These converters have a boost converter stage that raises the voltage of the battery pack to a higher level, followed by a buck converter stage that lowers the voltage to the level required by the vehicle's electrical system. In regenerative braking mode, the converter's direction is reversed, and the buck converter stage becomes a boost converter, allowing the battery to be charged. Another type of bidirectional converter is the four-quadrant converter, which uses four transistors to enable bi-directional power flow. These converters are more complex and expensive than two-stage converters, but they offer more precise control of power flow and are more efficient at high power levels.

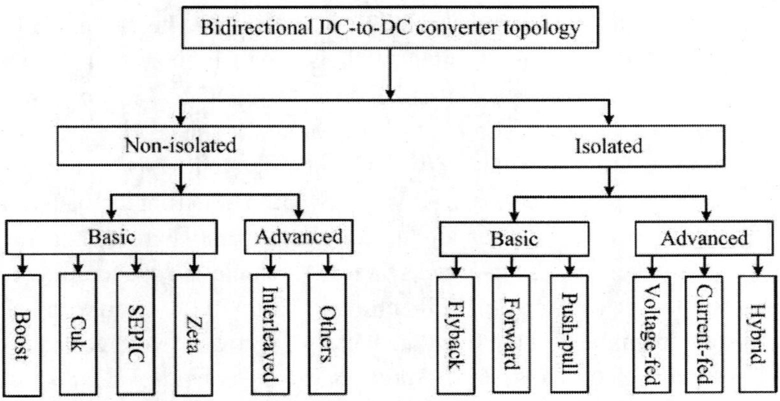

Figure 1.11. Classification of various bidirectional DC-to-DC converters.

In addition to their bi-directional capabilities, efficiency is a crucial factor for bidirectional DC-to-DC converters, and they must be designed to minimize power losses and heat generation. The efficiency of a converter is determined by its conversion ratio, switching frequency, and power losses in the converter components. Increasing the switching frequency of the converter increases its efficiency but also increases the cost and complexity of the circuit. There are also several design challenges associated with bidirectional DC-to-DC converters. These include the need for high voltage isolation between the input and output sides of the converter, the need for effective cooling of the converter components, and the management of EMI generated by the converter.

A dual-active-bridge (DAB) converter uses two active bridges to allow power flow in both directions. The DAB converters are a type of DC-to-DC converter used in EVs to convert the high voltage from the battery to the lower voltage required by various components (Zhao, Song, Liu, & Sun, 2014). The DAB converter consists of two full-bridge converters connected back-to-back, with a high-frequency (HF) transformer in the middle as shown in Figure 1.12. Eight switches (S_1 to S_8) used in DAB are insulated gate bipolar transistors (IGBTs). Switches are typically IGBTs or metal-oxide-semiconductor field-effect transistors (MOSFETs).

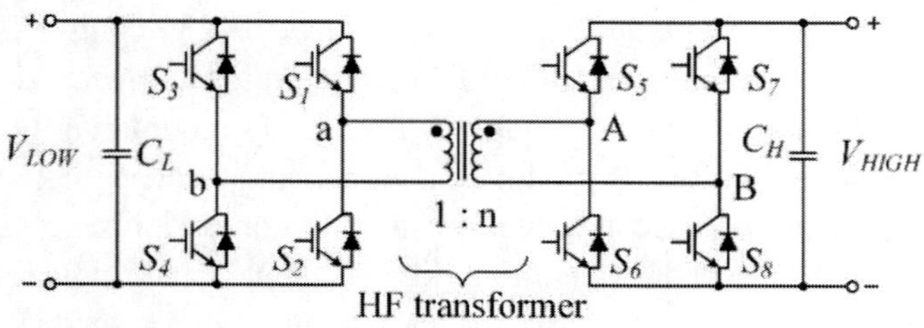

Figure 1.12. Dual active bridge converter.

The half-bridge and full-bridge concepts in DAB converters refer to the type of bridge configuration used in each converter. In a half-bridge converter, there are two switches on one side of the bridge and a center-tapped transformer on the other side. The input voltage is applied to the center tap of the transformer, and the switches are controlled to regulate the output voltage. In a full-bridge converter, there are four switches arranged in a bridge configuration, with the center-tapped transformer in the middle. The input voltage is applied across two diagonally opposite switches, and the other two switches are controlled to regulate the output voltage.

In a DAB converter, the two full-bridge converters are connected back-to-back with the transformer in the middle, creating a bidirectional converter. The input voltage is applied to one side of the DAB, and the output voltage is taken from the other side, with the transformer providing isolation between the input and output. The DAB converter is a high-efficiency converter that can handle high power levels, making it well-suited for use in EVs. It offers precise control over the output voltage and can be designed to operate over a wide range of input and output voltages. Overall, the half-bridge and full-bridge concepts in DAB converters are important design considerations for optimizing the performance and efficiency of the converter in EV applications.

1.2.3. Control techniques for DC-to-DC Converters

The performance of DC-to-DC converters can be significantly improved by employing suitable control techniques. Common control techniques used in DC-to-DC converters for EVs include (Ravi, Letha, Samuel, & Reddy, 2018), (Turksoy, Teke, & Alkaya, 2020), (Hossain, Rahim, & Selvaraj, 2018):

- *Pulse-width modulation (PWM) control:* PWM control is a widely used technique in DC-to-DC converters, allowing for precise regulation of output voltage or current by modulating the width of switching pulses. A modulation signal, and a carrier signal of higher frequency is used to generate suitable control pulse as depicted in Figure 1.13(a). It offers flexibility, efficiency, and ripple reduction by adjusting the duty cycle of the high-frequency switching signal. PWM control enables fast response to changing conditions and finds applications in voltage regulation, power conversion, and motor drives. However, it introduces switching losses, which affect efficiency and heat generation. EMI/EMC interference can also arise, necessitating proper filtering and shielding. Implementing PWM control requires careful design and control algorithm implementation to accurately generate switching signals based on feedback information. Overall, PWM control is versatile and effective but requires addressing challenges related to switching losses, EMI/EMC interference, and control complexity for optimal performance.

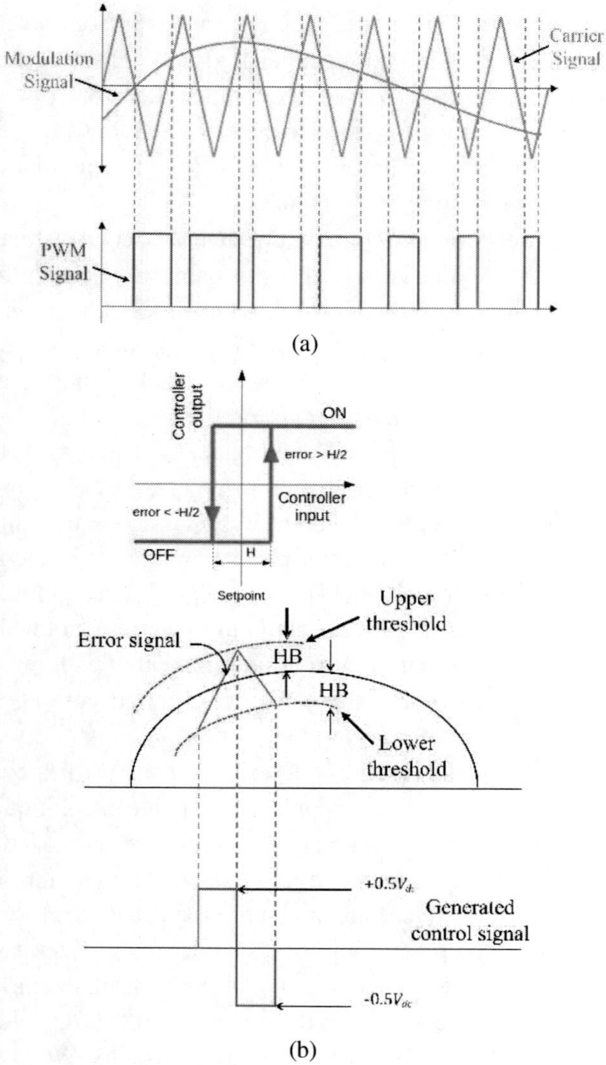

Figure 1.13. Control signal generation using (a) PWM and (b) Hysteresis control.

- *Voltage Mode Control:* It is a commonly used control technique in DC-to-DC converters that regulates the output voltage based on a feedback loop. It involves comparing the actual output voltage with a desired reference voltage and adjusting the duty cycle of the switching signal to maintain the desired output. Voltage mode control provides good stability and accuracy in regulating the output voltage, making it suitable for applications that require precise voltage regulation. The control loop continuously monitors the output voltage and adjusts the duty cycle accordingly, ensuring that any variations or disturbances are quickly compensated for. Voltage mode control offers simplicity in implementation and allows for efficient control of the converter's output voltage, making it a popular choice in various DC-to-DC converter designs.

- *Current Mode Control:* Current mode control regulates the output current of the converter by sensing and controlling the inductor current. Unlike voltage mode control, which focuses on regulating the output voltage, current mode control adjusts the duty cycle of the switching signal based on the feedback of the inductor current. By monitoring the inductor current, the control loop can respond quickly to load variations and disturbances, providing better stability and dynamic response compared to voltage mode control. Current mode control also offers inherent cycle-by-cycle current limiting, which enhances the converter's protection against overcurrent conditions. This control technique is particularly suitable for applications that require precise current regulation, such as motor drives and power supplies with strict current requirements. It allows for accurate and efficient control of the converter's output current, ensuring reliable operation under varying load conditions.
- *Hysteresis Control:* Hysteresis control, also known as bang-bang control or comparator control, is a control technique used in DC-to-DC converters to regulate the output voltage or current. It involves comparing the actual output value with a predetermined threshold or reference value and making a control decision based on the comparison result. In hysteresis control, a hysteresis band is defined around the desired setpoint, and the control action is triggered when the output value crosses the upper or lower threshold of the hysteresis band (HB) as illustrated in Figure 1.13(b). When the output value exceeds the upper threshold, the control action is activated to reduce the input power or increase the output power. Conversely, when the output value falls below the lower threshold, the control action is triggered to increase the input power or decrease the output power. This on/off control mechanism ensures that the output value stays within the hysteresis band, minimizing steady-state errors and providing robust regulation. Hysteresis control offers several advantages, including simplicity, fast response, inherent stability, and noise immunity. It is particularly suitable for applications that require

quick transient response and where precise regulation within a certain tolerance is not critical. However, hysteresis control can introduce higher output ripple and may require careful selection of the hysteresis band width to balance between stability and dynamic performance. Overall, hysteresis control provides a straightforward and effective approach for regulating DC-to-DC converters, offering a trade-off between simplicity and performance.

- *Sliding Mode Control (SMC):* SMC is a robust control technique used in DC-to-DC converters to achieve accurate regulation of the output voltage or current. It drives the system states to a predefined sliding surface through a switching control signal, ensuring precise regulation despite uncertainties and disturbances. SMC offers advantages such as robustness against parameter variations and disturbances, fast response, reduced sensitivity to modeling inaccuracies, and fault tolerance. However, it can exhibit chattering, rapid switching near the sliding surface, which can be mitigated using boundary layer techniques. Designing and tuning an SMC controller requires a good understanding of system dynamics, selection of appropriate sliding surfaces, and design of switching control laws. In summary, SMC provides accurate regulation and robustness in DC-to-DC converters, with fast response and reduced sensitivity, but attention is needed to mitigate chattering effects and ensure proper design and tuning of the controller.
- *Digital Control:* Digital control in DC-to-DC converters involves the use of digital signal processing techniques and algorithms to implement control strategies (Restrepo & González-Castaño, 2021). It offers advantages such as flexibility in adjusting control parameters, enhanced performance through advanced control strategies, real-time parameter estimation and adaptation, communication capabilities for remote monitoring and coordination, and advanced fault detection and diagnostics. However, digital control also presents challenges including sampling and quantization effects that can affect control performance, computational requirements for real-time execution, and implementation complexity in terms of hardware design, software development, and system integration. Addressing these challenges is crucial to ensure the successful implementation of digital control strategies and harness the benefits it offers in DC-to-DC converters.

- *Predictive Control:* Predictive control is an advanced control technique that is gaining popularity in DC-to-DC converters. It utilizes mathematical models of the system and predicts the future behavior of the converter based on the current and past measurements. By optimizing a cost function over a finite time horizon, predictive control determines the control actions that minimize the deviation from desired performance criteria. Predictive control in DC-to-DC converters provides enhanced performance through optimization of control actions over a finite time horizon, considering system dynamics and constraints. It effectively handles nonlinearities and constraints, adapts to changing conditions, and enables model-based fault detection. However, implementing predictive control requires addressing computational complexity, ensuring model accuracy, and mitigating parameter sensitivity. With ongoing research, predictive control holds promise for improving the performance and reliability of DC-to-DC converters in diverse applications.
- *Maximum power point tracking (MPPT) control:* MPPT control is a control technique used in DC-to-DC converters to optimize the power transfer from a renewable energy source, such as solar panels or wind turbines. The MPPT control continuously adjusts the operating point of the converter to track the maximum power point (MPP) of the energy source, where the power output is highest. By tracking the MPP, the converter can extract the maximum available power and improve the overall efficiency of the system. MPPT control techniques, such as perturb and observe (P&O) or incremental conductance, utilize feedback information from the energy source and adjust the converter's operating parameters, such as duty cycle or switching frequency, to maintain the system at the MPP. MPPT control offers benefits such as increased energy harvesting, improved efficiency, and better utilization of renewable energy sources. However, it also requires careful design, proper selection of control algorithms, and consideration of converter dynamics and limitations to ensure accurate and stable tracking of the MPP in varying environmental conditions.
- *PID control:* PID control, which stands for Proportional-Integral-Derivative control, is a widely employed control technique in various systems, including DC-to-DC converters. It combines three control actions - proportional, integral, and derivative - to regulate the output voltage or current. The proportional control adjusts the control signals

based on the error between the desired and actual values, the integral control eliminates steady-state error by accumulating past errors, and the derivative control minimizes the rate of change of the error to enhance stability. By summing these control actions, the PID controller dynamically adjusts the duty cycle of the converter's switching signals. Often a PI is also sufficient to control a system instead of PID. A detailed block diagram of a DC-DC converter using PI controller is depicted in Figure 1.14. PID control offers advantages such as accuracy, stability, dynamic response, robustness, and tunability. However, its performance depends on proper gain tuning, lacks adaptability to changing system dynamics, and may have limitations in highly nonlinear or wide-ranging applications. Nonetheless, PID control remains popular for its simplicity, effectiveness, and well-established theory, providing accurate regulation, stability, and fast response in DC-to-DC converters when properly implemented and tuned.

- *Soft-switching techniques:* Soft-switching techniques are utilized in DC-to-DC converters to reduce switching losses and improve efficiency by minimizing the switching stress on power semiconductor devices. These techniques, such as resonant converters or zero-voltage switching (ZVS) and zero-current switching (ZCS), aim to eliminate or significantly reduce the switching losses that occur during the transition of power semiconductor devices. By incorporating resonant tank circuits or additional circuit elements, soft-switching techniques allow for smooth switching transitions, enabling the devices to turn on or off when the voltage or current across them is close to zero. This minimizes the power dissipation and improves the overall efficiency of the converter. Soft-switching techniques offer advantages such as reduced power losses, lower EMI, and improved reliability of the power devices. However, the design and implementation of soft-switching converters require careful consideration of circuit complexity, control strategies, and component selection to achieve optimal performance while ensuring stable operation across different load and input conditions.

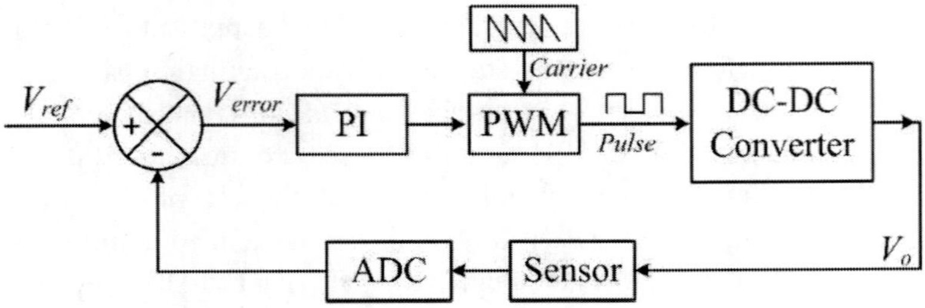

Figure 1.14. Generalized control block diagram of a DC-DC converter using a PI controller.

- *Adaptive control techniques:* Adaptive control techniques are employed in DC-to-DC converters to adaptively adjust control parameters based on real-time measurements or system conditions. These techniques aim to enhance the performance and robustness of the converter by continuously updating the control actions to account for uncertainties, variations in operating conditions, or changes in system dynamics. Adaptive control algorithms utilize adaptive mechanisms, such as parameter estimation, online identification, or model reference adaptation, to adjust the control parameters and optimize the control response. By adapting to changes in the system, adaptive control techniques can improve the accuracy of output regulation, enhance transient response, and provide robustness against uncertainties or disturbances. However, implementing adaptive control requires careful consideration of algorithm design, stability analysis, convergence properties, and the selection of suitable adaptation mechanisms to ensure the stability and performance of the control system.

The selection of the control technique depends on various factors, including the converter topology, desired performance specifications, complexity requirements, and cost considerations. Each technique has its advantages and trade-offs in terms of stability, response time, efficiency, and implementation complexity. The choice of the control technique should be based on the specific application requirements and design constraints to achieve optimal performance and reliability.

1.3. DC-to-AC Converters for Electric Vehicles

This section focuses on the crucial role of DC-to-AC converters, also known as inverters, in electric vehicles. These converters are responsible for converting the DC power from the vehicle's battery into AC power required for motor drive applications. We will explore the various types of DC-to-AC converters used in electric vehicles and their characteristics, including voltage source inverters, current source inverters, and impedance source inverters. Additionally, case studies highlighting the practical applications of these converters in electric vehicles will be presented.

1.3.1. Introduction to DC-to-AC Converters

DC-to-AC converters are fundamental components in electric vehicles that play a crucial role in converting the DC power from the vehicle's battery into AC power needed for motor drive applications. In this section, we will provide an in-depth introduction to DC-to-AC converters, exploring their operation principles, key components, and significance in the context of electric vehicles. At the heart of an inverter is the power electronic circuitry that enables the conversion of DC to AC power. This circuitry consists of various semiconductor devices, such as transistors or thyristors, which are controlled by a switching mechanism to produce the desired AC waveform. The switching mechanism can be implemented using PWM techniques, where the switches are rapidly turned on and off to control the output voltage and frequency.

The primary function of a DC-to-AC converter is to generate a high-quality AC waveform with variable frequency and amplitude to drive the electric motor in an EV. The motor requires AC power to produce the rotational motion that propels the vehicle. By converting the DC power from the battery to AC power, the inverter enables precise control of the motor's speed, torque, and direction. The working principle of a DC-to-AC converter involves two main stages: the DC input stage and the AC output stage. In the DC input stage, the DC power from the battery is conditioned and prepared for conversion. This stage typically includes components such as filters, input voltage regulators, and protection circuits to ensure stable and clean DC power supply to the converter.

The AC output stage is responsible for generating the desired AC waveform based on the motor drive requirements. This stage consists of semiconductor switching devices, such as power transistors or IGBTs, which are controlled by a pulse width modulation (PWM) technique. The PWM technique adjusts the switching frequency and duty cycle of the transistors to regulate the output voltage magnitude and frequency.

There are several types of DC-to-AC converters used in electric vehicles, each with its unique characteristics and suitability for different applications. The most common types include voltage source inverters (VSIs), current source inverters (CSIs), and impedance source inverters (ZSIs). A comparison between VSI, CSI and ZSI is illustrated in Table 1.2. The choice of DC-to-AC converter type depends on various factors, including the specific application requirements, system efficiency, cost considerations, and overall system integration. Designers must carefully evaluate these factors to select the most suitable converter topology for the electric vehicle system.

1.3.2. Voltage Source Inverters

VSIs are one of the most commonly used types of DC-to-AC converters in electric vehicles. They play a crucial role in converting the DC power from the vehicle's battery into high-quality AC power required for motor drive applications. In this section, we will explore the working principle, topologies, control strategies, and advantages of voltage source inverters in electric vehicles.

The working principle of a voltage source inverter involves converting the fixed or variable DC voltage from the battery into a variable frequency AC voltage, as depicted in Figure 1.15(a). This is achieved through the use of semiconductor switching devices, such as IGBTs, which are controlled by a PWM technique. The PWM control adjusts the switching frequency and duty cycle of the transistors to regulate the output voltage magnitude and frequency.

Voltage source inverters can be implemented in various topologies, including two-level inverters, three-level inverters, and multilevel inverters. Two-level inverters are the simplest and most commonly used topology, where the DC input is connected to the positive and negative terminals of the inverter. Three-level inverters provide an additional voltage level, resulting in reduced voltage stress on the switching devices and improved output waveform quality. Multilevel inverters, such as the cascaded H-bridge or flying capacitor inverters, offer further advantages in terms of increased output voltage levels and reduced harmonic distortion.

Table 1.2. A comparison table between VSI, CSI and ZSI

Parameter	VSI	CSI	ZSI
Input voltage range	Requires stable DC input voltage	Can tolerate fluctuations in DC input voltage	Requires stable DC input voltage
Output voltage control	Easy control of output voltage	Output voltage control can be more complex	Easy control of output voltage
Output current control	Output current control is more complex	Easy control of output current	Output current control is more complex
Protection requirements	Requires protection against overcurrent and overvoltage	Requires protection against overcurrent	Requires protection against overcurrent and overvoltage
Fault tolerance	Less fault-tolerant due to reliance on power devices	More fault-tolerant due to current source characteristics	More fault-tolerant due to impedance-based design
Efficiency	Generally higher efficiency	Efficiency can be lower due to the presence of current source	Efficiency can be lower due to the presence of impedance network
Cost	Generally lower cost due to the simplicity of the design	Cost can be higher due to the need for current source components	Cost can be higher due to the need for specialized impedance network
Harmonic distortion	Can introduce higher harmonic distortion in the output	Generally lower harmonic distortion in the output	Generally lower harmonic distortion in the output
Voltage stress	Higher voltage stress on power devices	Lower voltage stress on power devices	Lower voltage stress on power devices
Applicability	Commonly used in EV motor drives	Less commonly used in EV motor drives	Emerging technology in EV motor drives

Control strategies for voltage source inverters are crucial for achieving accurate and efficient motor drive operation in electric vehicles. One of the commonly used control techniques is PWM method. PWM controls the duty cycle of the switching devices to regulate the output voltage magnitude and frequency. By adjusting the switching frequency and duty cycle, the desired AC waveform can be synthesized, providing precise control of the motor speed and torque. Other control strategies, such as space vector modulation (SVM) or direct torque control (DTC), can also be employed to enhance the performance and efficiency of voltage source inverters in electric vehicle applications.

Voltage source inverters offer several advantages in electric vehicle systems. Firstly, they provide good voltage control, allowing for precise control of the motor drive operation. Secondly, they offer a high-quality output waveform with low harmonic distortion, ensuring smooth and efficient motor operation. Moreover, voltage source inverters can operate at variable frequencies, enabling flexibility in controlling the motor speed and torque.

Additionally, the use of PWM techniques in voltage source inverters allows for efficient use of battery power, reducing energy losses and improving overall system efficiency.

However, voltage source inverters also have certain challenges that need to be addressed. One of the major challenges is mitigating the switching losses and improving the overall converter efficiency. This can be achieved through the use of advanced semiconductor devices with low on-state and switching losses, as well as optimal control strategies that minimize power losses. Another challenge is managing the heat dissipation generated by the switching devices. Efficient thermal management techniques, such as heatsinks, cooling fans, or liquid cooling systems, are required to ensure the reliable and long-term operation of voltage source inverters. The importance of VSIs in EV application is as follows:

- VSIs are essential components in EVs as they convert the DC power from the battery into AC power required for motor drive applications.
- They provide precise control of the motor speed and torque, allowing for smooth and efficient operation of the vehicle.
- VSIs offer good voltage control, enabling accurate regulation of the output voltage magnitude and frequency.
- They produce high-quality output waveforms with low harmonic distortion, ensuring optimal motor performance and reducing electrical noise.
- The variable frequency operation of VSIs allows for flexible control of the motor speed, offering improved efficiency and responsiveness.
- VSIs are crucial for regenerative braking, as they facilitate the conversion of excess energy from the motor back into the battery, improving overall energy efficiency.
- They contribute to the overall efficiency of the EV system by efficiently utilizing the battery power and minimizing energy losses.
- VSIs play a vital role in achieving optimal power transfer and utilization in the EV drivetrain, enhancing the overall performance and range of the vehicle.
- These inverters are instrumental in enabling advanced control strategies such as PWM and SVM, enhancing motor control accuracy and efficiency.

- With the continuous advancements in VSI technology and control strategies, they continue to play a key role in the development of efficient and reliable EV systems.

1.3.3. Current Source Inverters

CSIs are a type of DC-to-AC converter used in electric vehicles to convert DC power from the battery into AC power required for motor drive applications. Unlike VSIs, which control the output voltage, CSIs control the output current, making them suitable for specific high-power applications in EV systems. In this section, we will explore the working principle, characteristics, control strategies, and advantages of current source inverters in electric vehicles.

A general topology of the CSI is shown in Figure 1.15(b). The working principle of a current source inverter involves controlling the output current instead of the output voltage. CSIs maintain a constant output current and vary the output voltage according to the load requirements. This is achieved by utilizing an inductor connected in series with the load, which acts as a current source. The inductor limits the rate of change of current and stabilizes the output current waveform. Current source inverters are typically used in applications that require high-power operation, such as traction drives in electric vehicles. They offer several advantages in these specific applications. One of the key advantages of CSIs is their inherent short-circuit protection. Since the output current is controlled, a short circuit in the motor or load does not result in excessive current flow, protecting the system from damage. This makes CSIs suitable for high-power applications where robust short-circuit protection is required.

CSIs also provide excellent linearity in current control, allowing for precise control of motor torque and speed. They exhibit good dynamic response and can handle rapid changes in load conditions, making them suitable for applications that require quick and accurate adjustments in motor drive operation. Additionally, the output current waveform of CSIs has low harmonic distortion, ensuring smooth motor operation and reducing electrical noise. Control strategies for current source inverters focus on regulating the output current and maintaining stability in the system. Various control techniques can be employed, including hysteresis current control, proportional-integral (PI) control, and model predictive control (MPC). These strategies ensure accurate current regulation, efficient motor operation, and system stability.

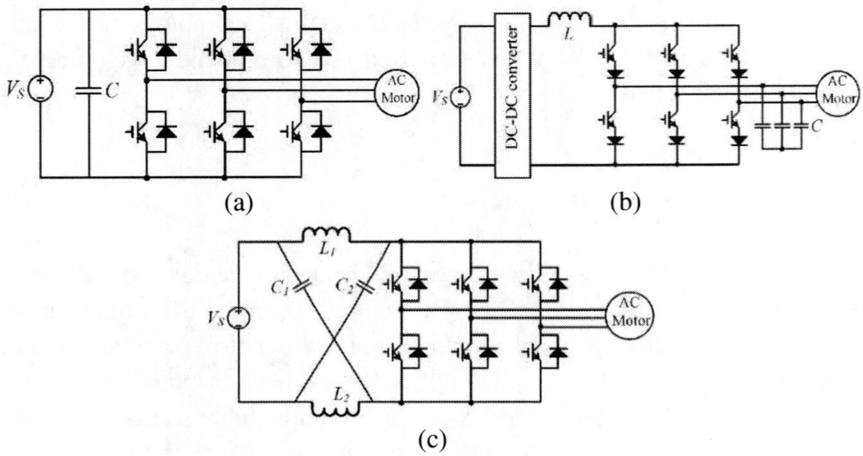

Figure 1.15. Typical inverter topology for EVs (a) VSI, (b) CSI, and (c) ZSI.

Despite their advantages, current source inverters also have certain limitations and challenges. One of the challenges is the requirement for a large inductor in the current path, which adds bulk and weight to the system. Additionally, CSIs have higher losses compared to VSIs due to the current-controlled operation, leading to lower overall system efficiency. Careful consideration is needed in the design and selection of components to minimize losses and optimize efficiency. Hence, current source inverters play a vital role in converting DC power into AC power in high-power applications of electric vehicles. They offer inherent short-circuit protection, precise current control, and good dynamic response, making them suitable for traction drives and other high-power motor applications. Control strategies ensure accurate current regulation and system stability. However, challenges related to size, weight, and efficiency need to be addressed for optimal performance in electric vehicle systems. With ongoing advancements in power electronics and control techniques, current source inverters continue to contribute to the development of efficient and reliable electric vehicle systems. The importance of CSIs is given below:

- CSIs are crucial components in electric vehicles for converting DC power from the battery into AC power required for high-power motor drive applications.
- CSIs provide inherent short-circuit protection, ensuring the safety and reliability of the system during motor operation.

- They offer precise current control, allowing for accurate regulation of motor torque and speed, enhancing the overall performance and efficiency of the EV.
- CSIs exhibit good dynamic response, enabling quick adjustments in motor drive operation to accommodate varying load conditions.
- The output current waveform of CSIs has low harmonic distortion, resulting in smooth motor operation and reduced electrical noise.
- CSIs are well-suited for high-power applications in EVs, such as traction drives, due to their ability to handle high currents and provide robust performance.
- Control strategies for CSIs ensure accurate current regulation and system stability, allowing for optimal motor drive performance.
- CSIs contribute to the overall efficiency of the EV system by providing effective power conversion and utilization.
- These inverters are particularly useful in applications that require high-power operation, where precise current control and robust short-circuit protection are essential.
- With ongoing advancements in power electronics and control techniques, CSIs continue to play a significant role in the development of efficient and reliable electric vehicle systems.

1.3.4. Impedance Source Inverters

ZSIs are a type of DC-to-AC converter used in EVs for converting DC power from the battery into AC power required for motor drive applications. ZSIs differ from conventional VSIs and CSIs in their unique impedance network, which provides certain advantages in terms of voltage boost capability, input current ripple reduction, and flexible voltage conversion. In this section, we will explore the working principle, characteristics, control strategies, and advantages of impedance source inverters in electric vehicles.

The working principle of an impedance source inverter involves the use of an impedance network, as shown in Figure 1.15(c). The impedance network typically consists of an inductor and a capacitor, is connected between the input DC source and the inverter. The impedance network provides a unique boost functionality that allows for voltage step-up and efficient voltage conversion. The inductor-capacitor network provides a continuous energy flow path, allowing for energy transfer between the DC and AC sides of the inverter.

ZSIs offer several advantages in electric vehicle applications. One of the key advantages is the inherent voltage boost capability. The impedance network allows for higher output voltage levels than the input voltage. It enables the use of lower voltage batteries while still achieving the desired high voltage levels required for motor drive applications. This voltage boost capability provides flexibility in system design, reduces the number of battery cells required, and can contribute to cost savings. Another advantage of ZSIs is the reduced input current ripple compared to traditional inverters. The impedance network smoothens the input current waveform, resulting in lower current ripple and improved power quality. This reduction in input current ripple can have a positive impact on battery life and overall system efficiency.

Control strategies for impedance source inverters focus on regulating the output voltage and achieving optimal voltage conversion. Various control techniques can be employed, including PWM control, MPPT control, and adaptive control algorithms. These strategies ensure accurate voltage regulation, efficient power conversion, and system stability.

Impedance source inverters offer flexibility in voltage conversion, allowing for bi-directional power flow. This capability is particularly beneficial in electric vehicles where energy can be transferred between the battery and other energy storage devices, such as supercapacitors or fuel cells. ZSIs enable efficient energy management and utilization, improving the overall energy efficiency of the EV system. Additionally, ZSIs have the potential for fault tolerance and resilience in the event of component failures. The impedance network can help isolate and mitigate the impact of faults, providing a certain level of robustness and reliability to the system.

However, impedance source inverters also have certain challenges that need to be addressed. The design and optimization of the impedance network require careful consideration to achieve the desired voltage boost capability and minimize losses. Efficient control strategies need to be implemented to ensure accurate voltage regulation and optimal power conversion.

In conclusion, impedance source inverters play a significant role in converting DC power to AC power in electric vehicles. They offer advantages such as voltage boost capability, reduced input current ripple, and flexible voltage conversion. Control strategies enable accurate voltage regulation and efficient power conversion. ZSIs provide flexibility in energy management and have the potential for fault tolerance. However, challenges related to design optimization and control need to be addressed for optimal performance in electric vehicle systems. With ongoing advancements in power electronics and control techniques, impedance source inverters continue to contribute to the development of efficient and reliable electric vehicle systems.

1.3.5. Case Studies of DC-to-AC Converters in Electric Vehicles

In this section, three case studies will be explored that highlight the applications and performance of DC-to-AC converters in EVs. These case studies demonstrate the versatility and effectiveness of these converters in different EV systems.

1.3.5.1. Electric Bus Powertrain

One of the prominent applications of DC-to-AC converters is in the powertrain of electric buses (Lajunen, 2014). Electric buses require efficient and reliable conversion of the battery's DC power to AC power to drive the electric motor. A case study conducted on an electric bus powertrain demonstrated the effectiveness of a three-phase VSI in converting the battery's DC voltage into three-phase AC power. The VSI used in the electric bus powertrain employed PWM control to regulate the output voltage and frequency. The control strategy enabled precise control of motor speed and torque, resulting in smooth and efficient bus operation. The case study evaluated the performance of the VSI in terms of power conversion efficiency, dynamic response, and overall system reliability.

The results showed that the VSI achieved high power conversion efficiency, exceeding 95%. The efficient conversion of DC power to AC power minimized energy losses and maximized the driving range of the electric bus. The VSI also demonstrated excellent dynamic response, allowing for quick adjustments in motor speed and torque in response to changing driving conditions. The reliability of the system was ensured through comprehensive fault protection mechanisms and robust control algorithms.

1.3.5.2. Electric Vehicle Battery Charging

DC-to-AC converters also play a crucial role in electric vehicle battery charging systems. A case study focused on an AC bi-directional charger showcased the effectiveness of the converter in facilitating efficient and flexible battery charging (Bai, et al., 2022). The AC bi-directional charger utilized a combination of a DC-to-AC converter and an AC-to-DC converter. The DC-to-AC converter was responsible for converting the AC grid power into high-frequency AC power, which was then fed to the AC-to-DC converter to convert it into DC power for battery charging. The bi-directional capability of the charger allowed energy to flow bidirectionally between the grid and the vehicle battery, enabling vehicle-to-grid capabilities.

The case study evaluated the performance of the converter in terms of charging efficiency, power factor correction, and grid integration. The DC-to-AC converter demonstrated high efficiency in converting grid power to AC power, minimizing losses and ensuring efficient charging of the vehicle battery. A schematic layout of an EV energy storage (e.g., battery) charging is illustrated in Figure 1.16. Power factor correction techniques were implemented to improve power quality and reduce the strain on the grid. The converter also showcased seamless grid integration, allowing for smart charging functionalities and V2G operations.

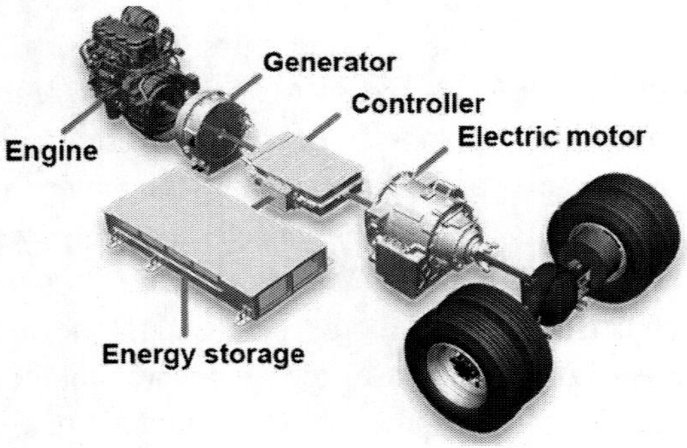

Figure 1.16. Layout of EV battery charging (Image source: BAE Systems).

1.3.5.3. Electric Vehicle Regenerative Braking

Regenerative braking is a key feature in electric vehicles that enables the recovery of kinetic energy during deceleration and braking. DC-to-AC converters play a crucial role in regenerative braking systems by converting the recovered energy from the electric motor into DC power for storage in the battery.

A case study focused on an electric vehicle with regenerative braking demonstrated the effectiveness of a bidirectional DC-to-AC converter in energy recovery (Fajri, Lee, Prabhala, & Ferdowsi, 2016). The bidirectional converter allowed for seamless switching between the motor drive mode and the regenerative braking mode, ensuring efficient energy transfer between the motor and the battery. The case study evaluated the performance of the converter in terms of energy recovery efficiency, control responsiveness, and battery charging capabilities. The bidirectional converter achieved high energy recovery efficiency, allowing for significant energy savings and increased driving range. The responsive control algorithms enabled smooth transition between motor drive and regenerative braking modes, ensuring seamless operation and optimal energy recovery. The converter also facilitated efficient battery charging during regenerative braking, ensuring effective utilization of the recovered energy.

Hence, the case studies highlight the versatility and effectiveness of DC-to-AC converters in various electric vehicle applications. From electric bus powertrains to battery charging systems and regenerative braking, these converters play a crucial role in ensuring efficient power conversion, energy management, and system performance in electric vehicles.

These case studies have shown that DC-to-AC converters offer several benefits and features that contribute to the success of electric vehicle systems:

1) *Efficient Power Conversion:* The converters demonstrated high power conversion efficiency, minimizing energy losses and maximizing the driving range of electric vehicles. Efficient conversion of DC power to AC power ensures optimal utilization of the available energy.
2) *Dynamic Response and Control:* The converters exhibited excellent dynamic response, allowing for quick adjustments in motor speed, torque, and charging rate. Precise control algorithms enable smooth and responsive operation, enhancing the overall driving experience and system performance.
3) *Flexibility and Bi-Directional Power Flow:* The converters enabled bi-directional power flow, allowing energy to be transferred between the battery and other components, such as the motor or the grid. This flexibility facilitates functions like regenerative braking and vehicle-to-grid capabilities, enhancing energy efficiency and grid integration.
4) *Reliability and Fault Tolerance:* The case studies emphasized the importance of robust control algorithms and fault protection mechanisms in ensuring system reliability. Comprehensive fault detection and mitigation strategies enhance the safety and durability of electric vehicle systems.
5) *Integration with Smart Grid:* The converters demonstrated seamless integration with the grid, enabling advanced functionalities such as bidirectional power flow, V2G capabilities, and smart charging. This integration promotes efficient grid utilization, load balancing, and

demand response, contributing to a more sustainable and resilient energy ecosystem.

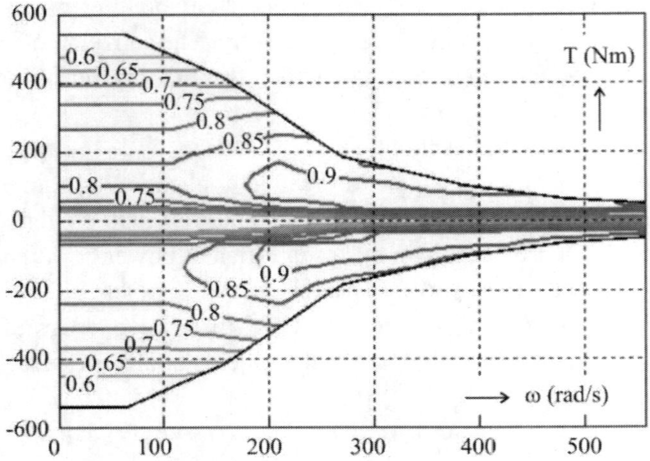

Figure 1.17. Efficiency map of an inverter combined with electric motor.

Figure 1.17 represents the efficiency characteristics of the entire EV system across a range of operating conditions. Motor speed (rad/s) is given on the horizontal axis and the motor torque (Nm) on the vertical axis. The efficiency is represented by contour lines on the graph. It provides insights into the efficiency of power conversion from the DC input to the AC output, considering various motor loads and operating points. The efficiency map shows how efficiently the inverter and motor system converts electrical energy to mechanical energy under different operating conditions. At each operating point, the efficiency is calculated as the ratio of output power (mechanical power delivered by the motor) to input power (electrical power supplied by the inverter). The efficiency of the system can vary depending on factors such as the motor design, inverter topology, switching frequency, control strategy, and losses associated with power conversion. At low motor loads, the efficiency may be lower due to losses in the inverter and motor windings. At high motor loads, the efficiency may be reduced due to increased losses in the motor core and bearing friction.

The efficiency map helps in selecting the optimal operating point for the system to maximize overall efficiency. It provides a visual representation of the regions where the system operates most efficiently, allowing designers and operators to make informed decisions regarding motor sizing, control strategies, and operational optimization.

These case studies also highlight some ongoing challenges and future research directions in the field of DC-to-AC converters for electric vehicles:

1) *Enhancing Efficiency:* While the converters demonstrated high efficiency, there is always room for improvement. Research efforts can focus on reducing losses and optimizing control strategies to enhance overall energy conversion efficiency.
2) *Advanced Control Techniques:* Developing advanced control techniques, such as predictive control algorithms and advanced modulation strategies, can further enhance the dynamic response, stability, and overall performance of DC-to-AC converters.
3) *Thermal Management:* Efficient thermal management is crucial for maintaining the performance and reliability of the converters. Research on innovative cooling techniques and thermal management strategies can help ensure optimal operating temperatures and extend the lifespan of the converter components.
4) *Standardization and Interoperability:* Developing standardized interfaces and protocols for DC-to-AC converters can facilitate interoperability and compatibility among different electric vehicle systems, enabling seamless integration and exchange of power between various components.
5) *Cost Reduction and Scalability:* Continued research and development efforts are needed to reduce the cost of DC-to-AC converters, making them more economically viable for mass adoption. Additionally, scalable designs and manufacturing techniques can further enhance their commercial viability and widespread implementation.

In summary, the case studies demonstrate the importance and effectiveness of DC-to-AC converters in electric vehicle applications. They provide valuable insights into their performance, benefits, and areas for improvement. With ongoing research and advancements in power electronics, control strategies, and system integration, DC-to-AC converters will continue to play a vital role in driving the transition to efficient and sustainable electric transportation.

1.4. AC-to-DC Converters for Electric Vehicles

The conversion of AC to DC is a fundamental process in electric vehicles that enables them to charge their batteries and power various vehicle systems. AC-to-DC converters, also known as rectifiers or onboard chargers, play a crucial role in converting the AC power from the charging infrastructure or the grid into the DC power required by the vehicle's battery and electrical systems. This section delves into the topologies of AC-to-DC converters, power factor correction techniques, and the charging infrastructure and standards associated with these converters.

AC-to-DC converters in EVs are responsible for efficiently and safely converting the AC power into the appropriate DC voltage and current levels required for battery charging. They are typically integrated within the vehicle's powertrain or as a separate charging unit, allowing the EV to be charged from various power sources, including home charging stations, public charging stations, and fast-charging networks. Broadly, AC-to-DC converters are divided into three types such as uncontrolled, half-controlled and fully controlled rectifiers as shown in Figure 1.2 as well. A comparative data of these rectifiers is given in Table 1.3.

Understanding the operation, characteristics, and standards of AC-to-DC converters is crucial for the development and deployment of efficient and reliable charging infrastructure for electric vehicles. This section aims to provide insights into the various topologies used in these converters, the techniques employed for power factor correction, and the standards and infrastructure necessary for seamless and standardized EV charging. By exploring these topics, we can gain a comprehensive understanding of the technologies and considerations involved in the AC-to-DC conversion process in electric vehicles.

1.4.1. Topologies of AC-to-DC Converters

AC-to-DC converters for EV charging can be implemented using various topologies, depending on factors such as power level, efficiency requirements, and system complexity. Some common topologies include (Yu, et al., 2022), (Maroti, Padmanaban, Bhaskar, Ramachandaramurthy, & Blaabjerg, 2022):

1) *Single-Phase Diode Rectifier:* The single-phase diode rectifier, also known as single-phase uncontrolled rectifier, is the simplest form of

AC-to-DC converter. It consists of a diode bridge connected to the input AC supply. The diodes conduct during the positive half-cycles of the input voltage, allowing current flow through the load and charging the EV battery. However, due to the unidirectional nature of the diodes, this topology results in a pulsating DC output with significant ripple voltage. Single-phase diode rectifiers are typically used in low-power applications where cost and simplicity are primary considerations.

2) *Single-Phase Full-Bridge Rectifier:* The single-phase full-bridge rectifier is an improvement over the diode rectifier as it utilizes four diodes arranged in a bridge configuration. This topology offers a full-wave rectification, resulting in a smoother DC output with reduced ripple compared to the diode rectifier. The full-bridge rectifier is widely used in EV charging applications due to its simplicity, cost-effectiveness, and adequate performance for medium-power levels.
3) *Three-Phase Diode Rectifier:* In three-phase diode rectifiers, three diodes are connected in each phase of the input AC supply, resulting in a six-pulse rectification. This topology provides higher power capability compared to single-phase rectifiers and offers a more balanced output current. Three-phase diode rectifiers are commonly used in high-power EV charging stations where a three-phase AC power supply is available.
4) *Single-Phase Controlled Rectifiers:* Controlled rectifiers use thyristors or other controllable switches to regulate the output DC voltage by adjusting the conduction angle. They offer improved power factor and reduced harmonic distortion compared to uncontrolled rectifiers.
5) *Three-Phase Controlled Rectifiers:* These converters utilize controlled switches, such as thyristors or IGBTs in place of diodes, to regulate the DC output voltage. They provide better power factor and lower harmonic distortion compared to uncontrolled rectifiers.
6) *Active Power Factor Correction (PFC) Converters:* PFC converters actively shape the input current waveform to achieve a near-unity power factor. They employ additional control circuitry to modulate the switching devices, resulting in reduced harmonic distortion and improved efficiency. There are various PFC converters available such as three-phase active rectifier, interleaved PFC boost rectifier and bridgeless PFC rectifiers.

7) *Multi-stage rectifiers:* Multi-stage rectifiers refer to rectification circuits that consist of multiple stages or levels of rectification e.g., cascaded rectifier, voltage multiplier rectifiers and interleaved multi-phase rectifiers. In these circuits, the AC input voltage is converted into a DC output voltage by passing through multiple rectifier stages, each performing a partial rectification. The purpose of using multi-stage rectifiers is to achieve higher voltage levels or improve the quality of the rectified output. The multi-stage conversion also involves indirect AC-to-DC converters, at the beginning AC-to-AC (e.g., matrix converter (Drabek, Peroutka, Pittermann, & Cedl, 2011)) then AC-to-DC converters, or AC-to-DC then DC-to-DC converters (e.g., dual active bridge converter, LLC resonant converter ((Gore), et al., 2022) and solid-state transformer concept (Zheng, Kandula, & Divan, 2023)). The dual active bridge (DAB) converter utilizes two sets of power switches and transformers to provide bidirectional power flow and galvanic isolation. It enables efficient AC-to-DC and DC-to-DC conversion, making it suitable for applications such as onboard vehicle chargers and DC fast charging stations. The DAB converter offers flexible voltage conversion, improved efficiency, and advanced control capabilities.

Each AC-to-DC converter topology has its own advantages and limitations, and the selection of the appropriate topology depends on factors such as power requirements, cost, efficiency, power factor correction, and system complexity. As technology advances, new converter topologies may emerge, offering improved performance and additional features. It is important to select the appropriate AC-to-DC converter topology based on specific application requirements, including power levels, efficiency targets, power factor correction, galvanic isolation, and size constraints. Additionally, considerations such as cost, complexity, control strategies, and compliance with industry standards and regulations play a significant role in the selection process.

As technology advances and the demand for electric vehicles continues to grow, research and development efforts are focused on improving the efficiency, power density, and cost-effectiveness of AC-to-DC converters. The aim is to develop innovative converter topologies and control techniques that can further enhance the performance and reliability of electric vehicle charging systems. Moreover, AC-to-DC converters are essential components in electric vehicles, responsible for converting AC power from the grid into the DC power required for charging the vehicle's battery. Various converter topologies, such as voltage source inverters, current source inverters, impedance source inverters, and other advanced topologies, are used to achieve efficient and reliable power conversion. The selection of the appropriate topology depends on factors such as power requirements, efficiency targets, power factor correction, and system complexity. Ongoing research and development in AC-to-DC converter technologies continue to drive advancements in the field, contributing to the growth and adoption of electric vehicles.

Table 1.3. Comparison of uncontrolled rectifiers, half-controlled rectifiers, and fully-controlled rectifiers

Comparison factor	Uncontrolled rectifiers	Half-controlled rectifiers	Fully-controlled rectifiers
Control capability	No control over output voltage or current	Control over output voltage	Control over output voltage and current
Output quality	Higher harmonic distortion in output	Improved harmonic distortion in output	Improved harmonic distortion in output
Power factor	Lower power factor due to non-linear load	Improved power factor compared to uncontrolled rectifiers	Further improved power factor compared to half-controlled rectifiers
Efficiency	Generally lower efficiency	Improved efficiency compared to uncontrolled rectifiers	Further improved efficiency compared to uncontrolled and half-controlled rectifiers
Controllability range	No control over the output voltage	Limited control over the output voltage	Wide control range over the output voltage
Applicability	Simple applications with no need for control or power factor correction	Applications requiring moderate control and power factor improvement	Applications requiring precise control and power factor correction

1.4.2. Power factor Correction Techniques

Power factor correction techniques are essential in AC-to-DC converters to ensure efficient power transfer from the grid to the EV battery and comply with grid regulations. These techniques aim to achieve a power factor close to unity, reducing the reactive power demand and minimizing harmonic distortion. Some common power factor correction techniques include:

1) *Passive Power Factor Correction:* Passive PFC techniques utilize passive components such as inductors, capacitors, and resistors to

improve the power factor. These components are added to the converter circuit to reshape the input current waveform, aligning it closer to the input voltage waveform. The commonly used passive PFC technique is the use of input inductors and capacitors to create a filter that reduces the harmonic content and improves power factor. Although passive PFC techniques are simple and cost-effective, they have limited effectiveness in achieving high power factor correction.

2) *Active Power Factor Correction:* Active PFC techniques involve the use of active components, such as power switches and control circuitry, to actively shape the input current waveform. The most commonly used active PFC technique is the boost converter. In this technique, the input AC voltage is rectified and then fed into a boost converter, which operates in continuous conduction mode. The boost converter actively controls the output voltage and regulates the input current to follow the input voltage waveform, resulting in a near-unity power factor. Active PFC techniques offer higher power factor correction compared to passive techniques and can achieve power factors close to 1.
3) *Bridgeless Power Factor Correction:* Bridgeless PFC techniques aim to address the limitations of traditional diode bridge rectifiers by eliminating the diode bridge altogether. In bridgeless PFC topologies, power switches such as MOSFETs or IGBTs are used to control the current flow, resulting in reduced conduction losses and improved efficiency. The bridgeless PFC technique offers higher power factor correction, lower total harmonic distortion (THD), and reduced EMI compared to diode-based rectifiers.
4) *Interleaved Power Factor Correction:* Interleaved PFC techniques utilize multiple parallel-connected power converter modules to share the input current, reducing the input current ripple and improving power factor correction. The interleaved technique divides the load current among multiple converter modules, allowing each module to handle a fraction of the total current. By interleaving the current from multiple modules, the overall input current ripple is reduced, resulting in improved power factor and lower THD. Interleaved PFC is particularly beneficial for high-power applications where the converter needs to handle a significant amount of current.
5) *Resonant Power Factor Correction:* Resonant PFC techniques utilize resonant tank circuits to achieve power factor correction ((Gore), et al., 2022). These techniques involve the use of resonant inductors and

capacitors in the converter circuit to achieve soft-switching operation, minimizing switching losses and improving overall efficiency. Resonant PFC techniques, such as LLC resonant converters, offer high power factor correction, reduced switching losses, and improved power density. They are particularly suitable for high-power applications where efficiency and size are critical factors.

6) *Digital Control Techniques:* Digital control techniques play a vital role in achieving effective power factor correction. Advanced control algorithms, such as digital signal processing (DSP) and microcontroller-based control, enable precise control of the converter operation and power factor correction. Digital control techniques allow adaptive control, real-time monitoring, and adjustment of the converter parameters to optimize power factor correction under varying operating conditions.

Figure 1.18 describes the PFC concept in AC-to-DC converters (Jenkins, 2018). It shows that The PFC controller can be designed to minimize harmonic distortion in the supply current, resulting in an input AC line current waveform that aligns with the sine input voltage waveform. Power factor correction techniques provide numerous benefits in electric vehicle charging systems. They improve power quality by reducing harmonic distortion and aligning input current with the voltage waveform, resulting in reduced voltage distortion and minimized stress on the electrical grid. Additionally, power factor correction enhances efficiency by reducing losses associated with harmonic currents, allowing converters to operate closer to optimal efficiency and improving energy conversion. This, in turn, enables efficient power transfer from the grid to the electric vehicle and maximizes the utilization of electrical infrastructure. By maintaining a good power factor, electric vehicle chargers operate in a grid-friendly manner, complying with regulations and standards for power quality and stability. Moreover, power factor correction techniques play a vital role in AC-to-DC converters for electric vehicles. They reduce demand for reactive power, leading to cost savings in electrical infrastructure and supporting the integration of renewable energy sources. Overall, power factor correction enhances system reliability, reduces power fluctuations, and contributes to the sustainability of electric vehicle charging systems. Ongoing research and development in power factor correction continue to drive advancements in converter design and control techniques.

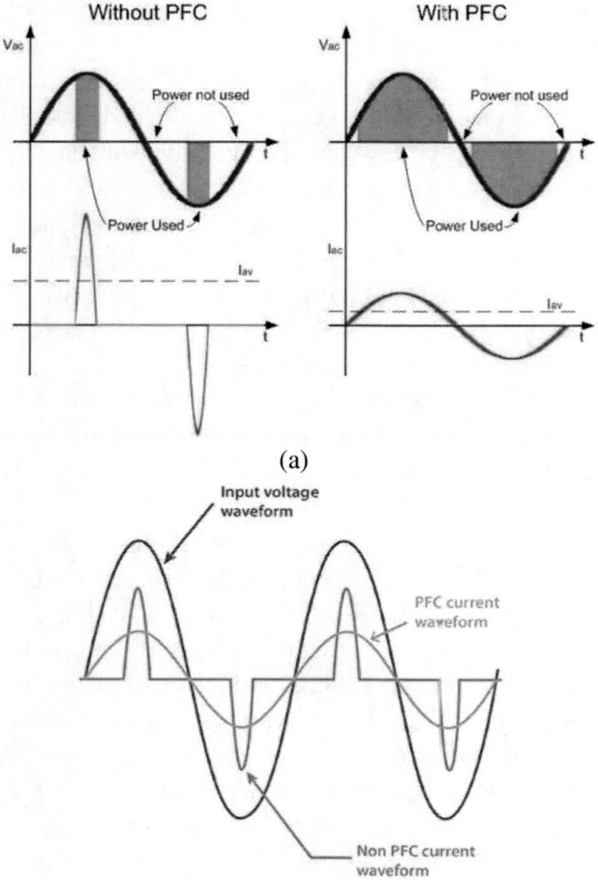

(a)

(b)

Figure 1.18. Represents the role of PFC in (a) power quality, and (b) power factor.

1.4.3. Charging Infrastructure and Standards for AC-to-DC Converters

AC-to-DC converters, also known as electric vehicle supply equipment (EVSE) or charging stations, are essential components of the charging infrastructure. Charging stations equipped with AC-to-DC converters provide charging capabilities for EVs, enabling them to recharge their batteries from the grid. Thus, AC-to-DC converters are an integral part of the charging infrastructure for electric vehicles. Standardization and compatibility are essential in ensuring seamless charging experiences and interoperability between different EVs and charging infrastructure. Several charging standards and protocols have been established to govern the AC-to-DC charging process (Acharige, et al., 2023), (Welcome to EV101: Electric Vehicle Charging Basics, n.d.). Some prominent standards include:

1) *SAE J1772:* Developed by the Society of Automotive Engineers (SAE), the J1772 standard defines the AC Level 1 and Level 2 charging connectors and communication protocols used in North America. It specifies the physical connector design, electrical requirements, and communication protocols between the EV and the charging station.
2) *IEC 61851:* The International Electrotechnical Commission (IEC) developed the IEC 61851 series of standards to ensure compatibility and interoperability of EV charging equipment globally. It covers AC charging (Mode 1, Mode 2, and Mode 3) as well as DC fast charging (Mode 4). These standards define the charging connector types, charging modes, safety requirements, and communication protocols.
3) *CHAdeMO:* CHAdeMO (Charge de Move) is a DC fast charging standard developed by the CHAdeMO Association, primarily used in Japan and supported by various EV manufacturers. It enables high-power DC charging with power levels ranging from 50 kW to 400 kW. CHAdeMO chargers use a unique connector design and communication protocol to facilitate fast charging.
4) *Combined Charging System (CCS):* CCS is a global standard developed by international organizations, including the CharIN (Charging Interface Initiative) consortium. It combines AC and DC charging capabilities into a single connector, accommodating both AC Level 1 and Level 2 charging as well as DC fast charging. CCS supports power levels up to 350 kW, enabling faster charging times for EVs.
5) *GB/T:* GB/T is a Chinese standard for EV charging developed by the Standardization Administration of China (SAC). It defines AC and DC charging modes and connectors used in China. The GB/T standard covers both low-power and high-power charging, providing compatibility with various EV models.

The establishment of these standards ensures that EVs can be charged at various charging stations globally, regardless of the AC-to-DC converter topology used. It promotes interoperability, safety, and a consistent user experience for EV owners. Charging infrastructure for AC-to-DC converters encompasses a range of charging solutions (Atanalian, Al Haddad, Zgheib, & Kanaan, 2021):

1) *Level 1 Charging:* Level 1 charging utilizes a standard household AC outlet with a power rating typically around 1.4 kW. It provides a slow charging rate and is suitable for overnight charging at home or in workplaces where longer charging durations are feasible.
2) *Level 2 Charging:* Level 2 charging employs dedicated charging stations with higher power ratings, typically ranging from 3.3 kW to 22 kW. These stations require professional installation and can deliver faster charging compared to Level 1, enabling quicker replenishment of the EV battery.
3) *Level 3 Charging:* Level 3 charging, also known as DC fast charging, utilizes high-power DC chargers capable of delivering power levels above 50 kW. These chargers employ AC-to-DC converters to directly convert grid AC power into high-power DC power, significantly reducing charging times. DC fast charging stations are typically found along highways, public charging stations, and commercial locations.

Charging infrastructure for AC-to-DC converters continues to evolve, with advancements in power electronics, communication protocols, and grid integration. Emerging technologies such as vehicle-to-grid (V2G) and smart charging aim to enhance the efficiency, reliability, and grid interaction of AC-to-DC charging systems.

There are four types of plugs commonly used for charging electric vehicles (EVs), two for AC charging and two for DC charging, as shown in Figure 1.19. Type 1 and Type 2 plugs are commonly used for AC charging, while CHAdeMO and CCS plugs are used for DC charging. Type 1 plugs are frequently used in American vehicles and are designed for single-phase charging with a maximum charging speed of up to 7.4 kW.

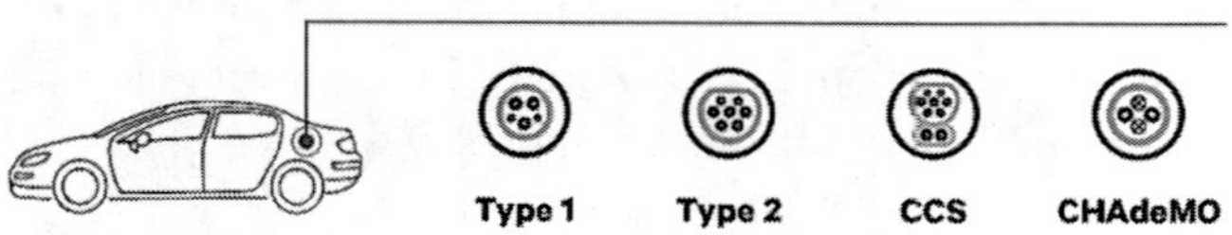

Source: Wallbox Chargers.

Figure 1.19. Various EV charging plug type.

In addition, AC-to-DC converters are vital components of the charging infrastructure for electric vehicles. They enable efficient conversion of AC power from the grid into DC power required for EV charging. Standardization and interoperability are crucial for ensuring compatibility between EVs and charging stations, leading to a seamless charging experience. As the adoption of electric vehicles continues to grow, the development of robust charging infrastructure and adherence to charging standards become increasingly important. Here are some additional aspects to consider:

a) Charging Station Types:
   - *Public Charging Stations:* These charging stations are typically located in public areas such as parking lots, shopping centers, and city streets. They offer convenience to EV owners who are on the go and require quick charging services.
   - *Residential Charging Stations:* Residential charging stations are installed in homes, apartments, or housing complexes. They provide a convenient and accessible charging solution for EV owners, allowing them to charge their vehicles overnight or during periods of non-use.
   - *Workplace Charging Stations:* Many workplaces are now equipped with charging stations to facilitate EV charging for employees. This supports sustainable transportation initiatives and encourages the adoption of electric vehicles.

b) Grid Integration:
   - *Smart Charging:* Smart charging systems use advanced communication and control technologies to optimize charging processes and manage the load on the grid. These systems can adjust the charging rate based on grid conditions, electricity demand, and user preferences. Smart charging helps to minimize peak loads, balance power distribution, and improve overall grid stability.
   - *Vehicle-to-Grid:* V2G technology enables bidirectional power flow between the EV battery and the grid. It allows EVs to not only consume electricity from the grid but also return excess energy back to the grid during peak demand periods. V2G supports grid stability, demand response, and energy management, contributing to a more flexible and sustainable energy ecosystem.

c) Interoperability and Roaming:
- *Interoperability:* Interoperability ensures that EVs can charge at any compatible charging station, regardless of the charging network or service provider. Common standards and protocols enable seamless communication and transaction processes between the EV and the charging infrastructure.
- *Roaming:* Roaming agreements between charging network operators allow EV drivers to access charging services across different regions or countries using a single charging card or mobile application. Roaming promotes convenience and accessibility, enabling EV owners to travel long distances without concerns about incompatible charging networks.

d) Standards and Regulations:
- *Safety Standards:* Charging infrastructure and AC-to-DC converters must comply with safety standards to ensure user and vehicle safety during the charging process. Standards such as IEC 60364 and IEC 61851-1 define safety requirements for electrical installations and EV charging systems.
- *Grid Regulations:* Grid operators and utility companies establish regulations to manage the impact of EV charging on the electrical grid. These regulations may include requirements for power factor correction, demand response capabilities, and grid connectivity.

The development and expansion of charging infrastructure, along with adherence to charging standards, are crucial for the widespread adoption of electric vehicles. It promotes user confidence, convenience, and accessibility, ultimately accelerating the transition to cleaner and more sustainable transportation. Overall, AC-to-DC converters play a vital role in the charging infrastructure for electric vehicles. Standardization, interoperability, and adherence to safety and grid regulations are essential for the seamless integration of AC-to-DC converters into the charging ecosystem. As technology continues to advance, the charging infrastructure will evolve to support higher power levels, faster charging speeds, and smarter grid integration, further driving the growth of electric mobility.

1.5. Requirements and Protection of Converters in Electric Vehicles

Electric vehicles rely on power electronic converters to efficiently convert and control electrical energy throughout the vehicle's operation. These converters play a crucial role in ensuring proper power delivery, voltage regulation, and overall system performance. However, due to the demanding operating conditions and dynamic nature of EVs, converters must meet specific requirements and be equipped with robust protection mechanisms. This section explores the requirements and protection considerations for converters in electric vehicles, highlighting the unique challenges and solutions associated with their operation (Wang, Dehghanian, Wang, & Mitolo, 2019), (Sanchez Sutil, Hernández, & Tobajas, 2015).

1) *Voltage and Power Requirements:* Converters in EVs must meet certain voltage and power requirements to ensure compatibility with the vehicle's electrical system and charging infrastructure. The voltage levels typically depend on the battery pack configuration and the specific voltage requirements of various vehicle components. Additionally, the power requirements of the electric vehicle, including the motor drive, auxiliary systems, and charging infrastructure, need to be taken into account. The converter must have the capability to handle the required power levels, providing efficient and reliable conversion between AC and DC power. By ensuring that the converter meets the voltage and power requirements, it can effectively support the electrical demands of the electric vehicle, enabling optimal performance and functionality. High-power converters are necessary to accommodate the increasing power demands of EVs, especially for fast charging applications. Compliance with industry standards and regulations, such as the voltage levels specified by the IEC and regional standards, is essential to ensure interoperability and safety.
2) *Efficiency and Thermal Management:* Efficiency and thermal management are crucial considerations in the design and operation of converters for electric vehicles. Efficiency refers to the ability of the converter to convert electrical energy with minimal losses. In the context of electric vehicles, high efficiency is desired to maximize the utilization of the vehicle's battery power, prolong driving range, and minimize energy consumption. Efficient converters help reduce heat

generation, which is essential for effective thermal management. Thermal management is vital to maintain the optimal operating temperature of the converter and associated components. Efficient heat dissipation prevents overheating, which can degrade the performance and lifespan of the converter. Thermal management techniques include the use of heat sinks, fans, liquid cooling systems, and thermal interface materials to dissipate heat effectively and maintain acceptable operating temperatures. Efficiency and thermal management are closely related since high temperatures can negatively impact efficiency. Excessive heat can increase the resistance in electronic components, leading to increased power losses and reduced efficiency. Conversely, an inefficient converter generates more heat, necessitating more robust thermal management strategies. To achieve high efficiency and effective thermal management, converter design should focus on minimizing losses through the selection of high-quality components, efficient topologies, and optimized control algorithms. Additionally, thermal management techniques should be implemented to dissipate heat efficiently, ensuring proper airflow and cooling to prevent component damage and ensure the longevity of the converter. Efficiency and thermal management directly impact the performance, reliability, and lifespan of converters in electric vehicles. By prioritizing efficiency and implementing effective thermal management strategies, converters can contribute to the overall energy efficiency of the vehicle, enhance driving range, and ensure the reliable operation of the electric vehicle's power system.

3) *Electromagnetic Interference (EMI):* Converters in EVs generate electrical noise and can interfere with other vehicle systems and external electronic devices. Proper design and implementation of EMI filters and shielding mechanisms are necessary to mitigate EMI issues (Wang, Li, Kang, & Wang, 2019). These measures ensure compliance with EMC standards and regulations, minimizing the risk of interference and maintaining the performance of other vehicle components.
4) *Protection Mechanisms:* Converters in EVs require robust protection mechanisms to safeguard against various operating conditions and potential faults. Some key protection features include:
   - *Overvoltage Protection:* Overvoltage protection is an essential feature in converters used in electric vehicles to safeguard the

system from excessive voltage levels that can occur due to various reasons. Sudden changes in the input voltage, faults within the system, or regenerative energy from braking can lead to voltage spikes that may exceed the safe operating limits of the converter and other electrical components. Overvoltage protection mechanisms are designed to detect these high voltage conditions and activate protective measures to prevent damage and ensure the safety and reliability of the system. These mechanisms typically involve voltage sensing circuits, feedback control, and protective devices such as transient voltage suppressors or voltage clamping circuits. By promptly responding to overvoltage events, the converter can limit the voltage to a safe level, preventing potential damage to sensitive components and ensuring the overall stability and longevity of the electric vehicle's power system.

- *Undervoltage Protection:* It is a crucial aspect of converters used in electric vehicles to prevent damage and ensure proper operation in the event of low voltage conditions. In electric vehicle systems, undervoltage situations can occur due to factors such as a depleted or faulty battery, voltage drops in the power distribution network, or faults in the electrical system. Undervoltage protection mechanisms are implemented to monitor the input voltage and activate protective measures when the voltage falls below a predefined threshold. These measures may include reducing power output, activating warning signals, or even initiating a safe shutdown of the converter. By detecting and responding to undervoltage conditions, the converter can prevent potential malfunctions, safeguard critical components, and maintain system integrity. Undervoltage protection is essential for the reliable and safe operation of converters in electric vehicles, ensuring that the power system functions within the specified voltage range and avoiding detrimental effects on the vehicle's performance and longevity.
- *Overcurrent Protection:* It prevents excessive current flow that can lead to component damage and system malfunction. In an electric vehicle system, overcurrent situations can occur due to factors such as short circuits, component failures, or abnormal operating conditions. Overcurrent protection mechanisms are designed to detect these high current events and activate

protective measures to limit the current to safe levels. These mechanisms typically involve current sensing circuits, feedback control, and protective devices such as fuses or circuit breakers. When an overcurrent condition is detected, the protection system responds by interrupting the current flow, isolating the faulty section, or reducing the power output. By effectively managing overcurrent situations, the converter can safeguard sensitive components, prevent damage to the system, and ensure the overall safety and reliability of the electric vehicle. Overcurrent protection is essential for maintaining the integrity and longevity of the converter and the entire electrical system, enabling smooth and efficient operation of the electric vehicle.

- *Short-Circuit Protection:* It is a critical feature in converters used in electric vehicles to prevent damage and ensure the safety and reliability of the system in the event of a short circuit fault. Short circuits occur when there is an unintended connection between two points in an electrical circuit, resulting in a significant increase in current flow. This can cause excessive heating, component damage, and even pose a fire hazard. Short-circuit protection mechanisms are designed to detect these faults and quickly interrupt the current flow to prevent further damage. Common short-circuit protection methods include the use of fuses, circuit breakers, or current-limiting devices that can detect the excessive current and open the circuit. By activating protective measures in the presence of a short circuit, the converter can isolate the fault, prevent current overload, and protect sensitive components within the electric vehicle system. Short-circuit protection is crucial for maintaining the overall safety, reliability, and performance of the converter, ensuring the smooth and uninterrupted operation of the electric vehicle.
- *Temperature Protection:* It is required to safeguard against excessive temperatures that can lead to component failure, system malfunction, and even safety hazards. In an electric vehicle, converters are subjected to varying operating conditions that can cause temperature fluctuations. Overheating can occur due to factors such as high ambient temperatures, extended periods of high load, or inadequate cooling. Temperature protection mechanisms are implemented to monitor the temperature of critical components and activate protective

measures when the temperature exceeds safe limits. These measures may include reducing power output, activating cooling systems, or initiating a safe shutdown of the converter. By effectively managing temperature levels, the converter can prevent thermal damage, ensure the longevity of components, and maintain the overall performance and reliability of the electric vehicle system. Temperature protection is essential for maintaining optimal operating conditions, avoiding thermal stress, and ensuring the safe and efficient operation of converters in EVs.

- *Fault Diagnostics and System Monitoring:* Fault diagnostics and system monitoring play a crucial role in ensuring the reliable and efficient operation of converters in electric vehicles. Faults and failures can occur in various components or subsystems of the converter, which can impact the overall performance and safety of the electric vehicle. Fault diagnostics involve the implementation of monitoring systems and algorithms that continuously analyze the converter's operation, detect abnormalities, and identify specific faults or failures. This includes monitoring key parameters such as voltage, current, temperature, and other relevant signals. System monitoring enables real-time analysis of the converter's health, allowing for the early detection and diagnosis of potential faults or malfunctions. By promptly identifying and isolating faults, system monitoring facilitates efficient troubleshooting, reduces downtime, and enables proactive maintenance. Additionally, fault diagnostics and system monitoring enable data logging and analysis, providing valuable insights into the converter's performance, helping to improve future designs and enhance overall system reliability.

5) *Reliability and Durability:* Converters in EVs must be designed to withstand the harsh operating conditions encountered in automotive environments. They are subjected to vibrations, temperature variations, moisture, and other environmental factors. Robust component selection, proper sealing, and rugged construction techniques are employed to enhance the reliability and durability of the converters. Additionally, rigorous testing and validation procedures, such as temperature cycling, humidity testing, and mechanical stress tests, are conducted to ensure the converters can

withstand the demanding operating conditions and have a long service life.

6) *Safety Considerations:* Safety is of paramount importance in electric vehicles, and converters must be designed with safety features to protect against potential hazards. This includes protection against electrical shock, fire hazards, and other risks. Safety mechanisms such as galvanic isolation, fault detection and isolation, and insulation coordination are implemented to ensure the safe operation of converters in EVs. Compliance with safety standards and regulations, such as ISO 26262 for functional safety, is essential to guarantee the highest level of safety for both occupants and service personnel.
7) *Fault Tolerance and Redundancy:* To enhance the reliability and availability of the power conversion system in EVs, fault tolerance and redundancy techniques may be employed. Fault tolerance refers to the ability of the converter to continue functioning and delivering power even in the presence of faults or failures. Redundant converters and control systems can ensure continued operation even in the event of a component failure or fault. Redundancy involves the inclusion of duplicate or backup components, circuits, or systems that can seamlessly take over the operation if a primary component or system fails. By incorporating fault tolerance and redundancy, converters can mitigate the impact of failures, improve system reliability, and minimize downtime. Redundancy can be implemented at various levels, such as redundant power modules, redundant control circuits, or even redundant converters operating in parallel. Fault tolerance and redundancy strategies provide robustness and resilience to the converter, allowing it to maintain continuous operation and support the critical functions of the electric vehicle system, enhancing safety and overall performance.
8) *Serviceability and Maintenance:* Ease of serviceability and maintenance is crucial for converters in electric vehicles. Serviceability and maintenance are essential aspects of converters used in electric vehicles to ensure their long-term reliability and functionality. Serviceability refers to the ease with which the converter can be serviced or maintained, including tasks such as inspections, repairs, and component replacements. Proper serviceability reduces downtime and enhances the overall availability of the electric vehicle. Design considerations such as accessible component layout, diagnostic capabilities, and modular designs

facilitate troubleshooting and repair activities. Maintenance involves routine activities performed to keep the converter in optimal working condition. Regular inspections, cleaning, and preventive maintenance tasks help identify potential issues before they escalate into major problems. These activities may include checking for loose connections, verifying insulation integrity, and monitoring key performance parameters. Well-designed converters include built-in diagnostic features, self-monitoring capabilities, and fault detection systems that aid in proactive maintenance planning.

9) *Cost Considerations:* Cost plays a significant role in the design and selection of converters for electric vehicles. Balancing performance, reliability, and cost-effectiveness is essential to make EVs more affordable and accessible to a wider range of consumers. Design optimizations, component selection, and manufacturing processes that streamline production and reduce material costs contribute to achieving cost-effective converter solutions.

In summary, converters in electric vehicles must meet specific requirements and be equipped with robust protection mechanisms to ensure efficient and reliable operation. Voltage and power requirements, efficiency, thermal management, EMI mitigation, protection mechanisms, reliability, safety considerations, fault tolerance, serviceability, and cost considerations are critical factors in the design and implementation of converters for electric vehicles. By addressing these considerations, manufacturers can develop high-performance converters that meet the demands of electric vehicles while ensuring safety, reliability, and compliance with industry standards and regulations.

1.6. Design Considerations

The design considerations play a crucial role in the development and implementation of power electronic converters for electric vehicles. This section explores various aspects that need to be carefully addressed during the design process to ensure optimal performance, reliability, efficiency, and cost-effectiveness of the converters. The design considerations encompass component selection, topology selection, control and modulation strategies, EMI/EMC mitigation techniques, efficiency and thermal considerations, as well as reliability, durability, and cost considerations. By making informed decisions in each of these areas, designers can create converters that meet the specific requirements of electric vehicles while adhering to industry standards and regulations. The selection of suitable components, such as power semiconductors, passive elements, and magnetics, is crucial for achieving desired performance characteristics and reliability. Choosing the appropriate converter topology is another critical aspect that determines the conversion efficiency, voltage regulation, and overall system complexity.

Control and modulation strategies define how the converter operates, including switching schemes, control algorithms, and feedback loops. These strategies are vital in achieving accurate and efficient power conversion, as well as maintaining stability and fault tolerance. EMI/EMC mitigation techniques are essential for reducing electromagnetic interference and ensuring compliance with electromagnetic compatibility standards, thus avoiding interference with other electronic systems. Efficiency and thermal considerations focus on maximizing power conversion efficiency while effectively managing heat dissipation. Optimal efficiency minimizes energy losses and extends the driving range of electric vehicles. Effective thermal management techniques, such as heat sinks, cooling systems, and thermal interface materials, ensure that the converter operates within acceptable temperature limits, preventing thermal damage and degradation. Reliability, durability, and cost considerations are paramount to ensure long-term operation and customer satisfaction. Robust design practices, fault tolerance mechanisms, and protective features contribute to the reliability and durability of the converter. Additionally, cost-effective design solutions, such as selecting cost-efficient components and optimizing manufacturing processes, enable the mass production and affordability of converters for electric vehicles.

This section investigates into each of these design considerations, providing insights, strategies, and best practices to guide the design process of power electronic converters for electric vehicles. By addressing these considerations holistically, designers can develop converters that meet performance requirements, regulatory standards, and market demands, fostering the advancement and adoption of electric vehicles.

1.6.1. Component Selection

In the design of power electronic converters for electric vehicles, component selection is a critical aspect that significantly influences the performance, reliability, and overall efficiency of the system. Choosing the right components, such as power semiconductors, passive elements, and magnetics, is essential to ensure optimal operation and longevity of the converter.

Power semiconductors, including MOSFETs, IGBTs, and diodes, play a vital role in power conversion within the converter. The selection of power semiconductors depends on factors such as voltage rating, current handling capability, switching speed, and conduction losses. Designers need to carefully evaluate these parameters to choose semiconductors that can handle the desired power levels, while also considering aspects such as cost, reliability, and availability. Advanced semiconductor technologies, such as wide bandgap (WBG) materials like silicon carbide (SiC) and gallium nitride (GaN), offer higher efficiency and power density, making them attractive options for high-performance electric vehicle converters.

Passive elements, including capacitors, inductors, and resistors, are also crucial in power electronic converter designs. Capacitors are used for energy storage and filtering purposes, while inductors help in energy transfer and magnetic field generation. The selection of passive elements depends on factors such as voltage rating, current handling capability, frequency response, and tolerance. The use of high-quality components with low equivalent series resistance (ESR) and equivalent series inductance (ESL) ensures minimal losses and improved converter efficiency. Additionally, designers must consider the size, weight, and cost of these passive elements to optimize the overall converter design.

Magnetics, such as transformers and inductors, are essential for voltage transformation and energy storage in the converter. The selection of magnetic components depends on factors such as magnetic material, core type, winding configuration, and thermal characteristics. Designers need to consider parameters such as magnetic flux density, core losses, and temperature rise to ensure efficient energy transfer and minimize losses. Proper magnetic component selection is crucial for achieving the desired voltage transformation ratios, reducing EMI, and maintaining high converter efficiency.

Apart from individual component selection, designers also need to consider the compatibility and integration of components within the overall converter system. Factors such as voltage and current ratings, switching speeds, thermal characteristics, and electrical interfaces need to be carefully evaluated to ensure harmonious operation and avoid component mismatches or failures.

Reliability is a significant consideration when selecting components for electric vehicle converters. Components with higher reliability, longer operational lifetimes, and lower failure rates contribute to the overall reliability of the converter system. Manufacturers' reputation, quality control measures, and reliability data should be considered during component selection.

Environmental considerations are also important in component selection for electric vehicle converters. Components that comply with environmental regulations, such as RoHS (Restriction of Hazardous Substances), contribute to the sustainability and eco-friendliness of the converter system.

In summary, component selection is a crucial step in the design of power electronic converters for electric vehicles. Careful evaluation of power semiconductors, passive elements, and magnetics based on factors such as voltage ratings, current handling capabilities, efficiency, reliability, cost, and environmental considerations is essential. By choosing the right components, designers can optimize the performance, reliability, and efficiency of the converter, ultimately contributing to the overall performance and success of electric vehicles.

1.6.2. Topology Selection

Topology selection is a crucial step in the design of power electronic converters for electric vehicles. The choice of converter topology determines the overall performance, efficiency, size, and complexity of the system. Different topologies have unique characteristics and are suitable for specific applications and operating conditions.

One of the commonly used topologies for DC-DC conversion in electric vehicles is the buck converter. The buck converter provides step-down voltage conversion and is suitable for applications where the input voltage is higher than the required output voltage. It offers high efficiency and simplicity, making it a popular choice for battery charging and voltage regulation in electric vehicles. Another widely used topology is the boost converter, which provides step-up voltage conversion. It is often employed in electric vehicle applications where the input voltage is lower than the required output voltage, such as in regenerative braking systems. The boost converter allows efficient energy transfer from the lower voltage source to a higher voltage storage system.

The buck-boost converter is a versatile topology that offers both step-up and step-down voltage conversion. It can handle a wide range of input and output voltages, making it suitable for electric vehicle applications where voltage variations occur, such as in hybrid or plug-in hybrid vehicles. The buck-boost converter enables efficient power transfer and voltage regulation under varying operating conditions.

In applications where bidirectional power flow is required, such as in electric vehicle charging systems, the full-bridge or H-bridge topology is commonly used. The full-bridge converter allows power to flow in both directions, enabling bidirectional power transfer between the grid and the vehicle's battery. This topology provides efficient and controllable charging and discharging capabilities.

The flyback converter is another topology that offers galvanic isolation, making it suitable for applications that require isolation between input and output voltages. This topology is commonly used in electric vehicle charging stations to provide isolation between the grid and the vehicle's battery system.

Apart from these basic topologies, more advanced topologies, such as the multi-level converters, resonant converters, and matrix converters, are also being explored for electric vehicle applications. These topologies offer benefits such as improved efficiency, reduced EMI, and enhanced power density. However, they are often more complex and require advanced control techniques. When selecting a topology for an electric vehicle converter, several factors need to be considered. These include the input and output voltage requirements, current rating, efficiency targets, power density requirements, size and weight constraints, cost considerations, and reliability requirements. Additionally, factors such as the availability of components, compatibility with other system components, and suitability for mass production should also be taken into account.

Simulation tools and modeling techniques can aid in the evaluation and selection of the most appropriate topology for a specific application. By considering the system requirements and constraints, designers can assess the advantages and disadvantages of different topologies and choose the one that best meets the performance and operational needs of the electric vehicle. In conclusion, topology selection is a critical aspect of designing power electronic converters for electric vehicles. The choice of topology determines the fundamental characteristics of the converter and influences its performance, efficiency, size, and complexity. Designers need to carefully evaluate the application requirements, operating conditions, and system constraints to select the optimal topology that maximizes performance, efficiency, and reliability while meeting the specific needs of electric vehicle applications.

1.6.3. Control and Modulation Strategies

Control and modulation strategies play a crucial role in the operation of power electronic converters for electric vehicles. These strategies ensure efficient energy conversion, accurate voltage and current regulation, and reliable operation of the converter system. Various control and modulation techniques are employed to achieve desired converter performance and meet the specific requirements of electric vehicle applications.

One of the key control strategies used in power electronic converters is PWM. PWM is a technique that controls the switching of power semiconductor devices to regulate the output voltage or current. By varying the duty cycle of the switching signal, the average output voltage or current can be adjusted. PWM offers several advantages, including precise control, low harmonic distortion, and high efficiency. It is widely used in voltage source inverters, current source inverters, and DC-DC converters for electric vehicles.

In voltage source inverters, PWM techniques such as carrier-based PWM and SVM are commonly employed. Carrier-based PWM involves comparing a high-frequency carrier waveform with a reference waveform to generate the switching signals for the power devices. This technique allows precise control of the output voltage magnitude and frequency. SVM, on the other hand, provides better utilization of the available DC link voltage and enables maximum voltage utilization in the inverter. It offers improved harmonic performance and better utilization of the switching devices.

For current source inverters, hysteresis current control is often used. This control strategy compares the actual current with predefined upper and lower limits, and based on the comparison, generates the switching signals to maintain the desired current level. Hysteresis current control offers fast dynamic response and inherent current limiting capability, making it suitable for applications such as motor drives in electric vehicles.

In DC-DC converters, control strategies such as voltage-mode control and current-mode control are commonly employed. Voltage-mode control regulates the output voltage by comparing it with a reference voltage and adjusting the duty cycle of the switching signal. Current-mode control, on the other hand, regulates the output current by sensing the inductor current and adjusting the duty cycle accordingly. Both control techniques provide accurate voltage or current regulation and can be adapted to different converter topologies.

In addition to PWM techniques, advanced control strategies such as predictive control, sliding mode control, and model predictive control are also being explored for power electronic converters in electric vehicles. These control techniques offer improved performance, faster response, and better dynamic behaviour. They utilize mathematical models and predictive algorithms to optimize the converter operation and achieve optimal control objectives.

Modulation strategies are closely related to control strategies and are used to generate the switching signals for the power devices in the converter. The choice of modulation strategy depends on the converter topology and the desired performance requirements. Some common modulation strategies include SPWM, space vector pulse width modulation (SVPWM), and selective harmonic elimination (SHE). These strategies aim to reduce harmonics, improve efficiency, and minimize switching losses in the converter system.

Moreover, advanced modulation techniques such as multi-carrier modulation, hybrid modulation, and advanced hybrid modulation are also being investigated for their potential benefits in electric vehicle applications. These techniques aim to optimize the harmonic content, minimize the total harmonic distortion, and improve overall converter efficiency. The selection of control and modulation strategies depends on various factors including the converter topology, performance requirements, system constraints, and cost considerations. Simulation tools and modeling techniques play a crucial role in evaluating the performance of different control and modulation strategies and selecting the most suitable approach for the specific electric vehicle application.

1.6.4. EMI/EMC Mitigation Techniques

EMI and EMC mitigation techniques are crucial considerations in the design of power electronic converters for electric vehicles. EMI refers to the unwanted electromagnetic emissions that can interfere with the proper functioning of other electronic devices, while EMC refers to the ability of a system or device to operate without experiencing or causing interference with other nearby devices.

The increasing use of power electronic converters in electric vehicles has led to a higher risk of EMI generation and potential interference with other vehicle systems, communication networks, and nearby electronic devices. Therefore, it is essential to implement effective mitigation techniques to ensure compliance with electromagnetic compatibility standards and regulations. There are several techniques employed to mitigate EMI and achieve EMC in power electronic converters (also shown in Figure 1.20) (Yazdani, Farzanehfard, & Faiz, 2011):

1) *Filtering:* Filtering is one of the most common techniques used to reduce conducted and radiated EMI. It involves the use of passive components such as capacitors, inductors, and ferrite beads to suppress high-frequency noise. Filters can be placed at input and output stages of the converter, as well as on signal lines and power lines.
2) *Shielding:* Shielding involves the use of conductive enclosures or shields to block electromagnetic fields and prevent EMI propagation. The shields can be made of materials like metal or conductive composites and are placed around sensitive components or circuitry. Proper grounding and bonding techniques are also crucial to ensure effective shielding.
3) *Grounding and Bonding:* Proper grounding and bonding techniques are essential for minimizing EMI. They involve establishing low impedance paths for the flow of electrical currents and preventing the build-up of ground loops. By ensuring a solid and low-impedance ground connection, the risk of noise coupling and EMI generation can be reduced.
4) *Layout and PCB Design:* Careful layout and PCB design play a significant role in EMI/EMC mitigation. Separating high and low current traces, minimizing loop areas, and using proper decoupling techniques can help reduce electromagnetic emissions and improve overall system performance.
5) *Snubber Circuits:* Snubber circuits are used to suppress voltage spikes and reduce high-frequency noise generated during switching events. These circuits typically consist of passive components such as resistors, capacitors, and diodes, and are connected across switching devices to dampen voltage transients and minimize EMI.
6) *Grounding Filters:* Grounding filters are used to filter out high-frequency noise present on the ground line. They help prevent noise

coupling between different sections of the converter and reduce the risk of common-mode EMI.

7) *Component Selection:* The choice of components can also impact EMI/EMC performance. Selecting components with low parasitic capacitance, inductance, and resistance can help reduce noise generation and improve overall system stability.
8) *Compliance Testing:* Conducting comprehensive EMI/EMC compliance testing is crucial to ensure that the converter meets the required electromagnetic compatibility standards and regulations. Testing involves evaluating the conducted and radiated emissions and verifying that the converter operates within acceptable limits.

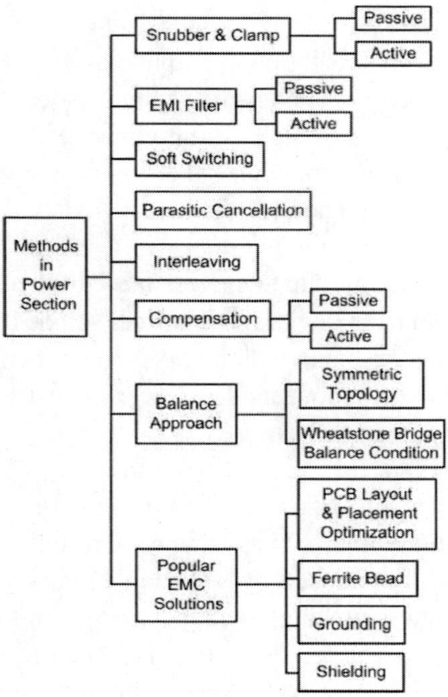

Figure 1.20. EMI/EMC mitigation techniques.

It is important to note that EMI/EMC mitigation should be considered throughout the entire design process, from the initial schematic design to the layout, component selection, and system integration. Simulation tools and modeling techniques can aid in predicting and analyzing the potential EMI issues, allowing for optimization and refinement of the design. By implementing effective EMI/EMC mitigation techniques, power electronic converters in electric vehicles can operate without interfering with other vehicle systems or nearby electronic devices, ensuring reliable and safe operation of the vehicle and maintaining compatibility with electromagnetic compatibility standards.

1.6.5. Efficiency and Thermal Considerations

Efficiency and thermal considerations are critical aspects in the design of power electronic converters for electric vehicles. These converters play a crucial role in the energy conversion process, and their efficiency and thermal management directly impact the overall performance, reliability, and lifespan of the system.

Efficiency refers to the ratio of output power to input power, and it determines how effectively the converter converts electrical energy from one form to another. In electric vehicles, high converter efficiency is desired to minimize energy losses and maximize the driving range. Factors that affect converter efficiency include conduction losses, switching losses, and control losses.

Conduction losses occur due to the resistance of the conducting elements, such as the semiconductor devices (transistors, diodes) and the conducting paths (copper traces, busbars) within the converter. These losses can be minimized by selecting components with lower on-resistance and using wider conductive paths to reduce resistance.

Switching losses occur during the transitions of the semiconductor devices when they switch on or off. These losses are primarily due to the energy dissipated in the internal capacitance and inductance of the devices. Switching losses can be reduced by selecting devices with lower switching losses, optimizing the switching frequency, and employing soft switching techniques such as zero-voltage switching or zero-current switching.

Control losses refer to the power dissipated in the control circuitry of the converter, including the gate drivers and control ICs. Minimizing control losses involves selecting efficient control algorithms, optimizing the control circuit design, and utilizing low-power consumption components.

Thermal considerations are crucial to ensure the safe and reliable operation of power electronic converters. As power is converted and transferred, heat is generated, and if not properly managed, it can lead to component overheating and degradation, which can significantly affect the converter's performance and reliability.

Thermal management techniques involve dissipating the heat generated by the converter to prevent temperature rise and maintain the components within their safe operating limits. Some common thermal management techniques include:

1) *Heat Sink:* Heat sinks are passive cooling devices that absorb and dissipate heat from high-temperature components, such as power semiconductors. They enhance the surface area for heat transfer and improve convective cooling. Proper heat sink design, including thermal conductivity, size, and mounting, is crucial for efficient heat dissipation.
2) *Thermal Interface Materials (TIM):* TIMs are used to improve the thermal coupling between components and heat sinks, reducing the thermal resistance and enhancing heat transfer. Examples of TIMs include thermal greases, thermal pads, and phase-change materials.
3) *Cooling Methods:* Active cooling methods, such as forced-air cooling or liquid cooling, can be employed to remove heat from the converter. Forced-air cooling uses fans or blowers to enhance convective cooling, while liquid cooling utilizes coolants circulating through channels or heat exchangers. Liquid cooling is more effective in dissipating higher heat loads but requires additional components and infrastructure.
4) *Thermal Modeling and Simulation:* Thermal modeling and simulation techniques can be employed to predict and analyze the temperature distribution and thermal behaviour of the converter under various operating conditions. This helps in optimizing the design, identifying potential hotspots, and validating thermal management strategies.

Efficiency and thermal considerations should be addressed in the early stages of converter design to optimize the system's overall performance and reliability. By selecting efficient components, optimizing control algorithms, implementing effective thermal management techniques, and conducting thermal simulations, power electronic converters can achieve high efficiency, minimize energy losses, and maintain safe operating temperatures, contributing to the overall efficiency and longevity of electric vehicle systems.

1.6.6. Reliability, Durability and Cost Considerations

Reliability, durability, and cost considerations are essential aspects in the design of power electronic converters for electric vehicles. These factors play a crucial role in ensuring the long-term performance, safety, and economic viability of the converter system.

Reliability refers to the ability of the converter to consistently perform its intended function without failure over a specified period of time. In the context of electric vehicles, reliability is crucial to ensure continuous and uninterrupted operation, as any converter failure can lead to vehicle downtime and inconvenience for the user. To enhance reliability, several design considerations need to be taken into account:

1) *Component Selection:* Choosing high-quality and reliable components is essential. This includes selecting reputable manufacturers, considering component ratings and specifications, and ensuring compatibility with the intended application and operating conditions. Components with established reliability metrics, such as mean time between failures (MTBF), can provide valuable insights into their expected lifespan.
2) *Robust Design:* The converter should be designed to withstand the harsh operating conditions encountered in electric vehicles, including temperature variations, vibration, and electrical noise. Adequate mechanical and electrical protection measures, such as proper enclosure design, secure component mounting, and isolation techniques, should be implemented to enhance durability and protect against environmental stresses.
3) *Testing and Validation:* Rigorous testing and validation procedures should be conducted during the design and manufacturing stages to identify and rectify any potential issues. This includes functional testing, stress testing under various operating conditions, and reliability testing to ensure the converter meets the specified performance and reliability requirements.

Durability refers to the ability of the converter to withstand mechanical and environmental stresses throughout its operational lifespan. In the context of electric vehicles, where the converter is subjected to constant vibrations, temperature fluctuations, and thermal cycling, durability is of utmost importance. Key considerations for ensuring durability include:

1) *Robust Construction:* The converter should be designed with robust construction techniques, utilizing sturdy materials, proper soldering techniques, and secure component mounting. This ensures that the converter can withstand mechanical shocks and vibrations without compromising its functionality.
2) *Thermal Management:* Effective thermal management techniques, as discussed in the previous section, play a crucial role in enhancing durability. By maintaining the components within their safe operating temperature limits, thermal stresses and fatigue can be minimized, prolonging the converter's lifespan.
3) *Environmental Protection:* The converter should be adequately protected against environmental factors such as moisture, dust, and chemical exposure. This can be achieved through proper sealing and conformal coating of sensitive components, as well as adherence to relevant industry standards and regulations for environmental protection.

Cost considerations are also important in the design of power electronic converters for electric vehicles. While ensuring reliability and durability, it is crucial to optimize the cost of the system to make it economically viable. Some key cost considerations include:

1) *Component Cost:* The selection of components should strike a balance between cost and quality. While it may be tempting to opt for lower-cost components, it is important to assess their reliability and performance to avoid compromising the overall system reliability.
2) *Manufacturing Cost:* Design choices and manufacturing processes should be optimized to minimize production costs. This includes considering factors such as component count, assembly complexity, and production efficiency.
3) *Serviceability and Maintenance:* The ease of servicing and maintenance can have a significant impact on the overall cost of the system. Designing the converter for easy access, modular components, and standardized maintenance procedures can help reduce downtime and minimize repair costs.

By considering reliability, durability, and cost throughout the design process, power electronic converters for electric vehicles can be optimized to deliver long-lasting and cost-effective solutions. This ensures the converters can meet the demanding requirements of electric vehicle applications while providing a reliable and economical solution for the end-users.

1.7. Applications

Power electronic converters play a vital role in various applications within electric vehicles. Application and interfacing of power electronic converters in EVs is illustrated in Figure 1.21. They enable efficient and reliable conversion of electrical power, allowing for optimal performance and functionality. In this section, some key applications such as battery charging, motor drive, regenerative braking, and on-board power supply are discussed. Each application highlights the significance of power electronic converters in the electrification of transportation and the advancement of electric vehicles (Abu Rub, Malinowski, & Al Haddad, 2014), (Rafin, Islam, & Mohammed, 2023).

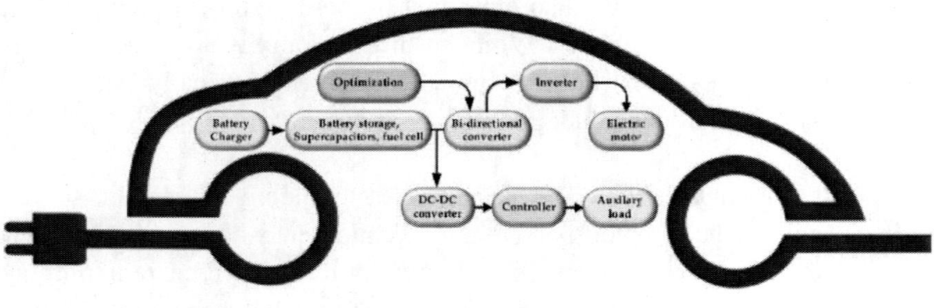

Figure 1.21. Application and interfacing of power electronic converters in EVs (Lipu, et al., 2022).

1.7.1. Battery Charging

Battery charging is a critical application in EVs that relies on power electronic converters to efficiently convert and regulate the electrical power supply. The charging process involves converting AC from the grid or charging station into DC to replenish the EV's battery pack. This conversion is essential for efficient and safe charging, as well as extending the battery life.

Power electronic converters used in EV battery charging systems employ various topologies and control strategies to meet the specific requirements of charging infrastructure and the EV battery pack. The most common charging standards include AC Level 1, AC Level 2, and DC fast charging.

AC Level 1 charging, also known as trickle charging, utilizes a simple power electronic converter to convert the AC power from a standard household outlet into DC power. It typically operates at a voltage of 120V with a low charging current, resulting in a slow charging rate. AC Level 1 charging is commonly used for overnight charging when the vehicle is parked for an extended period.

AC Level 2 charging is the most prevalent charging standard for residential, commercial, and public charging stations. It requires a power electronic converter that operates at higher voltage levels, typically 240V, allowing for faster charging rates. The converter efficiently converts the AC power from the grid into DC power, which is then supplied to the EV battery pack. AC Level 2 charging stations are commonly found in parking lots, shopping centers, and other public locations, offering convenient and relatively fast charging options for EV owners.

DC fast charging, also known as Level 3 charging, is designed for rapid charging on the go. It utilizes a more complex power electronic converter to directly convert the high-voltage DC power from the charging station into the appropriate voltage and current levels required by the EV battery pack. DC fast charging stations can charge an EV battery to 80% capacity in a significantly shorter time compared to AC charging. These stations are typically found along highways and major travel routes, providing quick charging options for long-distance travel.

The efficiency of power electronic converters used in EV battery charging is a crucial consideration. Higher efficiency translates to reduced energy losses during the conversion process, which enhances the overall charging efficiency and reduces the cost of charging. Power electronic converters employ advanced control techniques, such as PWM, to regulate the voltage and current output, ensuring efficient power transfer and minimizing losses.

Safety is another important aspect of EV battery charging. Power electronic converters incorporate protection mechanisms to prevent overvoltage, overcurrent, and overheating, safeguarding both the charging infrastructure and the EV battery pack. These protection features ensure safe and reliable charging, protecting against potential hazards and extending the lifespan of the battery.

Moreover, power electronic converters used in EV battery charging systems often incorporate communication protocols, such as the Open Charge Point Protocol (OCPP), to facilitate communication between the charging station and the EV. This communication enables features such as authentication, billing, and remote monitoring, enhancing the overall user experience and enabling efficient management of the charging infrastructure. In addition, power electronic converters enable the efficient conversion and regulation of electrical power, ensuring safe and reliable charging for EV owners. The different charging standards, such as AC Level 1, AC Level 2, and DC fast charging, require specific converter topologies and control strategies to meet the diverse charging needs of EVs. With ongoing advancements in power electronics technology, the efficiency, speed, and convenience of EV battery charging continue to improve, further driving the adoption of electric vehicles and contributing to a sustainable transportation future.

1.7.2. Motor Drive

The motor drive is a fundamental application of power electronic converters. Power electronic converters are responsible for controlling the electric motor that propels the vehicle, providing the necessary torque and speed control for optimal performance. The motor drive system consists of various components, including power converters, motor control algorithms, and feedback sensors, working together to convert electrical energy into mechanical motion.

Power electronic converters used in EV motor drives typically employ VSI or CSI to convert DC power from the vehicle's battery pack into AC power that drives the electric motor. The inverter's main function is to control the voltage, frequency, and waveform of the output AC power, allowing precise control of the motor's speed and torque.

In EV motor drives, precise and efficient control is crucial to ensure smooth acceleration, deceleration, and overall vehicle performance. Power electronic converters utilize advanced control strategies to regulate the output voltage and frequency of the inverter. These control techniques enable precise control of the motor's speed and torque, ensuring optimal performance and responsiveness.

Regenerative braking is another important aspect of EV motor drives. During braking, the motor can act as a generator, converting the kinetic energy of the vehicle into electrical energy. Power electronic converters play a critical role in the regenerative braking process by redirecting the generated electrical energy back to the battery pack for storage and reuse. This regenerative braking capability improves energy efficiency and extends the vehicle's range.

Efficiency is a key consideration in motor drives to maximize the vehicle's range and minimize energy losses. Power electronic converters in EV motor drives are designed to have high efficiency, converting electrical energy with minimal losses. This is achieved through careful component selection, advanced control algorithms, and optimized design techniques. Higher efficiency translates to improved overall vehicle performance, reduced energy consumption, and longer battery life.

Reliability and durability are essential considerations for power electronic converters in EV motor drives. These components must be designed to withstand harsh operating conditions, including temperature variations, vibrations, and electrical stresses. Robust design techniques, proper cooling systems, and robust protection mechanisms are implemented to ensure reliable and durable operation of the motor drive system.

Cost is also a significant factor in the design of power electronic converters for EV motor drives. As electric vehicle technology advances and becomes more widespread, cost reduction is crucial to drive mass adoption. Power electronic converters need to be designed with cost-effective components, manufacturing processes, and system integration methods, without compromising performance and reliability. They enable precise control of the motor's speed and torque, ensuring optimal vehicle performance and responsiveness. With advanced control strategies, regenerative braking capabilities, and a focus on efficiency, reliability, durability, and cost-effectiveness, power electronic converters contribute to the overall success of electric vehicles and their transition toward sustainable transportation solutions.

1.7.3. Regenerative Braking

Regenerative braking is a key feature in electric vehicles that allows the conversion of kinetic energy into electrical energy during braking. Unlike traditional vehicles that rely solely on friction-based braking systems, EVs utilize regenerative braking to recover energy that would otherwise be wasted as heat. This energy is then stored in the vehicle's battery pack for later use, effectively extending the vehicle's range and improving overall energy efficiency.

Regenerative braking in EVs is made possible through the use of power electronic converters. When the driver applies the brakes, the electric motor in the vehicle operates in reverse as a generator, converting the kinetic energy of the moving vehicle into electrical energy. The power electronic converters facilitate this process by converting the generated electrical energy from AC to DC and directing it to the battery pack for storage.

The regenerative braking system involves a complex control mechanism to seamlessly transition between the mechanical braking system and the regenerative braking mode. The power electronic converters, along with the motor control algorithms, play a critical role in regulating the power flow and ensuring a smooth transition between the two modes. This control mechanism is responsible for maintaining vehicle stability and providing a consistent braking experience for the driver.

One of the significant advantages of regenerative braking is its ability to improve energy efficiency in EVs. By recovering and storing energy that would otherwise be wasted, regenerative braking allows EVs to travel longer distances on a single charge. This energy efficiency not only reduces the need for frequent charging but also contributes to a more sustainable and environmentally friendly mode of transportation.

Furthermore, regenerative braking has a positive impact on the overall performance of the vehicle. The regenerative braking system provides additional braking force, which reduces wear and tear on the traditional mechanical braking system. This can lead to extended brake pad life and reduced maintenance costs for EV owners. Additionally, regenerative braking improves the driving experience by offering smoother and more controlled deceleration, enhancing driver comfort and safety.

However, it is important to note that regenerative braking may have limitations in certain driving conditions. For example, at high speeds or when the battery pack is fully charged, the ability to recover and store energy through regenerative braking may be limited. In such cases, traditional friction-based braking systems are still necessary to ensure safe and effective braking.

Overall, regenerative braking is a significant technological advancement in electric vehicles, made possible by power electronic converters. It allows for the recovery and utilization of energy that would otherwise be lost during braking, leading to increased energy efficiency, improved vehicle performance, and reduced environmental impact. As EV technology continues to evolve, further advancements in regenerative braking systems and power electronic converters are expected, driving the adoption and advancement of sustainable transportation solutions.

1.7.4. On-Board Power Supply

The on-board power supply system in electric vehicles plays a crucial role in providing electrical power to various vehicle components and systems (Pradhan, Keshmiri, & Emadi, 2023). It ensures the reliable and efficient operation of these components, ranging from lighting and infotainment systems to auxiliary power outlets and climate control. The on-board power supply system relies on power electronic converters to convert the high-voltage DC power from the battery pack into the required voltages and waveforms suitable for different vehicle loads.

One of the primary functions of the on-board power supply system is to regulate and distribute electrical power to the different subsystems and components in the vehicle. Power electronic converters, such as DC-DC converters and inverters, are utilized to step-down or step-up the voltage as needed, ensuring compatibility with the various electrical loads. These converters employ control strategies to maintain stable output voltages and provide the necessary power levels demanded by the vehicle's electrical systems.

The on-board power supply system also facilitates the charging of auxiliary batteries or devices within the vehicle. It may include outlets or USB ports that allow passengers to connect and charge their personal electronic devices, such as smartphones or tablets. Power electronic converters are responsible for regulating the charging process, providing the appropriate voltage and current levels to ensure safe and efficient charging.

Another important aspect of the on-board power supply system is its integration with the vehicle's energy management system. The power electronic converters, along with sophisticated control algorithms, enable intelligent energy management, optimizing the allocation of power based on priority and demand. This allows for efficient utilization of the available electrical energy and extends the vehicle's overall range.

Additionally, the on-board power supply system must comply with various safety standards and regulations to ensure the protection of vehicle occupants and prevent electrical hazards. Power electronic converters incorporate protection mechanisms, such as overcurrent protection, short-circuit protection, and thermal management, to safeguard against potential faults or failures. These measures help to maintain the reliability and durability of the on-board power supply system, ensuring its safe operation throughout the vehicle's lifetime.

The advancements in power electronic converters and on-board power supply systems have opened up new possibilities for integrating advanced features and functionalities into electric vehicles. For example, the on-board power supply system can support advanced driver-assistance systems (ADAS) by supplying power to sensors, cameras, and processing units. It can also facilitate the integration of electric heating and cooling systems, improving passenger comfort while minimizing the impact on the vehicle's driving range.

Furthermore, the on-board power supply system is crucial for enabling future innovations in electric vehicles, such as vehicle-to-grid technology. The V2G allows EVs to not only consume electrical energy but also feed excess energy back into the grid when needed. Power electronic converters and intelligent control systems play a vital role in managing the bidirectional power flow between the vehicle and the grid, enabling vehicle owners to actively participate in grid stabilization and demand response programs.

In summary, the on-board power supply system in electric vehicles is responsible for regulating and distributing electrical power to various vehicle components and subsystems. Power electronic converters play a key role in this system, ensuring efficient power conversion, intelligent energy management, and compliance with safety standards. The on-board power supply system enables the charging of auxiliary devices, supports advanced vehicle features, and paves the way for future innovations in electric vehicle technology.

1.8. Future Trends and Challenges

The field of power electronic converters for electric vehicles is continuously evolving, driven by the need for increased efficiency, improved performance, and enhanced functionalities. As electric vehicles become more widespread and the demand for sustainable transportation grows, several future trends and challenges emerge. This section explores some of these trends and challenges, highlighting the direction in which the technology is headed and the obstacles that need to be overcome (Trivedi & Sant, 2022), (Aghabali, et al., 2021), (Pradhan, Keshmiri, & Emadi, 2023).

1.8.1. Increased Power Density

One of the key future trends in power electronic converters for electric vehicles is the drive towards higher power density. Increased power density refers to the ability to deliver higher power levels within a smaller physical footprint. In the context of power electronic converters for electric vehicles, increasing power density is a crucial research direction and recommendation. By improving power density, converters can be made more compact and lightweight, allowing for better integration into the limited space available in electric vehicles. As electric vehicles strive for longer driving ranges and faster charging times, power converters need to become more compact and lightweight while maintaining high efficiency.

Achieving increased power density involves various aspects of converter design, such as optimizing component selection, improving thermal management techniques, and exploring advanced packaging technologies. For instance, advancements in semiconductor technology, such as the use of wide-bandgap materials like SiC and GaN, can enable higher switching frequencies and reduce the size and weight of power electronic components.

Additionally, improvements in magnetic components, such as inductors and transformers, can contribute to higher power density by reducing their size and losses. Advanced cooling techniques, such as liquid cooling or innovative heat sink designs, can enhance thermal management and allow for higher power handling capabilities. Increasing power density brings several benefits to electric vehicles. It allows for the design of more compact and lightweight power systems, which can improve overall vehicle efficiency and performance. Moreover, it can free up valuable space within the vehicle for other components or provide flexibility in vehicle design.

However, achieving higher power density is not without challenges. It requires careful consideration of thermal management to prevent excessive heat generation and ensure reliable operation. It also necessitates addressing potential issues related to EMI and EMC to maintain system reliability and compliance with regulations. To advance the research on increased power density, interdisciplinary collaboration between power electronics engineers, materials scientists, thermal management experts, and packaging specialists is crucial. Moreover, exploring innovative design approaches, advanced materials, and advanced manufacturing techniques will contribute to achieving the goal of increased power density in power electronic converters for electric vehicles.

1.8.2. Integration and System-Level Optimization

Future power electronic converters will likely see increased integration, with multiple functions and components combined into single units. This integration can lead to more efficient designs, reduced size, and improved reliability. Additionally, system-level optimization will become crucial, considering the interaction between power converters, energy storage systems, and other vehicle subsystems to achieve optimal overall performance and energy efficiency.

Integration and system-level optimization involve the seamless integration of power electronic converters into the overall electric vehicle system and the optimization of their performance in conjunction with other vehicle components. It goes beyond individual converter design and focuses on achieving optimal system efficiency, reliability, and functionality. The integration aspect emphasizes the need to consider the converter's interactions with other vehicle systems, such as the battery management system, motor controller, and energy storage systems. This requires a holistic approach to system design, where the converters are optimized to work synergistically with other components, minimizing losses, and maximizing overall system efficiency. System-level optimization also involves the development of advanced control strategies that coordinate the operation of different converters and system components. This includes intelligent power management algorithms, adaptive control techniques, and predictive control methods. By optimizing the coordination and interaction of various converters, the system can dynamically adapt to different driving conditions, improve energy efficiency, and enhance overall vehicle performance.

Furthermore, integration and system-level optimization can benefit from advancements in communication and networking technologies. By implementing intelligent communication protocols and data exchange between converters and other vehicle systems, real-time monitoring, diagnostics, and optimization can be achieved. This enables proactive maintenance, fault detection, and predictive analytics, leading to increased system reliability and reduced downtime. To pursue integration and system-level optimization, collaboration among different stakeholders, including vehicle manufacturers, power electronics suppliers, and research institutions, is essential. Furthermore, standardization efforts in terms of communication protocols, system interfaces, and interoperability will facilitate seamless integration and optimization across different vehicle platforms and manufacturers.

The ultimate goal of integration and system-level optimization is to create highly efficient and reliable electric vehicle power systems that maximize performance, range, and user experience. By leveraging the potential synergies and optimizing the interactions among different components, converters can be effectively integrated into the broader electric vehicle ecosystem, leading to enhanced overall system performance and driving the advancement of electric vehicle technology as a whole.

1.8.3. Bi-Directional Power Flow

Bi-directional power flow refers to the ability of power electronic converters to transfer power bidirectionally between different energy sources and loads. Currently, most EVs utilize converters primarily for power delivery from the battery to the motor during vehicle propulsion. However, future research and development efforts should focus on enabling bi-directional power flow capabilities in converters for various applications and benefits. As electric vehicles become an integral part of the smart grid, bidirectional power flow between the vehicle and the grid will gain prominence. This allows for V2G applications, where electric vehicles can not only consume energy from the grid but also feed excess energy back into it during peak demand periods. By utilizing the energy stored in EV batteries during periods of low demand, EVs can serve as flexible energy storage resources and contribute to grid stability and power management. This capability opens up possibilities for demand response, peak shaving, and grid support services, while also providing potential economic benefits for EV owners through revenue generation from selling surplus power back to the grid. Another application of bi-directional power flow is vehicle-to-home (V2H) integration, which enables EVs to serve as temporary power sources for homes during power outages or during peak energy demand periods. This capability not only provides backup power for essential household appliances but also allows for optimized energy management and increased self-consumption of renewable energy sources, such as solar power.

Furthermore, bi-directional power flow enables efficient energy management within the vehicle itself. It allows for energy regeneration during deceleration and braking, where the kinetic energy of the vehicle is converted back into electrical energy and stored in the battery. This regenerative braking capability improves overall energy efficiency and extends the vehicle's driving range. To enable bi-directional power flow, research efforts should focus on developing advanced control strategies, bidirectional power converter topologies, and energy management algorithms. These should ensure efficient power transfer, safe operation, and seamless integration with the grid and other power sources. Additionally, standardization of bi-directional power flow interfaces and protocols will be crucial to ensure interoperability and compatibility between different vehicle models and charging infrastructure.

Overall, bi-directional power flow in converters presents exciting opportunities for optimizing energy usage, grid integration, and enhancing the overall value proposition of electric vehicles. Future research should aim to advance this capability and explore its full potential in enabling a more sustainable and resilient energy ecosystem.

1.8.4. Wireless Charging

Wireless charging, also known as inductive charging or wireless power transfer, is an emerging technology that offers convenience and efficiency in charging EVs. It eliminates the need for physical cables and connectors, allowing for a seamless and effortless charging experience. Wireless charging offers convenience and eliminates the need for physical connectors. Power electronic converters will need to be designed to support wireless charging systems, allowing efficient transfer of power between the charging pad on the ground and the vehicle. This technology will require advancements in converter efficiency, alignment optimization, and standardization of charging protocols. Future research in wireless charging for EVs can focus on several aspects to further enhance its adoption and effectiveness.

One area of research is improving the efficiency and power transfer capability of wireless charging systems. Currently, wireless charging systems may have lower efficiency compared to traditional wired charging methods. Therefore, researchers can explore new designs and technologies to increase the efficiency of wireless power transfer, minimizing energy losses and optimizing charging times. This may involve advancements in coil design, resonant charging techniques, and intelligent control algorithms. Another important research direction is standardization and interoperability of wireless charging systems. Currently, there are different wireless charging technologies and standards available, which may lead to fragmentation and compatibility issues. Developing common standards and protocols for wireless charging can promote widespread adoption and interoperability among different vehicle models and charging infrastructure. Standardization efforts can also focus on safety aspects, such as ensuring EMC and alignment between charging pads and EVs.

Integration of wireless charging infrastructure into roadways and parking lots is another area of research. This concept, known as dynamic or in-motion wireless charging, aims to charge EVs while they are in motion, eliminating the need for frequent stops at charging stations. Research can focus on developing efficient power transfer mechanisms and infrastructure design to support dynamic charging. This can involve optimizing the charging infrastructure layout, managing power distribution, and ensuring safety measures for vehicles and pedestrians. Furthermore, research can explore the feasibility and implementation of bidirectional wireless charging, allowing EVs to not only receive power wirelessly but also transfer power back to the grid or other energy storage systems. This bi-directional capability can enable V2G integration and enhance the overall energy management and grid stability. Additionally, research efforts can be directed towards enhancing the user experience of wireless charging. This includes developing intuitive and user-friendly charging interfaces, integrating wireless charging systems with vehicle telematics and smart grid technologies, and addressing potential concerns such as electromagnetic radiation and the impact on battery health.

Overall, future research in wireless charging for EVs should focus on improving efficiency, standardization, integration, bidirectional capability, and user experience. Advancements in these areas can accelerate the adoption of wireless charging and contribute to the growth of the electric vehicle ecosystem.

1.8.5. Advanced Control Strategies

The development of advanced control strategies for power electronic converters is another significant future trend. These strategies aim to enhance the overall system performance, optimize energy management, and ensure stability and reliability. Potential research directions and recommendations for advanced control strategies in EV power converters include investigating the application of MPC techniques for improved dynamic response and optimal operation, exploring adaptive control techniques to adapt to changing conditions, and utilizing artificial intelligence and machine learning for intelligent control. Additionally, developing multi-objective control strategies, fault-tolerant control approaches, and coordinated control of multiple converters within the EV system are important areas to focus on. Real-time control and hardware implementation, robust control techniques, hybrid control approaches, and control strategies for emerging converter topologies also warrant attention. By addressing these research areas, advancements can be made in optimizing EV power converters and enhancing overall system performance and efficiency.

Hence, future research in advanced control strategies for EV power converters should focus on developing innovative and intelligent control techniques that can optimize system performance, ensure robust and reliable operation, and accommodate the increasing complexity and demands of modern electric vehicles.

1.8.6. Enhanced Fault Diagnosis and Prognostics

With the increasing complexity of power electronic converter systems, the ability to diagnose faults and predict failures becomes essential. Future converters will incorporate advanced fault detection and diagnostic techniques, enabling proactive maintenance and minimizing downtime. Prognostics algorithms will provide real-time health monitoring and predictive analytics to anticipate potential failures and take corrective actions.

Future research in this area can focus on the development of enhanced fault diagnosis and prognostic techniques that leverage advancements in data analysis, machine learning, and artificial intelligence. These techniques can be applied to various aspects of the converter system, including components such as semiconductors, capacitors, and inductors, as well as system-level faults such as short circuits, open circuits, and component degradation.

One research direction is the utilization of advanced signal processing and machine learning algorithms for fault detection and diagnosis. These algorithms can analyze the electrical signals and waveforms from the converter system to identify abnormal patterns or deviations indicative of faults. By training models on a comprehensive dataset of fault signatures, these algorithms can accurately classify and diagnose faults in real-time, allowing for prompt remedial actions. Another important aspect is the development of predictive maintenance techniques for power electronic converters. Prognostic algorithms can utilize historical operational data and real-time sensor measurements to estimate the remaining useful life of converter components and predict their failure probabilities. This information can be utilized to schedule maintenance activities, replace components before failure, and optimize the overall maintenance strategy, leading to improved reliability and reduced maintenance costs.

Furthermore, integrating advanced fault diagnosis and prognostics with the overall vehicle health management system can enable comprehensive monitoring and management of the converter system. This integration can facilitate the collection of real-time data from multiple sensors and systems, enabling a holistic view of the converter's health and performance. By combining data from various sources, such as temperature sensors, current sensors, and voltage sensors, advanced algorithms can provide accurate and reliable fault diagnosis and prognostics. In addition to algorithmic advancements, future research should also focus on the development of robust and reliable sensor technologies for fault detection and monitoring. This includes the exploration of novel sensor designs, such as smart sensors or integrated health monitoring systems, that can provide accurate and real-time information about the converter system's condition.

1.8.7. Cost Reduction

Cost reduction is a persistent challenge in the development of power electronic converters for electric vehicles. As the market expands and economies of scale are achieved, the cost of components and manufacturing processes is expected to decrease. Additionally, advancements in semiconductor technologies, improved converter designs, and efficient manufacturing techniques will contribute to cost reduction.

Furthermore, exploring advancements in manufacturing technologies can lead to cost reduction. For instance, the adoption of advanced manufacturing techniques like additive manufacturing (3D printing) can offer benefits such as reduced material waste, faster production cycles, and greater design flexibility. By leveraging these technologies, manufacturers can streamline production processes, reduce labor costs, and ultimately lower the overall cost of the converter. In addition, collaborative efforts between researchers, manufacturers, and policymakers are crucial for cost reduction. By fostering partnerships, it becomes possible to share knowledge, resources, and expertise, leading to more efficient research and development processes. Furthermore, policymakers can play a role by incentivizing cost reduction initiatives through grants, subsidies, and regulations that encourage the adoption of cost-effective converter technologies. It is also worth considering the potential for economies of scale. As the production volume of EVs and their components increases, manufacturers can benefit from economies of scale, which can drive down the cost of production. Therefore, efforts to promote and support the mass production of EVs can contribute to cost reduction in power electronic converters.

1.8.8. Thermal Management

Thermal management remains a critical challenge for power electronic converters in electric vehicles. As the power levels and operating temperatures of converters increase, effective thermal management becomes increasingly important to prevent overheating and optimize their performance. The high power densities and operating temperatures require effective cooling techniques to ensure reliable and efficient operation. Future converters will employ advanced thermal management solutions, including liquid cooling, heat pipes, and innovative packaging designs to dissipate heat effectively.

Advanced cooling techniques are an important area of research for EV power converters. This includes the development of methods such as liquid cooling, phase-change materials, and microchannel heat sinks to enhance heat dissipation and enable more efficient thermal management in compact converter designs. Additionally, the focus should be on thermal modeling and simulation tools that accurately predict the thermal behavior of converters under different conditions, considering factors like component placement, airflow, and heat transfer mechanisms. Implementing intelligent thermal control strategies that optimize cooling system performance based on real-time temperature measurements is another valuable approach. This includes adaptive control algorithms that adjust cooling mechanisms, fan speeds, and power distribution to maintain optimal operating temperatures and maximize converter efficiency. Furthermore, the investigation and development of new thermal interface materials, integrated cooling solutions, and comprehensive reliability testing under different thermal conditions are crucial for improving heat transfer, reducing thermal resistance, and ensuring the long-term reliability and durability of power electronic converters.

In addition to advanced cooling techniques, research efforts should also focus on packaging and enclosure design. Innovative designs that facilitate efficient heat dissipation and thermal management, such as using materials with high thermal conductivity, optimizing heat sink layouts, and employing effective ventilation strategies, can significantly enhance converter performance. Furthermore, the development of advanced thermal monitoring and diagnostic techniques is essential for real-time feedback on temperature distribution and component health status. This enables proactive maintenance and fault detection, preventing thermal-related failures and extending the converter's lifespan. It is also important to establish industry standards and guidelines for thermal management in power electronic converters for EVs. These standards would provide design recommendations, testing protocols, and performance criteria, ensuring consistent and reliable thermal management practices across different converter manufacturers. By addressing these research areas, significant advancements can be made in the thermal management of EV power converters, leading to improved efficiency, reliability, and overall performance.

1.8.9. Standardization and Interoperability

With the growing number of electric vehicle models and charging infrastructure providers, standardization and interoperability become crucial. Future trends include the establishment of common standards for power electronic converters, charging protocols, and communication interfaces. This will ensure compatibility between different vehicle models and charging infrastructure, promoting seamless integration and enhancing user convenience.

Firstly, there is a need for standardization of communication protocols and interfaces between the power electronic converters and other EV components, such as batteries, motors, and charging infrastructure. This would ensure interoperability and enable different components from various manufacturers to work together efficiently. Standardized communication protocols, such as Controller Area Network (CAN) or Ethernet-based protocols, can facilitate data exchange and control signals between the converter and other EV systems, allowing for plug-and-play functionality and easier integration.

Secondly, standardization efforts should focus on safety and performance requirements for power electronic converters. Establishing common safety standards and certification processes ensures that converters meet specific criteria for electrical safety, EMC, and environmental durability. These standards can address aspects like insulation requirements, fault protection mechanisms, thermal management, and EMI limits, ensuring the converters' safe operation and compatibility with other vehicle systems.

Furthermore, interoperability can be enhanced through the standardization of charging interfaces and infrastructure. For example, standardized charging connector designs, such as the widely adopted CCS or CHAdeMO connectors, enable EVs to charge at various charging stations, regardless of the charging network or service provider. Standardized charging protocols, such as the OCPP, enable communication between charging stations and back-end systems, facilitating seamless charging transactions and interoperability.

In conclusion, standardization and interoperability are critical factors for the widespread adoption and success of power electronic converters in EVs. Efforts should be directed towards establishing common standards for communication protocols, safety requirements, and charging interfaces. By promoting interoperability, EV manufacturers, component suppliers, and charging infrastructure providers can collaborate more effectively, ensure compatibility among different systems, and enhance the overall functionality and user experience of electric vehicles.

1.8.10. Environmental Impact and Sustainability

The environmental impact and sustainability of power electronic converters in EVs are of growing importance as the world seeks to reduce greenhouse gas emissions and transition to cleaner transportation. As the focus on sustainability and environmental consciousness increases, future trends in power electronic converters for electric vehicles will prioritize reducing their environmental impact. This includes minimizing the use of hazardous materials, improving energy efficiency, and adopting recyclable and environmentally friendly materials in the design and manufacturing process. Additionally, the recycling and disposal of power electronic converter components will be a critical aspect to address.

One key research direction is the development of more environmentally friendly materials and manufacturing processes for power electronic converters. This involves exploring alternative materials with lower environmental footprints, such as biodegradable or recyclable components, and reducing the use of hazardous substances in converter manufacturing. Additionally, advancements in manufacturing techniques, such as additive manufacturing or 3D printing, can contribute to sustainability by minimizing waste and energy consumption during production.

Efforts should also focus on optimizing the energy efficiency of power electronic converters. Higher conversion efficiencies reduce energy losses and contribute to the overall energy efficiency of EVs. Research can explore advanced semiconductor materials with improved electrical properties, innovative circuit topologies, and optimized control strategies to maximize converter efficiency. Additionally, incorporating energy management algorithms and intelligent power electronics can enhance the overall energy efficiency of the vehicle system.

Another research direction is the integration of renewable energy sources with power electronic converters in EVs. This involves exploring the integration of solar panels or other renewable energy harvesting technologies to directly power the converters or charge the vehicle's battery. By utilizing renewable energy, the dependence on grid electricity and the associated environmental impacts can be reduced.

Lifecycle analysis and eco-design principles should be applied to power electronic converters, considering their entire lifecycle from raw material extraction to end-of-life disposal. This includes assessing the environmental impact at each stage, implementing design strategies for recyclability, and promoting circular economy principles. Research can focus on developing converter designs that are easier to disassemble, reuse, or recycle, minimizing waste generation and maximizing resource efficiency.

Lastly, the adoption of sustainable and ethical sourcing practices for raw materials used in power electronic converters is crucial. Research can explore ways to ensure the responsible sourcing of materials, such as rare earth elements, and promote transparency and traceability in the supply chain to avoid environmental damage and social issues associated with mining and extraction.

In conclusion, future research and recommendations in the field of power electronic converters for EVs should prioritize environmental impact and sustainability. This includes exploring eco-friendly materials and manufacturing processes, optimizing energy efficiency, integrating renewable energy sources, considering lifecycle analysis, and promoting responsible sourcing practices. By addressing these aspects, power electronic converters can contribute to a more sustainable and environmentally friendly transportation system.

1.8.11. Grid Integration and Smart Charging

The integration of electric vehicles into the smart grid presents both opportunities and challenges. Future power electronic converters will need to support bidirectional power flow, grid communication, and smart charging capabilities. This enables V2G applications, demand response programs, and optimization of charging based on grid conditions and renewable energy availability. Future studies aim to develop charging strategies that are friendly to the electrical grid, taking into account factors like peak demand management, load balancing, and grid stability. By optimizing charging patterns and coordinating charging schedules, EVs can contribute to grid stability and alleviate strain during peak demand periods. Power converters will play a crucial role in managing the interaction between the electric vehicle, the charging infrastructure, and the grid to ensure efficient and reliable operation.

Smart charging systems can incorporate demand response mechanisms, aligning EV charging with periods of lower electricity demand or higher availability of renewable energy. By incentivizing off-peak charging and load shifting, the grid can optimize resource utilization and reduce the need for additional infrastructure investment. Another area of research is V2H integration, which allows EVs to provide power to homes or other buildings during power outages or as a backup power source. Efficient power conversion and control systems are being developed to seamlessly integrate EVs with residential or commercial buildings, increasing their resilience and reducing reliance on traditional grid infrastructure.

Grid communication and data analytics play a crucial role in smart charging systems. Reliable communication protocols, data management systems, and advanced analytics algorithms are being developed to optimize charging operations, consider user preferences, and support grid integration. Furthermore, power electronic converters in EVs can participate in ancillary markets and provide grid services such as frequency regulation, voltage support, and reactive power control. Future research aims to explore the technical and economic feasibility of these services, develop standardized interfaces, and address regulatory and market barriers to facilitate the participation of EVs in grid services.

By addressing these key points, research efforts are directed towards enhancing grid integration, improving the flexibility of charging systems, and maximizing the benefits of EVs in supporting grid stability and optimizing resource utilization.

1.8.12. Cybersecurity

Cybersecurity is a critical aspect of EV technology and grid integration. As EVs become increasingly connected and communicate with various networks, ensuring the security and protection of these systems becomes paramount. Future research in cybersecurity focuses on developing robust security measures to safeguard EVs and the associated charging infrastructure from potential cyber threats and attacks. With the increasing connectivity and integration of electric vehicles into the digital ecosystem, cybersecurity becomes a significant concern. Power electronic converters, being a vital component of the vehicle's electrical system, need to incorporate robust security measures to protect against cyber threats. Future trends will focus on implementing encryption protocols, secure communication interfaces, and intrusion detection systems to ensure the integrity and safety of the power converter system.

One area of research is the development of secure communication protocols and encryption techniques to protect data transmitted between EVs, charging stations, and grid networks. This includes authentication mechanisms to verify the identity of vehicles and charging infrastructure, ensuring that only authorized entities can access and interact with the system. By implementing strong encryption algorithms and secure communication channels, the confidentiality and integrity of data can be preserved, preventing unauthorized access or tampering. Another aspect of cybersecurity research involves detecting and mitigating potential vulnerabilities in EV systems. This includes identifying potential entry points for cyber attacks, such as compromised charging stations or malicious software embedded in the vehicle's software. Research aims to develop intrusion detection systems and real-time monitoring tools to detect and respond to cyber threats promptly. Additionally, the development of secure over-the-air (OTA) update mechanisms for EV software and firmware is crucial to ensure that security patches and updates can be deployed effectively to address newly discovered vulnerabilities.

Furthermore, research efforts focus on enhancing the resilience of EV systems against cyber attacks. This includes developing advanced anomaly detection algorithms to identify abnormal behavior or unauthorized access attempts. By integrating cybersecurity measures with existing safety systems, potential risks can be mitigated, and the overall system integrity can be maintained. Additionally, research aims to address the human factor in cybersecurity by promoting awareness and training programs for EV users, charging station operators, and other stakeholders to ensure best practices in cybersecurity are followed. Collaboration between automakers, charging infrastructure providers, cybersecurity experts, and regulatory bodies is crucial in developing standardized cybersecurity guidelines and regulations for the

EV industry. These guidelines can outline security requirements, best practices, and compliance frameworks to ensure a consistent and robust approach to cybersecurity across the ecosystem.

By addressing cybersecurity concerns in EV technology, researchers aim to enhance the trust and confidence in EV systems, encourage widespread adoption, and ensure the safety and security of EV users, charging infrastructure, and the grid as a whole.

1.8.13. Vehicle-to-X (V2X) Integration

V2X, also known as vehicle-to-everything, integration refers to the communication and interaction between vehicles and various entities in their surrounding environment, including other vehicles (V2V), infrastructure (V2I), pedestrians (V2P), and the grid (V2G). Future research focuses on exploring the potential of V2X integration and developing advanced technologies and protocols to enable efficient and secure communication between these entities.

One area of research is the development of V2V communication systems. This involves the exchange of information between vehicles, such as location, speed, and intention, to enable cooperative driving and enhance safety on the road. Research aims to improve the reliability and efficiency of V2V communication by exploring new wireless communication technologies, such as dedicated short-range communication (DSRC) or cellular vehicle-to-everything (C-V2X) communication, and developing robust protocols for data transmission and collision avoidance.

V2I integration focuses on the communication between vehicles and infrastructure elements, such as traffic lights, road signs, and smart traffic management systems. Research in this area aims to develop effective communication protocols and standards to enable seamless information exchange, enabling vehicles to receive real-time traffic updates, optimize route planning, and enhance traffic flow efficiency. This can lead to reduced congestion, improved energy efficiency, and enhanced overall transportation system performance.

V2P integration involves the interaction between vehicles and pedestrians or vulnerable road users. Research in this area aims to develop systems and technologies that improve pedestrian safety and enhance communication between vehicles and pedestrians. This includes the development of warning systems that alert drivers to the presence of pedestrians, as well as technologies that enable pedestrians to receive information from vehicles, such as safe crossing indications or notifications of approaching vehicles.

V2G integration focuses on the bi-directional interaction between electric vehicles and the electrical grid. Research in this area aims to develop technologies and control strategies that enable EVs to not only consume energy from the grid but also provide energy back to the grid. This includes the development of bidirectional power converters, advanced charging and discharging algorithms, and grid management systems that can leverage the energy storage capacity of EVs to support grid stability, renewable energy integration, and demand response programs.

Overall, V2X integration research aims to harness the potential of connected vehicles and advanced communication technologies to improve road safety, traffic efficiency, energy management, and overall transportation system performance. By enabling vehicles to communicate and interact with their surroundings, researchers aim to create a smarter, more interconnected transportation ecosystem that benefits both vehicle occupants and the wider community.

1.8.14. Advanced Materials and Packaging

Advanced materials and packaging play a crucial role in the development of efficient, reliable, and compact electronic systems for electric vehicles. The use of advanced materials and packaging techniques will be a future trend in power electronic converters for electric vehicles. Future research in this field focuses on exploring and developing new materials, packaging techniques, and designs that can meet the evolving demands of EV technology.

One area of research is the development of advanced semiconductor materials. Researchers are exploring new materials, such as wide-bandgap semiconductors like SiC and GaN, which offer improved efficiency, higher power density, and better thermal properties compared to traditional silicon-based semiconductors. These materials enable the design of more efficient power electronics systems, reducing power losses and increasing overall system performance.

Another area of research is focused on thermal management materials and techniques. As power levels increase in EVs, effective thermal management becomes crucial to ensure the reliable operation and longevity of electronic components. Researchers are investigating advanced thermal interface materials that have high thermal conductivity and low thermal resistance, enabling efficient heat transfer between components and heat sinks. Additionally, innovative cooling techniques, such as liquid cooling, phase-change materials, and microchannel heat sinks, are being explored to enhance heat dissipation capabilities and improve overall thermal management.

Packaging is another critical aspect of advanced materials and packaging research. Researchers are developing new packaging techniques and designs to enable higher component integration, improved reliability, and efficient thermal management. This includes the development of advanced packaging technologies, such as 3D packaging and embedded packaging, which allow for compact designs, shorter interconnects, and improved electrical and thermal performance. Additionally, the use of lightweight and robust materials for packaging, such as advanced polymers and composites, is being explored to reduce weight and improve overall system durability.

Furthermore, research in advanced materials and packaging aims to address challenges related to EMI and EMC. New materials and designs are being explored to minimize EMI, improve signal integrity, and ensure proper functioning of electronic systems in the presence of electromagnetic disturbances.

Overall, research in advanced materials and packaging for EVs aims to enhance the performance, reliability, and efficiency of electronic systems. By developing and implementing innovative materials, packaging techniques, and designs, researchers strive to overcome current limitations and enable the next generation of high-performance, compact, and reliable electronic systems for electric vehicles.

1.8.15. Continuous Improvement in Efficiency

Achieving higher efficiency in power electronic converters is an ongoing objective. Future trends will focus on reducing power losses, optimizing converter designs, and improving control strategies to maximize energy conversion efficiency. This includes minimizing switching losses, reducing conduction losses, and optimizing the converter's operating points based on load conditions. Advances in semiconductor technologies, control algorithms, and thermal management techniques will contribute to the continuous improvement in converter efficiency.

One area of research is the optimization of powertrain efficiency. This involves improving the efficiency of components such as electric motors, inverters, and power electronics. Researchers are exploring advanced motor designs, including permanent magnet motors and synchronous reluctance motors, which offer higher efficiency compared to traditional induction motors. Additionally, advancements in power electronics and control algorithms aim to minimize power losses during energy conversion, resulting in improved overall powertrain efficiency. Another aspect of efficiency improvement is the development of lightweight materials and vehicle structures. Researchers are investigating advanced materials, such as carbon fiber composites and aluminum alloys, to reduce the weight of EVs. By reducing vehicle weight, energy consumption is decreased, leading to increased efficiency and extended range. Additionally, aerodynamic enhancements, such as streamlined designs and active aerodynamics, are being explored to reduce drag and optimize vehicle efficiency. Energy management and optimization are also areas of research. Researchers are developing intelligent energy management systems that can optimize the distribution and utilization of electrical energy within an EV. This involves sophisticated algorithms and predictive models that consider factors such as driving conditions, traffic patterns, and battery characteristics to determine the most efficient use of energy. Smart charging systems are being developed to ensure optimal charging patterns, taking into account grid conditions and renewable energy availability to minimize energy waste.

Furthermore, advancements in battery technology are essential for improving efficiency. Researchers are working on developing high-energy-density batteries with improved charging and discharging efficiency. This includes advancements in battery chemistry, cell designs, and thermal management systems. By increasing the efficiency of energy storage and delivery, the overall efficiency of EVs can be significantly enhanced.

Lastly, research is focused on creating a more efficient charging infrastructure. This includes the development of fast-charging technologies that reduce charging times and minimize energy losses during the charging process. Additionally, the integration of renewable energy sources into the charging infrastructure helps to further enhance the efficiency and sustainability of EVs. Continuous improvement in efficiency is a vital aspect of EV research, aiming to maximize the range, reduce energy consumption, and enhance the overall sustainability of electric vehicles. By exploring advancements in powertrain technology, lightweight materials, energy management systems, battery technology, and charging infrastructure, researchers strive to push the boundaries of efficiency and accelerate the widespread adoption of electric vehicles.

In summary, the future of power electronic converters for electric vehicles holds exciting possibilities and challenges. The trends discussed above emphasize the need for higher power density, integration, advanced control strategies, enhanced fault diagnostics, cost reduction, and environmental sustainability. Overcoming these challenges will require collaborative efforts from researchers, manufacturers, policymakers, and stakeholders in the electric vehicle industry. By addressing these trends and challenges, power electronic converters will continue to play a pivotal role in enabling the widespread adoption of electric vehicles and driving the transition towards a sustainable transportation future.

Conclusion

Summary of the Chapter

In this chapter, ed the world of power electronic converters for electric vehicles are explored. Importance of these converters in the context of electric vehicles, emphasizing their role in facilitating efficient power conversion and management are highlighted. We then delved into the classification of converters, categorizing them based on their functionality and application.

Various types of converters, including DC-to-DC converters, DC-to-AC converters, and AC-to-DC converters are discussed. For each type, we explored different topologies, such as the SEPIC, Cuk, and Zeta converters, as well as voltage source inverters, current source inverters, and impedance source inverters. We examined their working principles, advantages, and limitations, providing a comprehensive understanding of their applications in electric vehicles.

Design considerations were a key focus of this chapter. Component selection, topology selection, control and modulation strategies, EMI/EMC mitigation techniques, efficiency and thermal considerations, and reliability, durability, and cost considerations are discussed extensively. These factors play a critical role in the design and development of power electronic converters, ensuring their optimal performance, reliability, and cost-effectiveness.

The chapter further explored the applications of power electronic converters in electric vehicles. Battery charging, motor drive, regenerative braking, and on-board power supply, showcasing how converters enable efficient and effective power conversion in each application are covered.

These applications are integral to the operation and functionality of electric vehicles, and power converters play a crucial role in their success.

Looking ahead, future trends and challenges in the field of power electronic converters for electric vehicles are explored and identified. The advancements in wide bandgap semiconductor materials, integration with the smart grid and V2X communication, cybersecurity measures, and the adoption of sustainable practices are also discussed. These trends present exciting opportunities for innovation and improvement in power converters, while the associated challenges will need to be addressed to ensure the continued growth and success of electric vehicles.

Overall, this chapter provided a comprehensive overview of power electronic converters for electric vehicles. Their types, applications, design considerations, and future trends and challenges are explored. The information presented in this chapter serves as a valuable resource for researchers, engineers, and industry professionals involved in the development and implementation of power electronic converters for electric vehicles. By understanding the intricacies of these converters and addressing the identified challenges, we can pave the way for a sustainable and efficient future of electric transportation.

Future Research Directions and Recommendations

The increasing demand for electric vehicles has driven the need for efficient, reliable, and cost-effective power electronic converters. As we move towards a more sustainable future, it is essential to focus on improving the performance and efficiency of these converters. There are several key research directions and recommendations that can further enhance the development and implementation of power electronic converters. These directions focus on addressing existing challenges, exploring new technologies, and improving the overall performance and efficiency of EV power systems. Here are top 15 future research directions and recommendations for power electronic converters in electric vehicles:

1) Investigation of advanced converter topologies such as multilevel converters, resonant converters, and hybrid converters.
2) Integration of power electronic converters with renewable energy sources for direct charging of EVs.

3) Research on energy storage integration with power electronic converters to improve energy utilization and vehicle efficiency.
4) Development of fault-tolerant and resilient converter designs to ensure reliable operation in the event of failures.
5) Exploration of advanced thermal management techniques to enhance the cooling and reliability of converters.
6) Standardization and interoperability efforts to establish common interfaces and protocols for converters in EVs.
7) Cost reduction strategies through innovative design, material selection, and manufacturing processes.
8) System-level optimization to consider the holistic performance of power systems, energy storage, and charging infrastructure.
9) Research on advanced control and modulation strategies to improve converter performance and efficiency.
10) Investigation of wide-bandgap semiconductor materials such as SiC and GaN for higher efficiency and power density.
11) Development of advanced power factor correction techniques for AC-to-DC converters in EV charging infrastructure.
12) Study of advanced EMI/EMC mitigation techniques to minimize electromagnetic interference and ensure compliance with regulatory standards.
13) Research on advanced fault detection and diagnostic methods for early detection and prevention of converter failures.
14) Exploration of innovative cooling methods such as liquid cooling or phase-change materials to enhance thermal management.
15) Investigation of advanced power semiconductor packaging technologies for improved reliability and thermal performance.

In conclusion, future research directions and recommendations in the field of power electronic converters for EVs are vast and varied. By addressing these research directions, researchers and industry professionals can contribute to the ongoing advancement and improvement of power electronic converter technologies, making EVs more efficient, reliable, cost-effective, and accessible for a sustainable transportation future. The future research directions and recommendations presented in this chapter can help guide the development of power electronic converters and address the challenges associated with their implementation in electric vehicles.

References

(Gore), S. D., Iqbal, A., Islam, S., Khan, I., Marzband, M., Rahman, S., & Al-Wahedi, A. M. (2022). Review on classification of resonant converters for electric vehicle application. *Energy Reports, 8*, 1091-1113.

Abu Rub, H., Malinowski, M., & Al Haddad, K. (2014). Electric and Plug-In Hybrid Electric Vehicles. In *Power Electronics for Renewable Energy Systems, Transportation and Industrial Applications* (pp. 387-421). Wiley-IEEE Press.

Acharige, S. S., Haque, M. E., Arif, M. T., Hosseinzadeh, N., Hasan, K. N., & Oo, A. M. (2023). Review of Electric Vehicle Charging Technologies, Standards, Architectures, and Converter Configurations. *IEEE Access, 11*, 41218-41255.

Aghabali, I., Bauman, J., Kollmeyer, P. J., Wang, Y., Bilgin, B., & Emadi, A. (2021). 800-V Electric Vehicle Powertrains: Review and Analysis of Benefits, Challenges, and Future Trends. *IEEE Transactions on Transportation Electrification, 7*(3), 927-948.

Atanalian, S., Al Haddad, K., Zgheib, R., & Kanaan, H. Y. (2021). A Review on Electric Vehicles Battery Chargers and AC/DC Converters for Fast Charging Stations. *IEEE 3rd International Multidisciplinary Conference on Engineering Technology (IMCET).* Beirut, Lebanon.

Babazadeh, Y., Sabahi, M., Babaei, E., & Kai, S. (2022). A New Continuous Input Current Nonisolated Bidirectional Interleaved Buck-Boost DC-DC Converter. *International Transactions on Electrical Energy Systems, 2022*, 1-19.

Bai, H. K., Costinett, D., Tolbert, L. M., Qin, R., Zhu, L., Liang, Z., & Huang, Y. (2022). Charging Electric Vehicle Batteries: Wired and Wireless Power Transfer: Exploring EV charging technologies. *IEEE Power Electronics Magazine, 9*(2), 14-29.

Drabek, P., Peroutka, Z., Pittermann, M., & Cedl, M. (2011). New Configuration of Traction Converter With Medium-Frequency Transformer Using Matrix Converters. *IEEE Transactions on Industrial Electronics, 58*(11), 5041-5048.

Fajri, P., Lee, S., Prabhala, V. A., & Ferdowsi, M. (2016). Modeling and Integration of Electric Vehicle Regenerative and Friction Braking for Motor/Dynamometer Test Bench Emulation. *IEEE Transactions on Vehicular Technology, 65*(6), 4264-4273.

Hossain, M., Rahim, N., & Selvaraj, J. a. (2018). Recent progress and development on power DC-DC converter topology, control, design and applications: A review. *Renewable and Sustainable Energy Reviews, 81*(1), 205-230.

İnci, M., Büyük, M., Demir, M. H., & İlbey, G. (2021). A review and research on fuel cell electric vehicles: Topologies, power electronic converters, energy management methods, technical challenges, marketing and future aspects. *Renewable and Sustainable Energy Reviews, 137*, 110648.

International Energy Agency. (2021). Retrieved from https://www.iea.org/reports/global-ev-outlook-2021.

Jenkins, J. (2018). *A closer look at power factor correction*. (Eaton) Retrieved from https://chargedevs.com/features/a-closer-look-at-power-factor-correction/.

Lajunen, A. (2014). Energy consumption and cost-benefit analysis of hybrid and electric city buses. *Transportation Research Part C: Emerging Technologies, 38*, 1-15.

Lipu, H., Shahadat, M., Miah, M. S., Ansari, S., Meraj, S. T., Hasan, K., . . . Hussain, A. (2022). Power Electronics Converter Technology Integrated Energy Storage

Management in Electric Vehicles: Emerging Trends, Analytical Assessment and Future Research Opportunities. *Electronics, 11*(4), 562.

Maroti, P. K., Padmanaban, S., Bhaskar, M. S., Ramachandaramurthy, V. K., & Blaabjerg, F. (2022). The state-of-the-art of power electronics converters configurations in electric vehicle technologies. *Power Electronic Devices and Components, 1*, 100001.

Pradhan, R., Keshmiri, N., & Emadi, A. (2023). On-Board Chargers for High-Voltage Electric Vehicle Powertrains: Future Trends and Challenges. *IEEE Open Journal of Power Electronics, 4*, 189-207.

Rafin, S. M., Islam, R., & Mohammed, O. A. (2023). Overview of Power Electronic Converters in Electric Vehicle Applications. *Fourth International Symposium on 3D Power Electronics Integration and Manufacturing (3D-PEIM).* Miami, FL, USA.

Ravi, D., Letha, S. S., Samuel, P., & Reddy, B. M. (2018). An Overview of Various DC-DC Converter Techniques used for Fuel Cell based Applications. *2018 International Conference on Power Energy, Environment and Intelligent Control (PEEIC).* Greater Noida, India.

Research and Markets. (2022). Retrieved from https://www.researchandmarkets.com/reports/5691643/global-electric-bus-market-by-vehicle-payload.

Restrepo, C., & González-Castaño, C. (2021). Digital control strategies of DC–DC converters in automotive hybrid powertrains. In *Modeling, Operation, and Analysis of DC Grids* (pp. 245-287). Academic Press.

Safayatullah, M., Elrais, M. T., Ghosh, S., Rezaii, R., & Batarseh, I. (2022). A Comprehensive Review of Power Converter Topologies and Control Methods for Electric Vehicle Fast Charging Applications. *IEEE Access, 10*, 40753-40793.

Sakka, M. A., Mierlo, J. V., & Gualous, H. (2011). DC/DC Converters for Electric Vehicles. In S. Soylu, *Electric Vehicles - Modelling and Simulations.* IntechOpen.

Sanchez Sutil, F., Hernández, J., & Tobajas, C. (2015). Overview of electrical protection requirements for integration of a smart DC node with bidirectional electric vehicle charging stations into existing AC and DC railway grids. *Electric Power Systems Research, 122*, 104-118.

Trivedi, S. S., & Sant, A. V. (2022). Comparative analysis of dual active bridge dc–dc converter employing Si, SiC and GaN MOSFETs for G2V and V2G operation. *Energy Reports, 8*(13), 1011-1019.

Turksoy, A., Teke, A., & Alkaya, A. (2020). A comprehensive overview of the dc-dc converter-based battery charge balancing methods in electric vehicles. *Renewable and Sustainable Energy Reviews, 133*, 110274.

Wang, B., Dehghanian, P., Wang, S., & Mitolo, M. (2019). Electrical Safety Considerations in Large-Scale Electric Vehicle Charging Stations. *IEEE Transactions on Industry Applications, 55*(6), 6603-6612.

Wang, J., Wang, B., Zhang, L., Wang, J., Shchurov, N., & Malozyomov, B. (2022). Review of bidirectional DC–DC converter topologies for hybrid energy storage system of new energy vehicles. *Green Energy and Intelligent Transportation, 1*(2), 100010.

Wang, Q., Li, W., Kang, J., & Wang, Y. (2019). Electromagnetic Safety Evaluation and Protection Methods for a Wireless Charging System in an Electric Vehicle. *IEEE Transactions on Electromagnetic Compatibility, 61*(6), 1913-1925.

Welcome to EV101: Electric Vehicle Charging Basics. (n.d.). (EVgo) Retrieved from https://www.evgo.com/ev-drivers/charging-basics/

Yazdani, M. R., Farzanehfard, H., & Faiz, J. (2011). Classification and Comparison of EMI Mitigation Techniques in Switching Power Converters – A Review. *Journal of Power Electronics, 11*(5), 767-777.

Yu, H., Niu, S., Shang, Y., Shao, Z., Jia, Y., & Jian, L. (2022). Electric vehicles integration and vehicle-to-grid operation in active distribution grids: A comprehensive review on power architectures, grid connection standards and typical applications. *Renewable and Sustainable Energy Reviews, 168*, 112812.

Zhao, B., Song, Q., Liu, W., & Sun, Y. (2014). Overview of Dual-Active-Bridge Isolated Bidirectional DC–DC Converter for High-Frequency-Link Power-Conversion System. *IEEE Transactions on Power Electronics, 29*(8), 4091-4106.

Zheng, L., Kandula, R. P., & Divan, D. (2023). Multiport Control With Partial Power Processing in Solid-State Transformer for PV, Storage, and Fast-Charging Electric Vehicle Integration. *IEEE Transactions on Power Electronics, 38*(2), 2606-2616.

Chapter 2

Battery Management System of EVs

Trilochan Penthia*

Department of Electrical Engineering, National Institute of Technology Patna, Bihar, India

Abstract

Energy storage systems are becoming increasingly important for electric vehicle (EV) charging infrastructure due to the significant benefits they can provide. By storing energy during off-peak hours, energy storage systems can reduce the strain on the electricity grid during peak hours when electricity demand is highest, helping to maintain grid stability. They can also help to reduce the cost of EV charging by allowing EV owners to charge their vehicles during periods when electricity prices are lower. Additionally, energy storage systems can play a key role in supporting the integration of renewable energy sources into the grid, as they can store excess energy generated by solar or wind power during times of low demand and supply it during periods of high demand. This can help to increase the utilization of renewable energy and reduce greenhouse gas emissions from the electricity sector. Overall, energy storage systems are critical for enabling the widespread adoption of EVs and supporting the transition to a more sustainable energy system. This book chapter provides an overview of energy storage systems (ESSs) and their importance for EV chargers. It discusses different types of ESSs, including battery-based, capacitor-based, flywheel-based, and hydrogen fuel cell-based ESSs, and outlines design considerations such as power and energy capacity, charging and discharging rates, scalability, efficiency, safety, reliability, and cost. The chapter also explores the integration of ESSs with EV chargers, including DC fast charging

* Corresponding Author's Email: trilochankoraput@gmail.com.

In: The Future of Hybrid-Electric Vehicles
Editors: Pawan Kumar Pathak and Anil Kumar Yadav
ISBN: 979-8-89113-515-4

systems, AC level 2 chargers, and bi-directional charging systems, and examines the benefits of ESSs for EV chargers, such as peak shaving, load levelling, renewable energy integration, resiliency and backup power, grid stabilization, and frequency regulation. Several case studies are presented, including Tesla's supercharger, Duke Energy's energy storage project in North Carolina, and Pacific Gas & Electric's smart grid program in California. The chapter concludes with a discussion of the challenges and opportunities associated with ESSs for EV chargers, including regulatory and policy barriers, technical and engineering challenges, and market and economic opportunities, and suggests future research directions. Overall, the chapter highlights the significant implications of ESSs for the future of EV charging and the transition to a more sustainable energy system.

Keywords: batteries, battery management system, electric vehicles, EV charging, BMS algorithms

2.1. Introduction

Electric vehicles (EVs) are becoming increasingly popular as a means of reducing greenhouse gas emissions and improving air quality. According to a report by the International Energy Agency, the global electric car fleet surpassed 10 million in 2020, up from 2 million in 2016, and the number is expected to reach 145 million by 2030 (Green Car Reports, 2022). This trend is expected to continue as more countries adopt policies and regulations to encourage the adoption of EVs. However, the success of EVs is heavily dependent on the performance and reliability of their battery systems. Estimates indicate that the global demand for lithium-ion batteries is expected to increase significantly, reaching 2000 gigawatt-hours (GWh) between 2022 and 2030, as depicted in Figure 2.1. The primary and continuing major market for lithium-ion batteries is the wide range of applications in the electric vehicle industry (Global EV sales by scenario, 2020-2030, 2022). The battery is the most expensive and critical component of an EV, and its performance is directly related to the vehicle's range, power, and durability. This is where the battery management system (BMS) plays a crucial role.

Research has shown that the performance and longevity of the battery pack in EVs are directly related to the design and operation of the BMS. A well-designed BMS can significantly improve the performance, safety, and lifespan of an EV battery. For example, a study by the University of Warwick found that a BMS with a more accurate state-of-charge estimation algorithm can increase the range of an EV by up to 5% (Greenwood, Marco, McMahon, & Sattar, 2020). Another study by the University of Michigan found that a BMS with a thermal management system can extend the lifespan of an EV battery by up to 20% (Weng, et al., 2021).

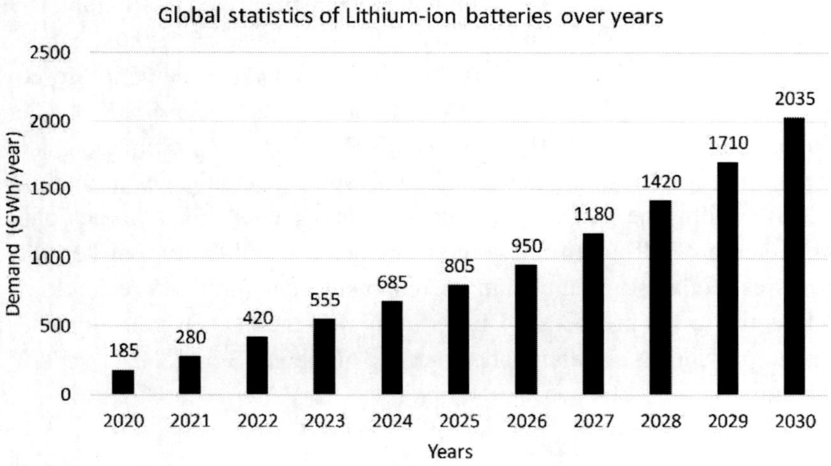

Figure 2.1. Demand and sales of Lithium-ion batteries globally (Zhao & Burke, 2022).

Moreover, research has shown that the adoption of EVs is closely tied to the availability and performance of their batteries. A survey by Deloitte found that the majority of consumers who consider purchasing an EV prioritize its range, battery life, and charging time over other factors such as price and brand (Recharging India's electric vehicle ambition by electrifying public transport, 2019). This suggests that the performance and durability of the battery are key factors that influence consumer behaviour and the adoption of EVs. As the EV market continues to grow, the development and deployment of advanced BMS technology are becoming increasingly important. Researchers and industry experts are exploring new approaches to BMS design, such as machine learning algorithms, that can improve the accuracy of battery state estimation and optimize battery performance (Li, Zhou, & Han, Advances in Battery Manufacturing, Service, and Management Systems, 2017). They are also developing new materials and manufacturing techniques that can enhance the safety, efficiency, and durability of the battery. However, the development and deployment of advanced BMS technology also pose significant challenges.

The design and operation of BMS must balance the trade-offs between performance, safety, cost, and complexity. Moreover, the standardization of BMS is a critical issue that needs to be addressed to ensure the compatibility and interoperability of BMS across different EV models and manufacturers.

This book chapter provides an overview of BMS technology in EVs, including its historical evolution, key components, functions and its current state-of-the-art design. It also highlights the importance of BMS in improving the performance, safety, and durability of EVs, as well as the challenges and opportunities associated with BMS development and deployment. Moreover, the chapter discusses different BMS algorithms used for State of Charge (SoC), State of Health (SoH), and State of Function (SoF) estimation, as well as fault detection and diagnosis. It also surveys the challenges in integrating and controlling the BMS with other vehicle systems and the types of tests conducted on BMS to ensure its performance and reliability. Furthermore, it discusses the latest research and development trends in BMS technology, as well as the future prospects of BMS in the EV market. Overall, this chapter aims to provide an in-depth understanding of the BMS in EVs and its critical role in the success of electric mobility.

2.1.1. Definition of Battery Management System

BMS is an electronic system designed to monitor, control, and protect the battery pack in an electric vehicle EV or other energy storage system. The BMS plays a critical role in ensuring the safe and efficient operation of the battery, as well as prolonging its lifespan and optimizing its performance. The BMS achieves this by monitoring and controlling the state of the battery, including its temperature, voltage, current, and SoC. The BMS consists of several key components, including sensors, voltage monitoring circuitry, current sensing circuitry, temperature sensing circuitry, and balancing circuitry. These components work together to monitor the battery's state and take corrective actions to prevent hazardous situations, such as overcharging, over-discharging, and overheating. A block diagram of BMS operation is illustrated in Figure 2.2.

Sensors are used to measure the battery's state, including its temperature, voltage, and current. Voltage monitoring circuitry measures the voltage of each cell in the battery pack, while current sensing circuitry measures the current flowing in and out of the battery. Temperature sensing circuitry measures the temperature of the battery, which is critical for maintaining its safety and preventing thermal runaway. Balancing circuitry is used to equalize the voltage of each cell in the battery pack, ensuring that the battery operates at its optimal performance and extending its lifespan. When a cell's voltage deviates from the others, the balancing circuitry redistributes the charge to ensure that all cells have the same voltage.

The BMS also performs several critical functions, including SoC estimation, thermal management, fault detection and diagnosis, and safety protection. SoC estimation is essential for determining the remaining range of the EV and preventing over-discharging, which can damage the battery. Thermal management is critical for preventing overheating, which can damage the battery and cause safety hazards. Fault detection and diagnosis are used to identify and correct any issues with the battery, while safety protection is essential for preventing hazardous situations such as short circuits, overcharging, and over-discharging.

The BMS is an essential component of an EV or any energy storage system, and its proper design and operation are critical for the success of electric mobility. A well-designed BMS can significantly improve the performance, safety, and lifespan of the battery, as well as the overall performance of the EV. Moreover, the BMS plays a critical role in the development of next-generation EVs and energy storage systems, where new technologies such as solid-state batteries and advanced sensors will require more sophisticated BMS designs.

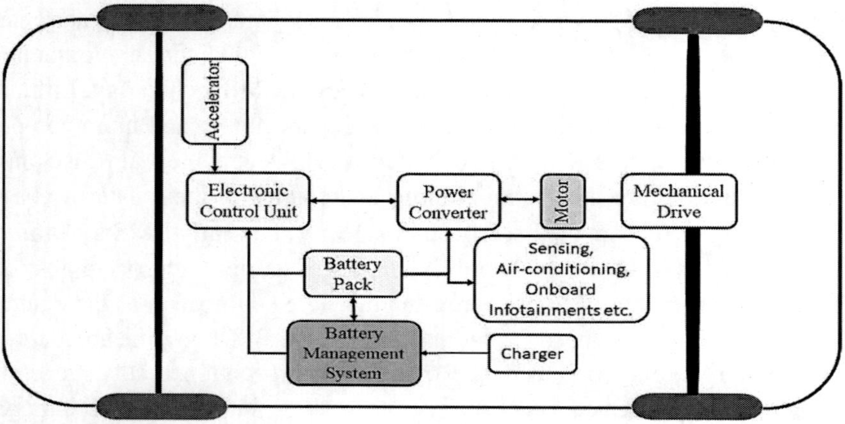

Source: MDPI.

Figure 2.2. Role of BMS in EVs.

2.1.2. Historical Overview of BMS in EVs

The BMS has been a critical component of EVs since their inception in the early 19th century (Chen, Xiong, Qiu, Zhao, & Chen, 2023). The first EVs were powered by lead-acid batteries, and their charging and discharging were controlled by primitive mechanical switches and relays. These early EVs had limited range and were often used as city taxis or delivery vehicles. One of the early pioneers of BMS technology was Alan Cocconi, the founder of AC Propulsion, a California-based electric vehicle manufacturer. In the mid-1990s, AC Propulsion developed a BMS for its electric sports car, the tzero (Lambert, 2017). The BMS was one of the first of its kind, and it incorporated advanced algorithms that allowed it to estimate the state of the battery with high accuracy. At that time, EVs were still a relatively new technology, and the lithium-ion battery technology was in its infancy (Liu, Placke, & Chau, 2022). The primary goal of BMS at that time was to monitor and control the battery's state to ensure safe and efficient operation of the battery. The tzero, with its advanced BMS, was a breakthrough in electric vehicle technology and became an inspiration for many electric vehicle manufacturers.

As battery technology improved, BMS became more sophisticated, incorporating electronic control systems that could monitor and manage the battery's state of charge and discharge. The development of nickel-cadmium batteries in the 1970s and 1980s led to the first modern BMS in EVs (The history and development of batteries: by Jose Alarco And Peter Talbot, The Conversation, 2015), which could monitor the battery's temperature and prevent overcharging and over-discharging. In the 1990s, the development of lithium-ion batteries led to significant advances in BMS technology. Lithium-ion batteries have a higher energy density and are lighter and more compact than lead-acid or nickel-cadmium batteries. However, they are also more complex and require precise monitoring and control of their state of charge and discharge to ensure their safety and longevity. The early BMS for lithium-ion batteries was based on simple voltage and current sensors that could measure the battery's state of charge and discharge. However, as the demand for EVs increased, the need for more sophisticated BMS became apparent. In 2004, Tesla Motors introduced the Roadster, the first commercially successful EV with a sophisticated BMS that could monitor and control the battery's temperature, voltage, and current.

The Tesla Roadster's BMS was based on an array of sensors and control circuits that could adjust the battery's charging and discharging rate to optimize its performance and safety (Tesla's Roadster, 2021), (Dorian, 2023).

The Roadster's BMS also had an active thermal management system that could regulate the battery's temperature, preventing it from overheating or freezing. Since then, BMS technology has continued to advance rapidly, with new developments in machine learning algorithms, artificial intelligence, and predictive analytics. These new technologies are allowing BMS to be more precise and efficient in managing battery performance and safety. Over the years, BMS technology has evolved significantly, driven by the increasing demand for more efficient, reliable, and safe electric vehicles. The early BMSs were relatively simple, consisting of only a few sensors and control circuits. However, with the introduction of lithium-ion batteries, which are more complex and require careful monitoring and management, BMSs have become more sophisticated and advanced.

One interesting story from the history of BMS in EVs is the case of General Motors EV1 (General Motors EV1, 2021). The EV1 was one of the first modern electric vehicles, introduced in the 1990s. However, it was controversial because General Motors only offered the EV1 on a lease, and they later recalled and destroyed most of the vehicles. One of the reasons for the recall was that the EV1's BMS had a design flaw that caused the battery to overcharge, leading to a fire risk. This incident highlights the critical importance of BMS in ensuring the safety and performance of EVs. Another interesting fact about BMS in EVs is that some BMS can communicate with the power grid to provide services such as frequency regulation and peak demand reduction. This technology, known as vehicle-to-grid (V2G) communication, allows EVs to be used as a flexible and reliable source of energy for the grid.

2.1.3. Importance of BMS for EVs

BMS technology has advanced rapidly in recent years, allowing EVs to have longer ranges, faster charging times, and better performance. BMS is essential for the safe, efficient, and reliable operation of an EV. It ensures that the battery operates within safe limits, optimizes its performance, and extends its lifespan. As EVs continue to grow in popularity, the importance of BMS will only increase, and it will continue to be a critical component in the development of future EV technologies. The importance of BMS for EVs can be summarized as follows (Chen, Xiong, Qiu, Zhao, & Chen, 2023), (Thomas, 2021):

- *Battery safety:* BMS ensures the safety of the battery by preventing overcharging, over-discharging, and overheating. This prevents damage to the battery and reduces the risk of fire or explosion.
- *Battery longevity:* BMS helps to extend the life of the battery by ensuring that it is not subjected to conditions that can reduce its performance or cause premature failure.
- *Performance optimization:* BMS optimizes the performance of the battery by regulating its charging and discharging rate, ensuring that it operates at the optimal temperature and preventing voltage sag.
- *Range estimation:* BMS provides accurate range estimation by monitoring the state of charge and predicting the remaining range based on driving conditions.
- *Energy efficiency:* BMS improves energy efficiency by reducing the energy lost during charging and discharging.
- *User experience:* BMS improves the user experience by providing real-time information about the battery's state of charge, remaining range, and charging status.
- *Grid integration:* BMS enables EVs to be used as a flexible source of energy for the grid through V2G communication, providing services such as frequency regulation and peak demand reduction.

2.2. BMS Architecture

The BMS architecture is an integral part of an EV's design, and it plays a crucial role in the vehicle's performance and safety. Understanding the architecture of the BMS is essential for engineers and designers involved in the development of EVs, as it impacts the vehicle's performance, range, and safety (Turksoy, Teke, & Alkaya, 2020). It allows them to design a BMS that is tailored to the specific needs of the vehicle and ensures that the battery operates efficiently and safely. The BMS architecture is composed of various functional blocks, each responsible for a specific aspect of the battery's management. By the end of this section, readers will have a comprehensive understanding of the different functional blocks of the BMS and the communication protocols used in EVs, as well as the different BMS architectures used in the industry (Saiteja & Ashok, 2022). They will have a better understanding of the various components that make up the BMS and how they work together to manage the battery pack's operation.

2.2.1. Functional Blocks of a BMS

BMS is comprised of several functional blocks. The functional blocks of a BMS work together to ensure the safe and efficient operation of the battery pack in an EV. They allow the BMS to monitor and control the battery's operation, communicate with other vehicle systems, protect the battery from external factors, and balance the cells in the battery pack. Various functionalities of a BMS are illustrated in Figure 2.3. In this section, various functional blocks of a BMS are discussed as follows:

a) *Sensing Block:* The sensing block is responsible for monitoring the battery pack's voltage, current, and temperature. It comprises various sensors, such as voltage sensors, current sensors, and temperature sensors, that provide accurate and reliable data to the BMS. The sensing block ensures that the battery operates within safe limits and prevents overcharging or over-discharging.
b) *Control Block*: The control block receives the data from the sensing block and processes it to determine the SoC, SoH, and state of power (SoP) of the battery. It also determines the battery's charging and discharging rates based on the vehicle's power demand and ensures that the battery operates within safe limits. The control block comprises a microprocessor, algorithms, and software that allow it to make accurate and timely decisions.
c) *Communication Block:* The communication block is responsible for communicating with other vehicle systems, such as the powertrain and the infotainment system, to ensure that the battery's operation is synchronized with the vehicle's overall performance. The communication block uses various communication protocols, such as the controller area network, to exchange data between the BMS and other systems.
d) *Protection Block*: The protection block ensures that the battery operates within safe limits and protects it from external factors that may damage it, such as high temperatures, overcharging, over-discharging, and short circuits. It comprises various protection mechanisms, such as fuses, circuit breakers, and overvoltage protection circuits.
e) *Balancing Block*: The balancing block ensures that the individual cells in the battery pack are charged and discharged equally, which prevents overcharging or over-discharging of any cell. It comprises

various balancing circuits, such as passive balancing and active balancing, that equalize the cells' voltage levels.

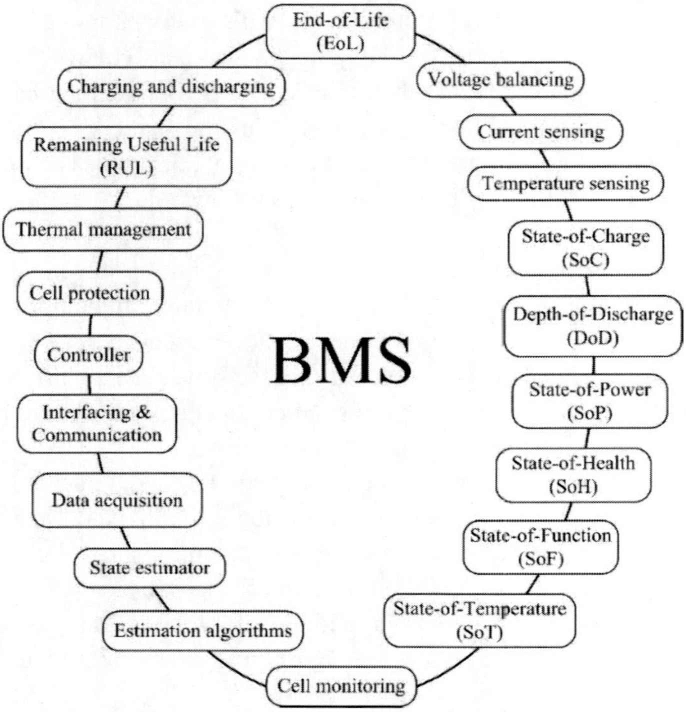

Figure 2.3. Pictorial depiction of various functionalities of BMS.

2.2.2. BMS Communication with Other Vehicle Systems

The BMS communicates with other vehicle systems to ensure that the battery is functioning correctly and to prevent any potential safety issues. It communicates depending on the vehicle's architecture and the communication protocol used. In this article, we will discuss the communication protocols used by the BMS to interact with other vehicle systems. The BMS communicates with other systems in the vehicle through a variety of protocols, including Controller Area Network (CAN), Local Interconnect Network (LIN), ethernet and wireless communication as shown in Figure 2.4. CAN is the most widely used protocol for communication between the BMS and other vehicle systems (Bae, Zhiguo, Koo, Kim, & Jang, 2013). It is a reliable and robust protocol that allows for high-speed communication between the BMS and other systems, such as the vehicle's engine control unit, transmission control unit, and charging system. The BMS uses CAN to provide the information about battery's SoC, SoH, SoP, and charging/discharging rates. Wireless communication is also used in some EVs to exchange data between the BMS and other vehicle systems. Wireless communication protocols, such as Bluetooth and Wi-Fi, are used to transfer data between the BMS and the driver's smartphone or other wireless devices.

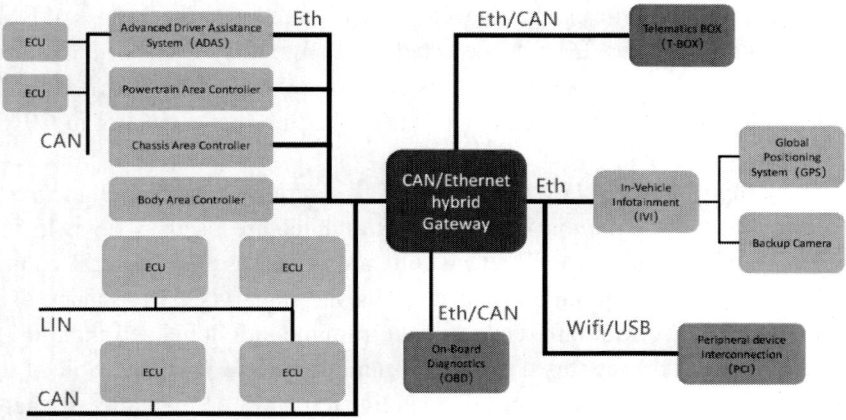

Source: MDPI.

Figure 2.4. BMS communication types and domain controller.

Ethernet is a high-speed communication protocol used in some EVs, enabling the BMS to exchange extensive data with other vehicle systems, like infotainment and driver assistance. It supports advanced features like autonomous driving and predictive maintenance. BMS communicates with the infotainment system to display battery performance information and provide driver recommendations. It also communicates with the thermal management system to maintain a safe battery temperature. The BMS exchanges data with the charging system to adjust the charging rate based on the battery state of charge. Additionally, the BMS may communicate with other systems, such as regenerative braking, to optimize battery operation.

Furthermore, the BMS plays a critical role in ensuring the safe and efficient operation of an electric vehicle's battery. It communicates with other vehicle systems, such as the engine control unit, infotainment system, thermal management system, charging system, and regenerative braking system, to ensure that the battery is functioning correctly and to prevent any potential safety issues. The BMS communicates with these systems using a variety of protocols, including CAN, LIN, and Ethernet, depending on the specific requirements of each system.

2.2.3. Types of BMS Architectures Used in EVs

BMS architecture depends on several factors, and it is an important consideration in EV design and development. There are several types of BMS architectures used in EVs as depicted in Figure 2.5, each with its own advantages and disadvantages. In this section, we will discuss the different types of BMS architectures used in EVs (Zhang, Cai, Zhang, Gu, & Liu, 2016).

a) *Centralized BMS Architecture*: The centralized BMS architecture is the most common type of BMS architecture used in EVs. In this architecture, all the battery cells are connected to a central control unit, which monitors and controls the battery's performance. The central control unit receives data from the individual cell monitoring units and uses this data to determine the SoC, SoH, and SoP of the battery. The advantages of a centralized BMS architecture are that it is relatively simple to implement and provides accurate battery monitoring and control.
b) *Distributed BMS Architecture*: The distributed BMS architecture is an alternative to the centralized BMS architecture. In this architecture, each battery cell has its monitoring unit, which communicates with neighbouring cells to monitor and control the battery's performance. The monitoring units communicate with each other using a communication bus, such as CAN or Ethernet. The advantages of a distributed BMS architecture are that it provides accurate cell-level monitoring and control and is more tolerant to failures than a centralized BMS architecture.
c) *Modular BMS Architecture*: The modular BMS architecture is a combination of centralized and distributed BMS architectures. In this architecture, the battery is divided into modules, each of which has its own monitoring unit. The module monitoring units communicate with a central control unit, which monitors and controls the battery's performance. The advantages of a modular BMS architecture are that

it provides accurate cell-level monitoring and control and is more flexible than a centralized BMS architecture.

d) *Hybrid BMS Architecture*: The hybrid BMS architecture is a combination of centralized and distributed BMS architectures. In this architecture, a central control unit monitors and controls the battery's performance, while distributed monitoring units provide additional cell-level monitoring and control. The distributed monitoring units communicate with the central control unit using a communication bus, such as CAN or Ethernet. The advantages of a hybrid BMS architecture are that it provides accurate cell-level monitoring and control and is more tolerant to failures than a centralized BMS architecture.

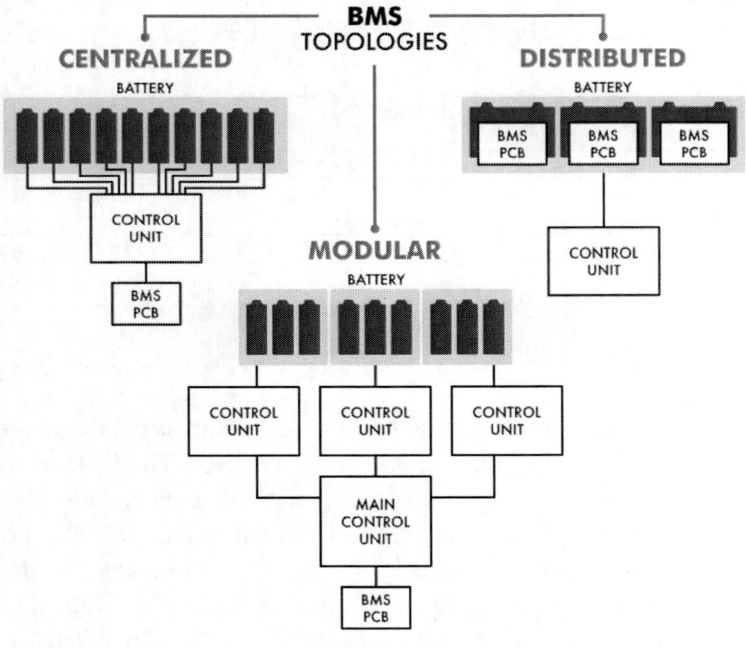

Figure 2.5. Typical BMS architecture used in EVs (Solovev & Petrova, 2021).

The choice of BMS architecture depends on several factors, such as the battery type, the number of battery cells, and the EV's performance requirements. For example, a distributed BMS architecture may be more suitable for EVs that require high power output and have a large number of battery cells, while a centralized BMS architecture may be more suitable for EVs that require high energy density and have a small number of battery cells.

Additionally, the choice of BMS architecture affects the cost, reliability, and safety of the EV. It is, therefore, an important consideration in EV design and development. A comparative analysis between the BMS architectures is presented in Table 2.1.

Table 2.1. A comparison between various BMS architectures

Parameters	Centralized	Distributed	Modular	Hybrid
Control	Central control unit	Distributed control units	Modular control units	Combination of control units
Communication	Centralized	Decentralized	Inter-module	Hybrid communication approach
Cooling	Easy	Difficult	Difficult	Variable
Scalability	Limited	High	Moderate	Flexible
Fault Tolerance	Vulnerable to the single point of failure	Redundant and fault-tolerant	Enhanced fault tolerance	Fault-tolerant approach
System Complexity	Lower	Higher	Moderate	Variable
Wiring	Centralized	Distributed	Modular	Combination of wiring approaches
Cost	Lower	Higher	Moderate	Variable
Implementation	Simpler	Complex	Flexible	Combination of implementations
Maintenance	Simpler	Difficult	Flexible	Variable approach

2.3. BMS Components

BMS is a complex system that includes several components to monitor and control the battery's performance in electric vehicles. The BMS constantly monitors the state of the battery and manages its charging and discharging to prevent overcharging, undercharging, and thermal runaway. It also provides crucial information about the battery to other vehicle systems, such as the motor controller, to optimize the overall performance of the vehicle. In this section, we will discuss the various components of a BMS, including the types of batteries used in EVs, the sensors used in the BMS, and the balancing circuitry that ensures the cells in the battery pack are evenly charged.

2.3.1. Types of Batteries Used in EVs

EVs rely on batteries as their primary source of energy, making the battery pack a critical component of the vehicle. The BMS plays a crucial role in ensuring the safety and efficiency of the battery pack. The BMS must be designed to work with the specific type of battery used in the EV (Waseem, Ahmad, Parveen, & Suhaib, 2023). There are several types of batteries used in EVs, each with its unique characteristics and advantages. Lithium-ion, nickel-metal hydride (NiMH), lead-acid and solid-state batteries are some of the most common types of batteries used in EVs.

a) *Lithium-Ion Batteries:* Lithium-ion batteries are the most common type of battery used in EVs (Xiong, et al., 2019). They are lightweight and have a high energy density, which makes them ideal for use in vehicles. Lithium-ion batteries also have a long lifespan and can be charged quickly. They offer a high energy density, which means they can store a lot of energy in a relatively small space. Lithium-ion batteries are also lightweight and can be charged quickly, making them ideal for use in EVs. One example of an EV that uses lithium-ion batteries is the Tesla Model S. The Model S has a battery pack made up of thousands of individual lithium-ion battery cells. The BMS in the Model S constantly monitors the state of each battery cell and manages its charging and discharging to ensure the safe and efficient operation of the battery pack. One advantage of lithium-ion batteries is their ability to be fast-charged. This means that EVs with lithium-ion batteries can be charged quickly, allowing drivers to get back on the road more quickly than with other types of batteries. For example, the Porsche Taycan, which has a 93.4 kWh lithium-ion battery pack, can be charged from 5% to 80% in just 22.5 minutes using a 350 kW DC fast charger. However, lithium-ion batteries also have some limitations. They can be expensive, which can increase the cost of EVs. They also have a limited lifespan, and their performance can degrade over time, which can reduce the range of the EV. The BMS in an EV with lithium-ion batteries must be designed to manage these limitations and ensure the safe and efficient operation of the battery pack over its lifespan.
b) *Nickel-Metal Hydride (NiMH) Batteries:* NiMH batteries were once the most common type of battery used in hybrid vehicles. They are heavier than lithium-ion batteries and have a lower energy density,

which means they are not as efficient as lithium-ion batteries. NiMH batteries also have a shorter lifespan than lithium-ion batteries. NiMH batteries are also less prone to overheating than lithium-ion batteries, which can make them safer in some applications. For example, an EV that used NiMH batteries is the Toyota RAV4 EV. The RAV4 EV, which was produced between 1997 and 2003, had a battery pack made up of 24 NiMH battery modules. Each module contained 6 individual battery cells, for a total of 144 battery cells in the pack. The BMS in the RAV4 EV monitored the state of each battery module and managed the charging and discharging of the battery pack to ensure the safe and efficient operation of the vehicle. While NiMH batteries have some advantages, such as their lower cost and longer lifespan, they also have some limitations. They have a lower energy density than lithium-ion batteries, which means they cannot store as much energy in the same space. This can reduce the range of the EV. NiMH batteries also have a slower charging rate than lithium-ion batteries, which can make charging times longer. Overall, NiMH batteries have been largely replaced by lithium-ion batteries in modern EVs due to their lower energy density and slower charging rates. However, NiMH batteries may still be used in some niche applications where their lower cost and longer lifespan are more important than their lower energy density and slower charging rates. The BMS in an EV with NiMH batteries must be designed to manage the limitations of these batteries and ensure the safe and efficient operation of the vehicle.

c) *Lead-Acid Batteries:* Lead-acid batteries are the oldest type of rechargeable battery and have been used in EVs since their inception, although they are less common in modern EVs compared to lithium-ion batteries. They are inexpensive, but they are heavy and have a low energy density, which makes them less efficient than other types of batteries. Lead-acid batteries are used in low-speed EVs like electric scooters, golf carts, and neighborhood electric vehicles (NEVs). They offer advantages such as cost-effectiveness, availability, ease of maintenance, and improved safety compared to some other battery chemistries. However, they have drawbacks, including their weight and size, limited energy density resulting in shorter driving ranges, limited cycle life, and slower charging times. While lead-acid batteries are gradually being phased out in mainstream EV applications in favor of advanced battery technologies like lithium-ion, they remain relevant in specific niche markets where cost-effectiveness and suitability for low-speed EVs are important considerations.

d) *Solid-State Batteries:* Solid-state batteries are a relatively new type of battery that uses a solid electrolyte instead of a liquid electrolyte. They are still in development, but they have the potential to be much lighter and have a higher energy density than lithium-ion batteries. For instance, an EV utilizing solid-state batteries is the Fisker Ocean. Fisker Inc., an electric vehicle startup, plans to equip the Ocean, an all-electric SUV, with advanced solid-state batteries. The solid-state batteries in the Fisker Ocean are expected to provide several benefits compared to traditional lithium-ion batteries. These include higher energy density, enabling a longer driving range per charge, and faster charging times. Solid-state batteries can be designed to have improved thermal stability, reducing the risk of thermal runaway and enhancing overall safety. Solid-state batteries have the potential to revolutionize the EV industry by addressing some of the limitations of traditional lithium-ion batteries.

Table 2.2. Comparison of features in different batteries

Feature	Li-Ion	NiMH	Lead-Acid	Solid-State
Energy Density	High	Medium	Low	High
Specific Power	High	Medium	Low	Medium
Cycle Life	High	Medium	Medium to High	High
Self-Discharge Rate	Low	High	High	Low
Charging Time	Short	Moderate	Long	Moderate
Temperature Sensitivity	Moderate to High	Moderate	Low	Low to Moderate
Memory Effect	None	Yes	None	None
Environmental Impact	Potential concern	Potential concern	Hazardous	Potential concern
Cost	High	Medium	Low to Medium	High
Application Range	Wider	Limited	Limited	Wider

The type of battery used in an EV depends on several factors, including the desired range of the vehicle, the weight of the battery pack, and the cost of the battery. Figure 2.6 represents a comparison between different battery technologies. Table 2.2 shows a comparison table highlighting seven key differences between lithium-ion batteries, NiMH batteries, lead-acid batteries, and solid-state batteries. Characteristic graphs of battery volume vs. vehicle range, and specific power vs. specific energy of a battery cell are shown in Figure 2.6(a) and Figure 2.6(b) respectively. The BMS must be designed to work with the specific type of battery used in the EV to ensure its safe and efficient operation. EVs use various battery types for energy storage. Lithium-ion batteries are the most common, with high energy density, long cycle life, and good power output. Different chemistries like LiFePO4, NMC, and LiPo have unique advantages. NiMH batteries are used in hybrids, while solid-state batteries show promise for the future with improved safety and energy density. Emerging technologies like Li-S and Li-Air offer even higher energy densities. Battery choice considers cost, energy density, weight, and safety, driving continuous innovation to improve EV performance and range.

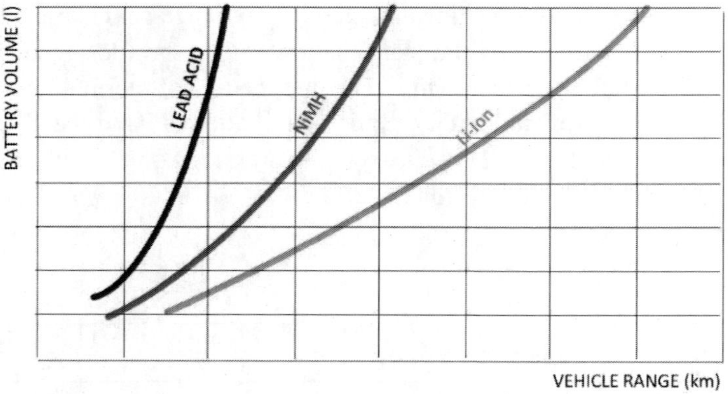

(a)

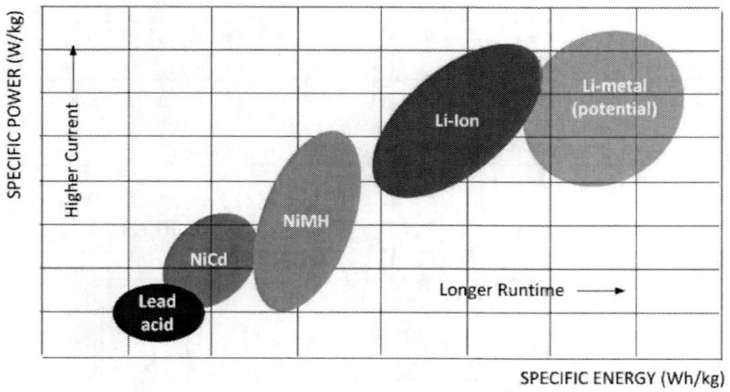

(b)

Figure 2.6. Graphs show the chemistry characteristics of a battery cell (a) battery volume vs. vehicle range, and (b) power vs. energy (Johnson & Ekler, 2020).

2.3.2. Sensors used in BMS

Sensors play a crucial role in the BMS of electric vehicles. They provide real-time data about the battery's condition, performance, and environment, allowing the BMS to make informed decisions and ensure optimal operation. Here are some of the key sensors used in BMS:

a) *Voltage Sensors:* Voltage sensors are used to measure the voltage levels of individual battery cells or modules. They provide critical information about the SoC and SoH of the battery. By monitoring voltage, the BMS can detect cell imbalances, overvoltage, under voltage, and abnormal voltage fluctuations.
b) *Current Sensors:* Current sensors measure the flow of electric current in and out of the battery pack. They help determine the battery's charging and discharging rates, as well as the overall power flow. By accurately measuring current, the BMS can ensure proper battery usage, prevent overcharging or over-discharging, and estimate the remaining battery capacity.
c) *Temperature Sensors:* Temperature sensors monitor the temperature of the battery cells and the surrounding environment. They play a critical role in maintaining the battery's thermal management. By monitoring temperature, the BMS can prevent overheating, optimize charging and discharging rates based on temperature conditions, and trigger safety mechanisms to protect the battery from extreme temperature conditions.
d) *Pressure Sensors:* In some advanced battery systems, pressure sensors are employed to monitor the internal pressure of the battery cells. They help detect any abnormalities or malfunctions that may lead to swelling or leakage. By monitoring pressure, the BMS can identify potential safety hazards and take appropriate actions to mitigate risks.
e) *Humidity Sensors:* Humidity sensors are used to measure the moisture content or relative humidity within the battery pack. They assist in identifying any moisture ingress, which can be detrimental to the battery's performance and longevity. By monitoring humidity levels, the BMS can detect moisture-related issues and prevent potential damage to the battery cells.
f) *Gas Sensors:* Gas sensors are employed to detect the presence of hazardous gases, such as hydrogen or carbon monoxide, which may

indicate a battery malfunction or abnormal conditions. These sensors help ensure the safety of the battery pack and the surrounding environment.

These sensors, along with others like accelerometers and strain gauges, provide essential data to the BMS, enabling it to monitor the battery's vital parameters and respond accordingly. The BMS processes the information from these sensors in real-time, making intelligent decisions for efficient battery usage, performance optimization, and safety enhancements. By integrating a variety of sensors, the BMS can accurately assess the battery's condition, detect anomalies, and take corrective actions. This allows for effective battery management, extending the battery's lifespan, improving efficiency, and ensuring the safety and reliability of EV operations.

2.3.3. Battery Cell Voltage Monitoring

Battery cell voltage monitoring is a critical aspect of the BMS in EVs. It involves continuously measuring and monitoring the voltage levels of individual battery cells or modules within the battery pack. This information is essential for assessing the SoC, SoH, and overall performance of the battery. Battery cell voltage monitoring serves several important purposes within the BMS:

1) *SoC Estimation:* Voltage measurement indicates the battery's remaining charge. By monitoring the voltage levels of each cell, the BMS can estimate the SoC of the entire battery pack. This information is crucial for determining the available energy and predicting the driving range of the EV.
2) *Cell Balancing:* Battery cell voltage monitoring helps identify any imbalances in the voltage levels across the cells within a battery pack. Cells within a pack can have slight variations in capacity or performance, leading to uneven voltage levels. The BMS uses this information to implement cell balancing algorithms or techniques to equalize the charge across cells, ensuring optimal performance and maximizing the overall capacity of the battery pack.
3) *Fault Detection and Diagnosis:* Abnormal voltage levels can indicate cell degradation, faulty cells, or other issues within the battery pack. By monitoring voltage levels, the BMS can detect potential faults or

malfunctions and trigger appropriate safety measures. This helps prevent catastrophic failures, extend the battery's lifespan, and ensure safe operation of the EV.

4) *Overvoltage and Undervoltage Protection:* Voltage monitoring enables the BMS to protect the battery cells from overvoltage or undervoltage conditions. If the voltage exceeds safe limits, the BMS can initiate actions such as reducing the charging rate or activating safety mechanisms to prevent damage to the cells. Similarly, if the voltage drops too low, the BMS can implement measures to avoid over-discharging, which can lead to reduced battery performance and potential damage.
5) *Performance Monitoring and Prognostics:* Voltage measurements provide insights into the overall performance and health of the battery cells. By analyzing voltage trends over time, the BMS can detect changes in cell behaviour, identify capacity degradation, and predict remaining useful life. This information helps in scheduling maintenance, optimizing battery usage, and maximizing the lifespan of the battery pack.

Battery cell voltage monitoring is typically achieved through dedicated voltage sensors or monitoring integrated circuits connected to each cell. These sensors provide accurate and real-time voltage measurements, which are processed by the BMS to make informed decisions about battery operation and management. By continuously monitoring battery cell voltages, the BMS can ensure the efficient and safe operation of the battery pack in an EV. It enables SoC estimation, cell balancing, fault detection, and protection mechanisms, while also providing valuable data for performance monitoring and prognostics. This comprehensive monitoring and control of cell voltages contribute to optimal battery performance, extended lifespan, and enhanced safety in electric vehicles.

2.3.4. Current and Temperature Sensing

Current and temperature sensing are essential components of the BMS in EVs. These sensors provide crucial data about the current flow and temperature conditions within the battery pack, enabling the BMS to monitor and manage the battery's performance, safety, and longevity.

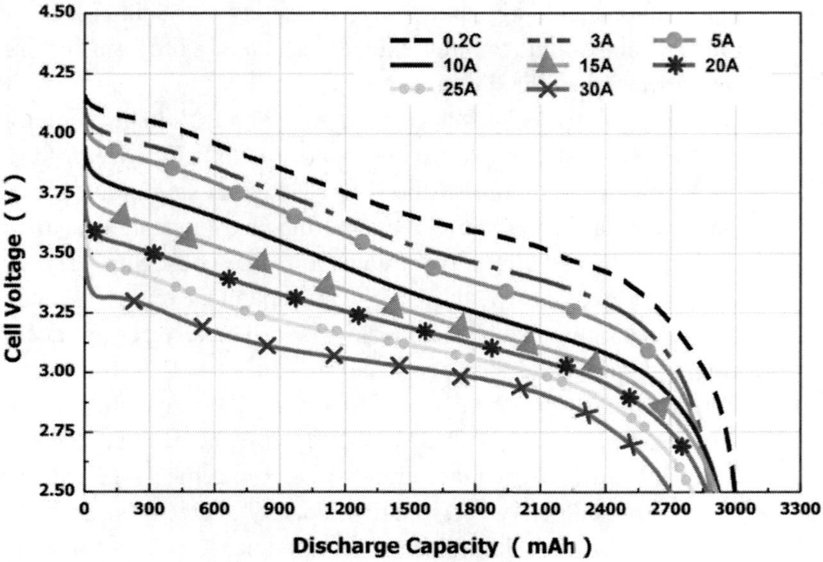

Figure 2.7. Discharging profile of a Li-ion cell. *(Image credit: LG Chem* (Discharge charateristics of Lithium-ion batteries, 2020)*).*

Current sensing involves measuring the flow of electric current into and out of the battery pack. This information is vital for accurately estimating the battery's state of charge, monitoring charging and discharging rates, and ensuring optimal utilization of the battery's capacity. A discharging characteristic of a Li-ion cell is shown in Figure 2.7. The figure shows how a cell voltage drops with discharge capacity. Current sensors, such as shunt resistors or Hall effect sensors, are integrated into the BMS to measure the current with high precision. By continuously monitoring the current, the BMS can detect abnormal charging or discharging patterns, prevent overcharging or over-discharging, and optimize the battery's operational efficiency.

Temperature sensing is equally important in the BMS as it helps monitor the thermal conditions of the battery cells and the surrounding environment. Lithium-ion batteries, in particular, are sensitive to temperature variations, and operating them within the recommended temperature range is crucial for their performance, safety, and lifespan. Temperature sensors, such as thermistors or RTDs (Resistance Temperature Detectors), are placed strategically within the battery pack to measure the temperature accurately. The BMS uses this information to implement thermal management strategies, such as controlling cooling systems or adjusting charging rates, to maintain the battery cells within the optimal temperature range. Monitoring temperature conditions also enables the BMS to detect overheating or extreme temperature events, triggering safety measures to prevent thermal runaway and ensure the longevity of the battery.

Current and temperature sensing in the BMS work together to monitor and control the battery's electrical and thermal behavior. By combining data from both sensors, the BMS can optimize battery performance, ensure safety, and extend battery lifespan. For example, it adjusts charging rates during high current flow to prevent overheating. Temperature monitoring during discharge enables thermal management to maintain safe operating ranges. The BMS uses this data for diagnostics, predicting battery life, scheduling maintenance, and optimizing EV operation. Accurate monitoring of current and temperature is critical for effective battery management in EVs.

2.3.5. Balancing Circuitry

Balancing circuitry is a vital component of the BMS in electric vehicles. It plays a crucial role in equalizing the charge across individual battery cells or modules within a battery pack, ensuring optimal performance, extended battery life, and improved safety. Battery cells in a pack can have slight variations in capacity or performance, leading to imbalances in charge levels. Over time, these imbalances can result in some cells being overcharged while others are undercharged, leading to reduced overall capacity, increased aging, and potential safety hazards. Balancing circuitry aims to address these imbalances by redistributing the charge among the cells, bringing them to a similar state of charge.

Balancing circuitry in the BMS consists of modules connected to each battery cell or module. Passive balancing uses resistors to dissipate excess charge, while active balancing employs electronic circuits to transfer charge actively between cells. When the BMS detects voltage differences beyond a threshold, it activates balancing to equalize cell charge levels. This prevents overcharging and reduces stress on cells, improving overall performance and battery lifespan. Balancing occurs during both charging and discharging, preventing damage and ensuring uniform discharge rates among cells.

The balancing process is controlled by the BMS, which monitors the voltage levels of individual cells and determines when balancing is needed. The BMS calculates the optimal amount of charge transfer required for each cell to achieve balance. It continuously evaluates the state of charge of the cells and adjusts the balancing process accordingly. Moreover, the balancing circuitry ensures that all battery cells within a pack are charged and discharged uniformly, preventing imbalances and improving overall battery performance. By redistributing the charge among cells, balancing circuitry extends the battery pack's lifespan, enhances energy efficiency, and enhances the safety and reliability of the EV.

2.4. BMS Algorithms

The BMS uses various algorithms to estimate the battery's state, including SoC, SoH, and SoF (M. S. Hossain Lipu a, et al., 2021), (Abdelaal, Mukhopadhyay, & Rehman, 2022). SoC estimation is used to determine the remaining energy in the battery, while SoH estimation is used to determine the battery's lifespan and remaining capacity. SoF estimation is used to determine the battery's ability to provide power, which is critical for maintaining the performance and range of the EV.

2.4.1. State of Charge (SoC) Estimation Algorithms

State of charge estimation is a fundamental aspect of BMS in EVs. It refers to the estimation of the remaining energy or capacity of the battery pack, providing crucial information about how much charge is available for use (Cheng, Divakar, Wu, Ding, & Ho, 2011). Accurate SoC estimation is essential for optimal utilization of the battery, range prediction, and ensuring the safe and efficient operation of the EV. SoC estimation algorithms employ various techniques and measurements to estimate the battery's state of charge, and different algorithms have been developed and implemented in EVs.

2.4.1.1. Coulomb Counting Method

The Coulomb Counting method is a commonly used technique for SoC estimation in battery management systems. It is based on the principle of measuring the cumulative amount of charge flowing into or out of the battery over time. By integrating the current measurement, the method calculates the state of charge of the battery, providing an estimate of the remaining energy or capacity.

The Coulomb Counting method for SoC estimation relies on measuring and integrating the current flowing in and out of the battery over time to track accumulated charge. It assumes a constant battery capacity and accurate current measurement, but several factors can affect its accuracy, such as aging, self-discharge, and temperature effects. To improve accuracy, adjustments for current measurement errors, battery capacity based on age or temperature, and calibration for integration inaccuracies are considered. However, the Coulomb Counting method has limitations due to uncertainties in current measurement and changes in battery performance over time, and it does not account for other factors affecting SoC, like temperature-related voltage changes.

To address these limitations, the Coulomb Counting method is often combined with other SoC estimation techniques, such as voltage-based methods or advanced algorithms like the Kalman filter. By integrating multiple estimation methods, the BMS can achieve higher accuracy and reliability in SoC estimation, leading to effective battery management and optimization of EV performance. While the Coulomb Counting method provides a practical approach to estimate SoC, its accuracy is enhanced when used in conjunction with other techniques and properly calibrated to achieve more reliable SoC estimates in EV batteries.

2.4.1.2. Open-Circuit Voltage (OCV) Method

The Open-Circuit Voltage (OCV) method is a widely used SoC estimation technique in BMSs, relying on the relationship between a battery's open-circuit voltage and its state of charge. It utilizes a predefined voltage vs. SoC table or mathematical model, developed through calibration, to estimate SoC based on the measured open-circuit voltage. The OCV method is less affected by current measurement errors compared to Coulomb Counting and provides a quick and straightforward estimation without complex current integration. However, its accuracy can be influenced by temperature variations and battery aging effects. Temperature compensation techniques are employed to adjust the OCV estimation based on the measured temperature, and periodic calibration is necessary to account for battery aging effects and maintain accuracy.

The OCV method for SoC estimation has limitations as it assumes a constant voltage vs. SoC relationship and neglects other factors like voltage hysteresis and internal resistance. To improve accuracy, it is often combined with other estimation techniques, like current integration or advanced algorithms like the extended Kalman filter. Integrating multiple approaches enhances SoC estimation for better battery management and EV performance. Despite its limitations, the OCV method is convenient and relatively accurate, especially when used with temperature compensation and periodic calibration, making it a valuable tool in the BMS for effective battery management and reliable range estimation in EVs.

2.4.1.3. Kalman Filter and Extended Kalman Filter (EKF)

The Kalman Filter and Extended Kalman Filter (EKF) are widely used estimation algorithms in BMS for SoC and SoH estimation. They aim to improve accuracy and reliability by considering uncertainties and noise in the system. The Kalman Filter is recursive, combining sensor measurements with a mathematical model for optimal state estimation. It operates through prediction and updates steps, where the prediction uses the system's dynamic model to estimate the current state based on previous data and control inputs, and the update incorporates sensor measurements to correct the predicted state. The EKF extends the Kalman Filter to handle nonlinear systems by linearizing the nonlinear equations using Taylor series expansion, allowing for more accurate estimation in the presence of nonlinear dynamics.

In BMS for EVs, the Kalman Filter and EKF use sensor measurements like voltage, current, temperature, and impedance, along with a battery model, to estimate the battery's state in real-time. They account for uncertainties, noise, and variations due to temperature changes and aging effects. The algorithms adaptively weigh the influence of measurements and model predictions using covariance matrices, reducing the impact of noise and improving accuracy. Both algorithms require a dynamic battery model that captures battery behavior under various conditions, based on electrical equivalent circuit models or physics-based models. When combined with other techniques like Coulomb Counting or Open-Circuit Voltage, the Kalman Filter and EKF significantly enhance BMS performance and reliability in EVs.

2.4.1.4. Model-Based Methods

Model-based methods are employed in BMS for EVs to estimate parameters like SoC and SoH using mathematical battery models. These methods utilize battery models representing electrochemical and electrical behavior to simulate battery performance under different conditions. By leveraging these models, model-based methods provide accurate and dynamic estimation, contributing to optimized battery management, enhanced EV efficiency, and reliability. The BMS uses the battery model to predict battery behavior based on input variables like current, voltage, temperature, and time. BMS compares these predictions with sensor measurements to estimate battery parameters. The models can be based on electrical equivalent circuit models, electrochemical models, or physics-based models, capturing complex battery behavior, including diffusion, reaction kinetics, and ion transport. Model-based methods can handle nonlinearities, temperature effects, and aging, and enable proactive battery management strategies by predicting battery behavior and degradation. However, they require high-fidelity models, which may be complex and computationally intensive, and necessitate periodic calibration to maintain accuracy.

Overall, model-based methods provide accurate and dynamic SoC and SoH estimation in BMS for EVs, enabling optimized battery management and improved EV performance. Their ability to handle various battery behaviors and anticipate degradation makes them valuable tools for ensuring reliable and efficient battery operation. However, the complexity and computational demands of high-fidelity battery models, as well as the need for periodic calibration, should be considered when implementing these methods.

2.4.1.5. Artificial Intelligence (AI) and Machine Learning (ML) Techniques

AI and ML techniques are increasingly utilized in electric vehicle BMS to enhance accuracy, efficiency, and adaptability. These techniques leverage algorithms and data analysis to optimize battery performance, improve state estimation, and make intelligent decisions without explicit programming. They excel at handling complex and nonlinear relationships in battery systems, making them well-suited for tasks like battery modeling, state estimation, and control. One common application of AI and ML in BMS is SoC estimation, where ML algorithms learn from historical battery data to build models for real-time SoC estimation, adapting to changing battery behavior and accounting for factors like aging effects and temperature variations. Additionally, predictive analytics can be used for battery health management, identifying patterns and correlations to predict battery degradation and enabling proactive maintenance strategies. AI and ML also optimize charging and discharging strategies by considering energy prices, user preferences, and usage patterns to develop intelligent algorithms for cost minimization and battery lifespan enhancement. Anomaly detection and fault diagnosis are other applications, that allow continuous monitoring and early identification of abnormal behaviors or system faults, contributing to improved safety and reliability of EV batteries. Challenges include the need for large datasets, computational resources, and interpretability of ML models to ensure transparency in the BMS.

Furthermore, AI and ML techniques provide powerful tools for optimizing EV battery performance, accurate state estimation, and intelligent decision-making in BMS. Their application holds great potential for improving the efficiency, reliability, and longevity of EV batteries, promoting the advancement of sustainable transportation.

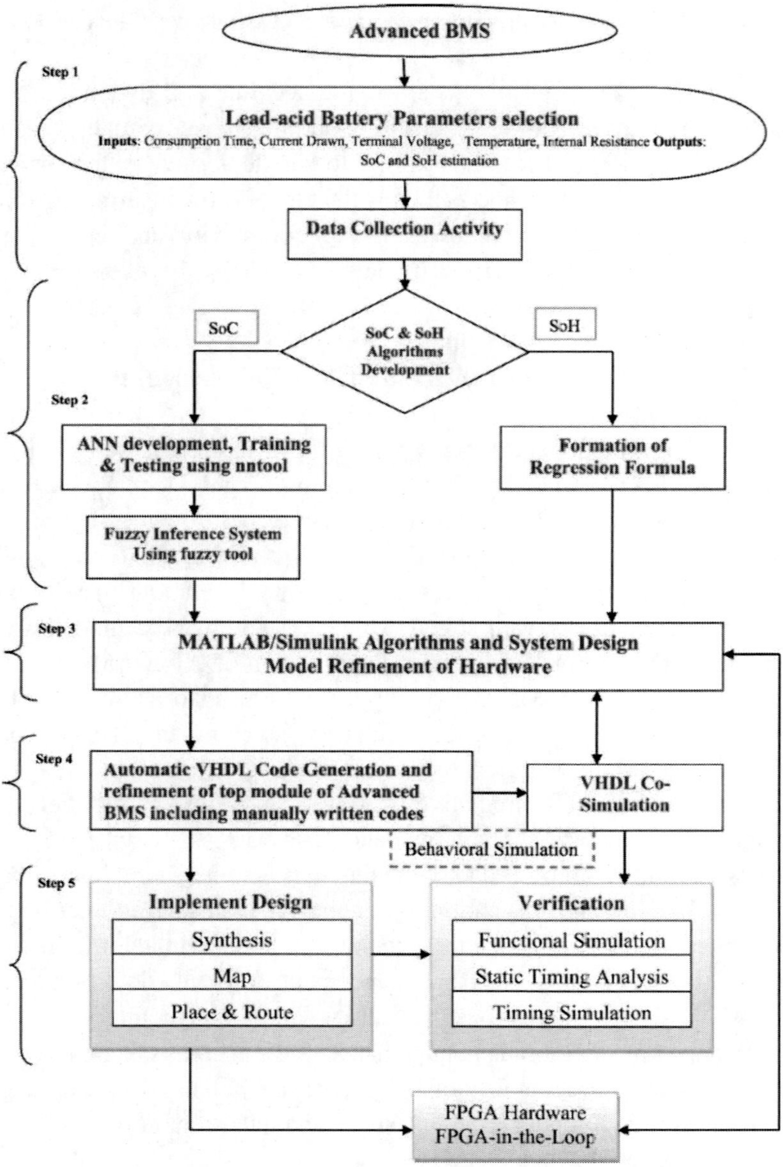

Figure 2.8. A flowchart for development of a BMS algorithm using MATLAB (Kumar, Khare, & Chaturvedi, 2018).

Each SoC estimation algorithm has its advantages and limitations, and the choice of algorithm depends on factors such as the accuracy requirements, available sensor measurements, computational resources, and the specific characteristics of the battery pack. A flowchart for the SoC estimation algorithm of a lead-acid battery is illustrated in Figure 2.8. In practice, a combination of different algorithms or hybrid approaches may be used to achieve more accurate and robust SoC estimation.

Accurate SoC estimation is crucial for several aspects of EV operation:

- *Range Estimation:* SoC estimation enables the prediction of the remaining driving range, allowing drivers to plan their journeys and make informed decisions about charging.
- *Battery Performance Optimization:* By knowing the battery's SoC, the BMS can optimize the battery's operating conditions, such as adjusting charging or discharging rates, to enhance overall performance and efficiency.
- *Battery Protection:* SoC estimation helps prevent overcharging or over-discharging, which can degrade the battery and reduce its lifespan. It also helps in implementing safety measures to prevent extreme SoC conditions.
- *SoH Estimation:* SoC estimation is often used in conjunction with other battery parameters to estimate the SoH of the battery. By tracking the battery's SoC over time, along with factors such as voltage, current, temperature, and capacity measurements, the BMS can assess the battery's degradation, identify potential issues, and provide insights into the overall health and remaining lifespan of the battery. Accurate SoC estimation is essential for reliable and precise SoH estimation, enabling proactive maintenance and replacement strategies for maximizing battery life and optimizing cost-efficiency.

2.4.2. State of Health (SoH) estimation algorithms

SoH estimation is a crucial aspect of battery management systems as it determines the overall health and degradation of the battery. Accurate SoH estimation enables effective battery management, optimized performance, and informed decision-making regarding battery replacement or maintenance. Figure 2.8 shows the SoC and SoH estimation process using a flowchart.

Several estimation algorithms have been developed to assess the SoH of EV batteries based on various factors and measurements.

a) *Capacity-Based Methods:* These algorithms estimate SoH by analyzing the capacity degradation of the battery over time. They typically compare the current capacity of the battery with its initial or reference capacity to determine the extent of degradation. The capacity can be obtained through various techniques such as full charge-discharge cycles, pulse discharge tests, or impedance spectroscopy. Capacity-based methods are straightforward and widely used, but they may not capture all degradation mechanisms and may be affected by factors such as operating conditions and usage patterns.
b) *Resistance-Based Methods:* These algorithms estimate SoH by analyzing the internal resistance of the battery, which is an indicator of its health. By monitoring the resistance changes over time, these methods can identify degradation and provide an estimate of the remaining useful life of the battery. Resistance-based methods can be implemented using techniques such as electrochemical impedance spectroscopy or incremental capacity analysis. However, they may not capture all types of degradation and can be influenced by temperature and operating conditions.
c) *Model-Based Methods:* These algorithms employ mathematical models of the battery to estimate SoH. The models consider various parameters, such as internal resistance, capacity fade, and aging effects, to simulate the behaviour of the battery over time. Model-based methods require accurate battery models and typically use techniques like state estimation, filtering, or optimization algorithms to adapt the model parameters based on measurements. These methods can provide accurate SoH estimation but require computational resources and accurate model calibration.
d) *Machine Learning (ML) Methods:* ML algorithms, such as neural networks, support vector machines, or random forests, can be used to estimate SoH based on historical battery data. These algorithms learn patterns and correlations from the data and develop models that can predict the SoH based on input variables such as voltage, current, temperature, and usage patterns. ML methods are flexible and can capture complex relationships, but they require large datasets for training and validation.

e) *Hybrid Approaches:* Hybrid approaches combine multiple estimation techniques to improve the accuracy and robustness of SoH estimation. For example, a hybrid approach may utilize capacity-based methods in combination with resistance-based methods or incorporate ML algorithms for data-driven estimation in conjunction with model-based methods. By leveraging the strengths of different approaches, hybrid methods aim to overcome limitations and provide more accurate and reliable SoH estimation.

SoH estimation in battery management systems is a complex task, and there is no one-size-fits-all algorithm for all battery chemistries and operating conditions. The choice of estimation method depends on factors like battery type, available data, computational resources, and accuracy requirements. BMS designers must carefully evaluate and select the most suitable SoH estimation algorithm based on the specific application and battery characteristics. Accurate SoH estimation is crucial for effective battery management in EVs, enabling health monitoring, predicting remaining lifespan, and optimizing battery usage. It provides valuable insights into battery degradation and performance, contributing to extended battery lifespan, improved reliability, and efficient utilization of EV batteries.

2.4.3. State of Function (SoF) Estimation Algorithms

SoF estimation algorithms play a critical role in BMS for electric vehicles by providing an assessment of the battery's ability to perform its intended functions. SoF estimation goes beyond measuring the battery's SoC, SoH, and power capability to evaluate its overall functional capability and reliability. These algorithms help ensure the safe and reliable operation of the battery system throughout its service life.

SoF estimation algorithms consider various factors to assess the functional performance of the battery. These factors include capacity fade, power fade, internal resistance changes, aging effects, and environmental conditions. By monitoring and analyzing these parameters, SoF estimation algorithms provide valuable insights into the battery's ability to deliver the required power and energy under different operating conditions. Several methods and techniques are used for SoF estimation in BMS. Here are some commonly employed approaches:

a) *Voltage-Based Methods:* Voltage-based algorithms estimate the SoF by analyzing the battery voltage behaviour during charge and discharge cycles. These methods rely on voltage measurements at different SoC to establish voltage profiles and calibrate voltage models. By comparing the measured voltage with the expected voltage, these algorithms can determine the functional state of the battery. Voltage-based methods are simple and widely used but may not capture all aspects of battery performance and degradation.
b) *Current-Based Methods:* Current-based algorithms estimate SoF by analyzing the current behaviour during charge and discharge cycles. These methods utilize current measurements and models to track the functional state of the battery. By comparing the actual current with the predicted or expected current, these algorithms can assess the performance and health of the battery. Current-based methods are sensitive to variations in current and may require accurate current sensing for reliable estimation.
c) *Model-Based Methods:* Model-based SoF estimation algorithms leverage mathematical models of the battery to estimate its functional performance. These models consider various parameters, such as internal resistance, capacity fade, and aging effects, to simulate the battery behaviour under different operating conditions. By comparing the model predictions with the actual measurements, these algorithms can estimate the SoF of the battery. Model-based methods require accurate battery models and often employ techniques such as state estimation, filtering, or optimization algorithms to adapt the model parameters based on real-time measurements.
d) *Data-Driven Methods:* Data-driven SoF estimation algorithms utilize historical battery data to develop models that can predict the battery's functional performance. Machine learning techniques, such as neural networks, support vector machines, or random forests, are commonly used in data-driven methods. These algorithms learn from historical data, including voltage, current, temperature, and usage patterns, to establish correlations between the inputs and the battery's functional performance. Data-driven methods are flexible and can capture complex relationships but require large datasets for training and validation.
e) *Adaptive Methods:* Adaptive SoF estimation algorithms dynamically adjust the estimation parameters and models based on real-time battery behaviour. These algorithms continuously update the SoF

estimation based on the measured data, allowing for adaptation to changing battery conditions and aging effects. Adaptive methods can be based on various techniques, such as recursive least squares estimation, adaptive filtering, or online optimization algorithms. By adapting to the evolving battery characteristics, adaptive methods enhance the accuracy of SoF estimation.

f) *Hybrid Approaches:* Hybrid approaches combine multiple estimation techniques to improve the accuracy and robustness of SoF estimation. These approaches may integrate model-based methods with data-driven methods or employ a combination of adaptive and model-based techniques. Hybrid algorithms aim to leverage the strengths of different methods to overcome limitations and provide more accurate and reliable SoF estimation.

SoF estimation algorithms are crucial for EVs as they enable monitoring and predicting the battery's functional performance. They provide valuable information for optimizing battery usage, managing load profiles, and ensuring safe and reliable operation. By assessing the battery's ability to deliver power and energy under various conditions, SoF estimation algorithms contribute to extending the battery's lifespan, maximizing performance, and preventing unexpected failures. The complexity of SoF estimation requires careful evaluation and selection of the most suitable algorithm based on battery chemistry, available data, computational resources, and desired accuracy. Accurate SoF estimation algorithms enable effective battery management, enhanced safety, and improved overall EV performance. By providing insights into the battery's functional capability and reliability, these algorithms contribute to the optimization and utilization of the battery system throughout its service life.

2.4.4. Fault Detection and Diagnosis Algorithms

Fault detection and diagnosis algorithms play a crucial role in the BMS. Various algorithms are depicted in Figure 2.9 to Figure 2.11. These algorithms are designed to identify and diagnose faults or abnormalities in the battery system, enabling timely intervention and maintenance (Hu, et al., 2020). By detecting and diagnosing faults accurately, the BMS can ensure the safe and reliable operation of the battery, extend its lifespan, and prevent potential failures.

Fault detection algorithms continuously monitor various parameters of the battery system, such as voltage, current, temperature, and impedance, to identify deviations from normal behaviour. These algorithms compare the measured values with predefined thresholds or statistical models to detect anomalies that may indicate the presence of a fault. The detected anomalies can range from simple deviations in voltage or temperature to more complex patterns that require advanced data analysis techniques. Once a fault is detected, fault diagnosis algorithms come into play to identify the specific type and location of the fault within the battery system. Fault diagnosis involves analyzing the collected data, identifying patterns or signatures associated with different fault types, and matching them with a fault database or expert knowledge. By accurately diagnosing the faults, the BMS can provide valuable information for maintenance actions, such as isolating faulty cells, triggering alarms, or initiating safety protocols. Various fault detection and diagnosis algorithms are utilized in BMS for EVs. Some commonly used approaches are presented here.

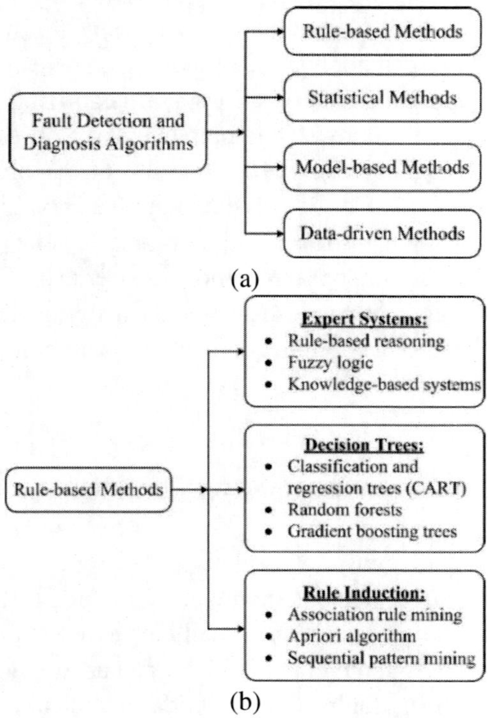

Figure 2.9. Various fault diagnosis methods (a) classification, and (b) widely used rule-based methods.

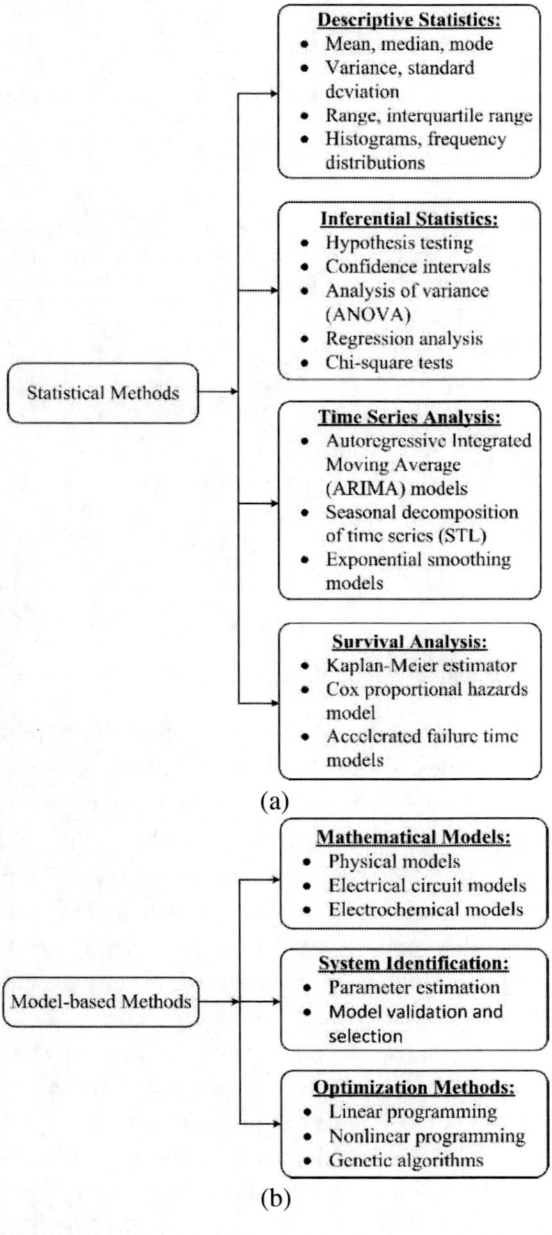

Figure 2.10. Fault diagnosis methods (a) statistical and (b) model-based methods.

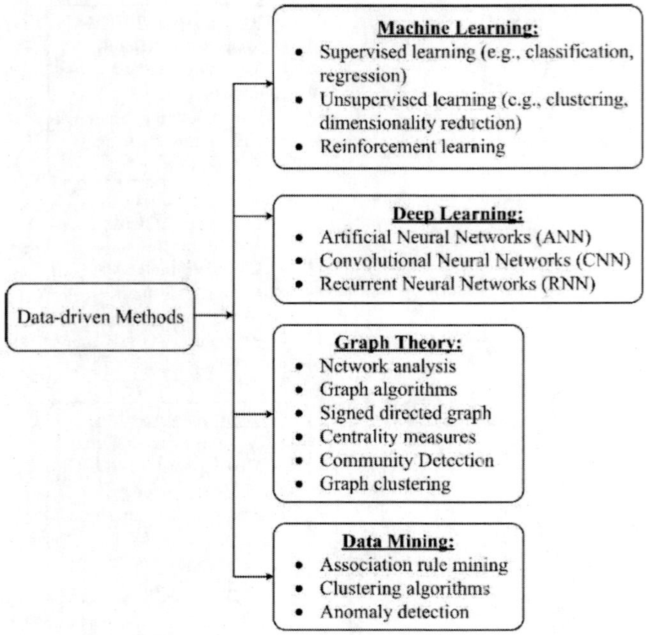

Figure 2.11. Various data-driven fault diagnosis methods.

a) *Rule-based Methods:* Rule-based algorithms rely on a set of predefined rules or thresholds to detect and diagnose faults. These rules are typically established based on expert knowledge and experience. For example, if the battery voltage exceeds a certain threshold, it may indicate an overcharge condition. Rule-based methods are straightforward to implement and interpret, but they may lack flexibility and struggle to detect complex faults or adapt to changing operating conditions. Several rule-based algorithms are shown in Figure 2.9 (b).

b) *Statistical Methods:* Statistical algorithms utilize statistical models to detect anomalies and diagnose faults. These methods employ techniques such as statistical process control, hypothesis testing, or pattern recognition (as depicted in Figure 2.10 (a)) to analyze the measured data and identify deviations from expected behaviour. Statistical methods can handle a wide range of fault types and provide quantitative insights into the severity or probability of the fault. However, they require historical data for training and may have limitations in detecting previously unseen faults.

c) *Model-based Methods:* Model-based algorithms utilize mathematical models of the battery system to detect and diagnose faults. Various model-based algorithms are shown in Figure 2.10 (b). These models capture the behaviour of the battery under normal and faulty conditions. By comparing the measured data with the model predictions, discrepancies can indicate the presence of faults. Model-based methods can handle various fault types and provide insights into the root causes of the faults. However, they require accurate battery models and may be sensitive to modeling uncertainties.

d) *Data-driven Methods:* Figure 2.11 illustrates various data-driven algorithms for fault detection and diagnosis. Data-driven methods refer to a category of techniques used in BMS that rely on analyzing and learning from data to make predictions, classifications, or decisions. These methods leverage the power of machine learning algorithms and statistical models to extract meaningful information and patterns from large volumes of battery-related data, such as voltage, current, temperature, and other sensor measurements. By training on historical data, data-driven methods can identify trends, correlations, and anomalies, enabling accurate state estimation, fault detection, and performance optimization. These methods are valuable in BMS development and deployment as they can adapt to varying battery chemistries, configurations, and operating conditions, making them flexible and versatile tools for efficient battery management in electric vehicles and other energy storage applications.

This comprehensive classification covers a wide range of methods used in battery management systems. Statistical methods offer a foundation for data analysis and pattern recognition. Rule-based methods utilize predefined rules and expert knowledge for decision-making. Model-based methods involve mathematical models to simulate and predict battery behavior. Data-driven methods use machine learning and deep learning algorithms to learn patterns directly from data. Graph theory is useful for analyzing relationships and interactions within the battery system. Each category provides unique insights and tools for effective battery management and optimization. Integrating multiple algorithms and techniques is common to enhance the accuracy and reliability of fault detection and diagnosis in BMS. Different methods can be combined or used together to address specific challenges in BMS development and optimization.

2.5. BMS Integration and Control

Integration and control are critical aspects of the BMS in electric vehicles. The BMS must seamlessly integrate with other vehicle systems and employ effective control algorithms to ensure optimal performance, safety, and efficiency (M. S. Hossain Lipu a, et al., 2021). Integration of the BMS with other vehicle systems involves coordinating with the powertrain system to optimize energy usage and others are the main objectives (Ebbs Picken, Silva, & Amon, 2023). The algorithms are meant for managing the charging system for efficient and safe charging. Communicating with the vehicle's network architecture, collaborating with safety systems to protect the battery and electrical components, and integrating with telematics systems for remote monitoring and data-driven maintenance decisions are also the objectives of the algorithms.

The BMS relies on various control algorithms to manage the battery system as already discussed in the previous section of this chapter. These algorithms regulate the SoC to maintain optimal energy levels, monitor the SoH to estimate battery life and performance, manage power limitation and distribution based on operational requirements, control thermal management for temperature regulation, and implement safety and protection protocols to detect and respond to abnormal conditions.

However, there are challenges associated with BMS integration and control. The complexity of coordinating multiple systems, ensuring compatibility and reliable communication protocols pose challenges. Scalability and flexibility are essential to accommodate different battery chemistries, sizes, and configurations. Standardization is also a challenge, as universal standards are necessary for interoperability and data exchange among different EV models. Addressing these challenges is crucial to achieve seamless integration, efficient control, and optimal performance of the BMS, ensuring the overall success of EVs in terms of reliability, safety, and user experience.

2.5.1. Integration of BMS with Other Vehicle Systems

The integration of the BMS with other vehicle systems is a critical aspect of electric vehicle design and operation. The BMS needs to seamlessly interface with various vehicle systems to ensure efficient operation, effective communication, and optimal performance. Let's explore some key areas of integration:

1) *Powertrain System:* The powertrain system is a critical component of an electric vehicle that encompasses all the systems and components responsible for generating and delivering power to propel the vehicle. The integration of the BMS with the powertrain system is crucial for the efficient operation and optimal performance of the EV. The BMS plays a key role in the powertrain system by managing and monitoring the battery's SoC and SoH. It communicates with the powertrain controller or motor controller to provide real-time information about the battery's capacity, voltage, temperature, and other relevant parameters. This information allows the powertrain system to optimize its performance based on the available battery capacity and limitations. The integration of the BMS with the powertrain system is crucial for controlling the power output and torque of the electric motor. The BMS provides the motor controller with information on the battery's energy availability, allowing the powertrain to adjust the motor's performance for efficient energy utilization and desired driving characteristics. This integration also involves coordinating with other components, like the motor drive inverter and regenerative braking system. The BMS communicates with the inverter to regulate power flow and interacts with the regenerative braking system to optimize energy recovery during braking. Additionally, the BMS enables features such as thermal management by monitoring battery temperature and controlling cooling or heating systems to ensure optimal battery performance, longevity, and safety.
2) *Charging System:* Integration with the charging system is essential for safe and efficient charging of the battery. The BMS communicates with the onboard charger or external charging infrastructure to monitor and manage the charging process. It ensures that the battery is charged within safe voltage and temperature limits, controls the charging current, and provides feedback to the charging system for optimal charging performance. The BMS also enables the implementation of various charging protocols and standards, such as AC charging (e.g., Level 1 and Level 2 charging) and DC fast charging (e.g., CHAdeMO, CCS, and Tesla Supercharger), as illustrated in Figure 2.12. It communicates with the charging station or charger to negotiate the charging parameters, including current,

voltage, and charging rate. The BMS plays a crucial role in ensuring the safe and optimized charging of the battery. It prevents overcharging or undercharging, preserving the battery's performance and lifespan. The integration of the BMS with the charging system enables advanced features like smart charging, allowing optimization of charging times and taking advantage of lower electricity rates. The BMS monitors and manages the charging process, continuously measuring parameters to detect abnormal conditions and protect the battery from potential damage.

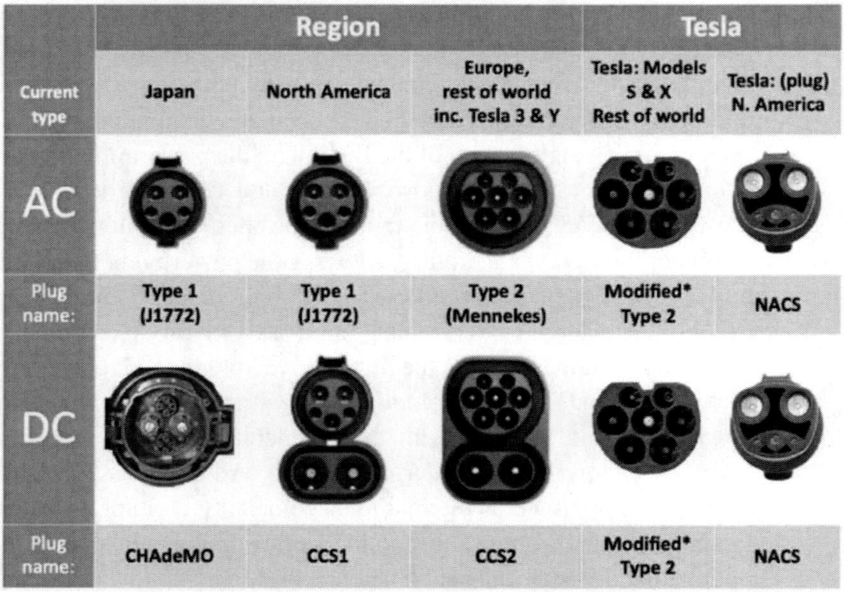

Figure 2.12. Comparison of charging plugs around the world (Gaton, 2022).

3) *Vehicle Communication Network:* The BMS communicates with the vehicle's communication network, such as the controller area network, to exchange information with other vehicle systems. This integration allows for seamless data sharing and coordination between the BMS and systems like the instrument cluster, infotainment system, and vehicle diagnostics. The BMS sends battery-related data, including SoC, SoH, and temperature, to the instrument cluster for display to the driver and receives commands from other systems, enabling effective battery management and optimized vehicle performance. Additionally, the BMS

communicates diagnostic information and fault codes to the onboard diagnostics system, enhancing the vehicle's ability to detect and address battery-related issues and ensuring the EV's safety and reliability.

4) *Safety Systems:* Integration with safety systems is crucial to ensure the protection of the battery and the overall vehicle. The BMS provides important battery-related information to the safety systems, allowing them to make informed decisions and take appropriate actions. For example, the BMS can communicate the battery's SoC and SoH to systems such as the Electronic Stability Control (ESC) and Anti-Lock Braking System (ABS). This information helps these systems optimize vehicle dynamics and braking performance based on the available battery power and condition, enhancing overall safety. Furthermore, BMS plays a crucial role in the thermal management and protection of the battery pack. It continuously monitors individual battery cell temperatures and collaborates with the vehicle's thermal management system to prevent overheating and thermal runaway. In case of collisions or abnormal conditions, the BMS can trigger safety measures such as airbag deployment, isolating the battery, or shutting down high-voltage circuits to protect occupants. The BMS also integrates with high-voltage safety systems, allowing it to communicate with contactors or disconnect switches to isolate the battery for technician and emergency responder safety. Overall, the BMS's integration with safety systems in EVs enhances battery management and ensures overall vehicle safety, providing critical protection for occupants and other road users.

5) *Telematics and Remote Monitoring:* Integration with telematics systems enables remote monitoring and data-driven management of the battery and EV performance. The BMS can send real-time data, including SoC, SoH, and charging status, to a cloud-based platform for remote monitoring. This allows fleet operators and vehicle owners to track battery health, analyze performance data, and make informed decisions regarding maintenance and optimization. Telematics and remote monitoring systems also enable features such as remote diagnostics, over-the-air updates, and predictive maintenance. EV owners can check the battery's status, receive charging notifications, and schedule sessions remotely. Manufacturers can diagnose faults, update BMS software, and optimize battery management strategies based on real-time data. Telematics systems also provide vehicle

tracking, remote locking, and emergency assistance features, enhancing user experience and security. Overall, integrating the BMS with telematics enables real-time monitoring and remote management, enhancing EV convenience and efficiency.

Seamless integration of the BMS with these vehicle systems ensures effective communication, coordination, and optimization of EV operation. It enables the BMS to gather relevant data, provide necessary control signals, and ensure the overall functionality and safety of the EV. Furthermore, the integration of the BMS with other vehicle systems plays a vital role in the efficient operation and performance of electric vehicles. By seamlessly integrating with the powertrain system, charging system, vehicle communication network, safety systems, and telematics, the BMS can optimize energy usage, facilitate safe charging, enable data exchange, enhance safety measures, and provide remote monitoring capabilities. This integration ensures the smooth coordination and functionality of the BMS within the broader EV architecture, contributing to a reliable and user-friendly electric vehicle experience.

2.5.2. BMS Control Algorithms

BMS control algorithms are a crucial component in electric vehicles. These algorithms are responsible for monitoring, controlling, and optimizing the operation of the battery pack to ensure its safety, performance, and longevity. Various BMS algorithms are already elaborated in section 2.4. They make use of various data inputs from sensors, as well as advanced mathematical and computational techniques, to make intelligent decisions and take appropriate actions. One of the primary functions of BMS control algorithms is state estimation, which involves estimating key parameters such as SoC, SoH, SoP, and SoF of the battery pack. Accurate state estimation is essential for effective battery management and reliable performance of the EV. The algorithms utilize inputs from voltage sensors, current sensors, temperature sensors, and other relevant data sources to estimate the battery's current state. These estimates are then used for various control purposes, such as determining optimal charging and discharging strategies.

Figure 2.13 represents a generalized block diagram of the BMS control. It typically consists of several interconnected units that work together to ensure safe and efficient battery operation. The basic control block diagram of a BMS includes a charger unit responsible for supplying power to the battery, a battery unit that stores electrical energy, a converter unit that regulates the voltage and current flow, a load unit that consumes the battery's power output, a data acquisition unit that collects and analyzes battery-related information, and a controller unit that makes decisions and implements control strategies based on the acquired data. The BMS plays a crucial role in maximizing battery life, protecting against overcharging or discharging, and maintaining optimal performance throughout the battery's lifespan.

BMS control algorithms are crucial for managing the charging and discharging of the battery pack in electric vehicles. They ensure safe and efficient battery operation by setting appropriate charging parameters based on the battery's state and characteristics, preventing overcharging or undercharging. During discharging, the algorithms monitor the current draw and manage the discharge rate to avoid excessive strain on the battery. Additionally, BMS control algorithms address cell imbalances within the battery pack, equalizing the state of charge among individual cells to improve overall efficiency, capacity utilization, and battery pack lifespan. Various balancing techniques, such as passive, active, or hybrid balancing, can be employed based on specific EV system requirements.

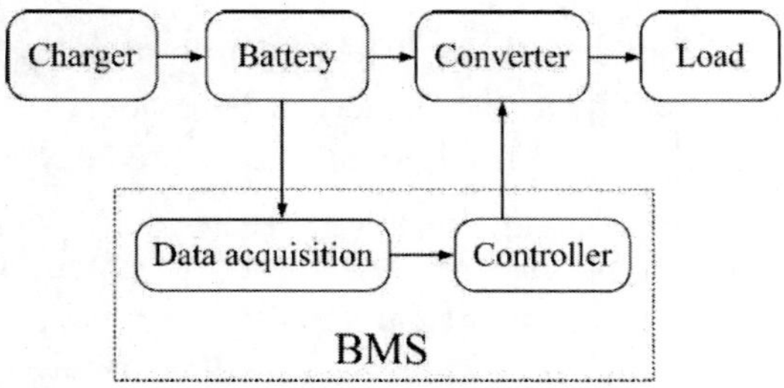

Figure 2.13. A basic control block diagram of a battery management system.

Safety is a paramount concern, and BMS control algorithms play a critical role in ensuring battery safety. These algorithms continuously monitor parameters like temperature, voltage, and current during operation, activating protective measures and fault detection mechanisms in case of any abnormalities. These safety mechanisms may include adjusting charging or discharging rates, initiating thermal management systems, or implementing emergency shutdowns to prevent hazardous conditions or damage to the battery. Advanced BMS control algorithms take advantage of predictive analytics and machine learning to optimize battery performance and lifespan.

They analyze historical data, environmental conditions, driving patterns, and other factors to make predictions and adapt control strategies accordingly. These algorithms can optimize energy consumption, predict charging needs, and suggest ideal charging schedules based on user preferences, contributing to an efficient and user-friendly electric vehicle experience. Overall, BMS control algorithms are vital for the safe and efficient operation of EV batteries, enhancing their performance, reliability, and longevity, and ultimately improving the overall efficiency and user satisfaction of electric vehicles.

Here are some commonly used BMS control algorithms list:

1) SoC Estimation Algorithms:
   - Coulomb Counting Method: This algorithm estimates SoC by integrating the current flowing in and out of the battery over time.
   - Open-Circuit Voltage Method: This algorithm utilizes the battery's terminal voltage to estimate SoC based on the voltage-SoC relationship.
   - Kalman Filter and Extended Kalman Filter: These algorithms use a recursive estimation approach that combines system dynamics and measurement data to estimate SoC with improved accuracy and robustness.
   - Model-Based Methods: These algorithms utilize battery models, which represent the battery's behaviour and dynamics, to estimate SoC by matching the model predictions with measured data.
   - Artificial Intelligence (AI) and Machine Learning (ML) Techniques: AI and ML algorithms, such as neural networks and support vector machines, can learn the complex relationships between battery parameters and SoC from training data, enabling accurate estimation.

2) SoH Estimation Algorithms:
   - Capacity Estimation: These algorithms estimate the available capacity of the battery by monitoring its voltage response during charging and discharging cycles.
   - Impedance-Based Methods: These algorithms analyze the battery's impedance spectrum to estimate SoH, considering factors like internal resistance and frequency-dependent behaviour.

- Aging Models: These algorithms utilize mathematical models that simulate the degradation mechanisms of the battery over time to estimate SoH based on degradation patterns and historical data.
- Machine Learning Approaches: ML algorithms can learn the correlation between battery operating conditions, usage patterns, and degradation, enabling accurate SoH estimation.

3) SoF Estimation Algorithms:
    - Resistance Estimation: These algorithms estimate the internal resistance of the battery, which is an indicator of its performance and health, by analyzing voltage and current measurements during load conditions.
    - Electrochemical Impedance Spectroscopy (EIS): EIS techniques measure the battery's impedance response to small amplitude sinusoidal signals at different frequencies, providing insights into its electrochemical behaviour and performance.
    - Signal Processing Techniques: Advanced signal processing algorithms, such as Fourier analysis or wavelet transform, can extract features from the battery's voltage and current signals to estimate its SoF.
    - Machine Learning-Based Approaches: ML algorithms can learn the complex relationships between battery parameters, behaviour, and SoF by analyzing historical data and patterns.

Each control algorithm has its advantages and limitations, and the choice of algorithm depends on factors such as the battery chemistry, available sensor data, computational requirements, and desired accuracy. BMS designers and researchers continue to explore and develop new control algorithms to improve the accuracy, efficiency, and safety of BMS operations in electric vehicles. There are other popular control techniques available that are also used for the BMS such as:

1) Proportional-Integral-Derivative (PID) Control:
    - PID control is a classic control algorithm widely used in various applications, including BMS.
    - It uses feedback from the system to continuously adjust the control output based on proportional, integral, and derivative terms.

- The proportional term provides an immediate response to errors, the integral term eliminates steady-state errors, and the derivative term improves the response to changes in error.
- PID control is effective in regulating battery charging and discharging currents to maintain desired setpoints.

2) State-Space Control:
   - State-space control is a mathematical modeling and control approach that represents a dynamic system in terms of its state variables.
   - It involves constructing a state-space model of the battery system, which includes the state variables, input variables, and output variables.
   - Control algorithms based on state-space models can effectively handle nonlinearities, uncertainties, and time-varying dynamics of the battery system.
   - State-space control allows for sophisticated control strategies such as optimal control, model predictive control, and adaptive control.

3) Fuzzy Logic Control:
   - Fuzzy logic control is a control algorithm that emulates human reasoning based on fuzzy set theory.
   - It utilizes linguistic variables and fuzzy rules to make control decisions.
   - Fuzzy logic control is well-suited for handling uncertainties and imprecise inputs in the battery system.
   - By defining fuzzy rules and membership functions, the algorithm can handle complex control scenarios and adapt to varying conditions.

4) Model Predictive Control (MPC):
   - MPC is an advanced control technique that utilizes a predictive model of the system to optimize control decisions.
   - It solves an optimization problem over a finite time horizon to determine the control actions that optimize a defined objective function.

- MPC considers system constraints, control objectives, and future predictions to generate optimal control inputs.
- MPC is particularly effective in managing complex dynamics, constraints, and nonlinearity in the battery system.

5) Adaptive Control:
   - Adaptive control algorithms continuously adjust the control parameters based on the changing characteristics of the battery system.
   - They adapt to variations in battery behaviour, aging effects, and uncertainties to maintain the desired performance.
   - Adaptive control algorithms employ adaptive laws and parameter estimation techniques to update the control parameters in real-time.
   - They can improve control accuracy and robustness in the presence of changing system conditions.

6) Neural Network Control:
   - Neural network control algorithms leverage the power of artificial neural networks to learn and approximate complex control mappings.
   - They use training data to train a neural network model that can predict control actions based on the system state.
   - Neural network control algorithms are effective in handling nonlinearities and can adapt to changes in the battery system over time.
   - They can provide accurate control even in the presence of uncertainties and disturbances.

These control algorithms can be combined or customized based on the specific requirements of the battery system and the desired performance objectives. The selection of the appropriate control algorithm depends on factors such as system dynamics, complexity, computational resources, and real-time requirements.

2.5.3. Control Challenges

While the battery management system control algorithms play a crucial role in ensuring the optimal performance and safety of the battery pack in EVs, they also face several challenges that need to be addressed. These challenges arise due to the complex nature of the battery system and the dynamic operating conditions of the EV. Understanding and overcoming these challenges is essential for the effective implementation of BMS control algorithms. Some of the key challenges include:

1) *Variability and Uncertainty:* Batteries exhibit inherent variability in their electrical characteristics, including capacity, resistance, and voltage response. Additionally, environmental factors such as temperature variations further contribute to this variability. BMS control algorithms need to account for these uncertainties and adapt their strategies to ensure accurate state estimation and control under varying conditions.
2) *Nonlinear Behaviour:* Battery dynamics are nonlinear, meaning that their behaviour is not directly proportional to inputs or state variables. Nonlinearity poses challenges for control algorithms, as traditional linear control techniques may not be sufficient. Advanced control strategies, such as adaptive control or model predictive control, may be required to effectively manage the nonlinearities of the battery system.
3) *State Estimation Accuracy:* Accurate estimation of the battery's state, including SoC, SoH, and SoF, is crucial for efficient battery management. However, estimating these states accurately can be challenging due to factors such as measurement noise, model inaccuracies, and degradation effects. BMS control algorithms must employ sophisticated estimation techniques, such as Kalman filters or machine learning approaches, to enhance state estimation accuracy.
4) *Thermal Management:* Battery temperature has a significant impact on performance, efficiency, and lifespan. Controlling and maintaining the temperature within a safe operating range is crucial. However, thermal dynamics are complex, influenced by factors such as external environment, current flow, and cell-to-cell temperature variations. BMS control algorithms need to integrate thermal management strategies, such as active cooling or heating, to ensure proper temperature regulation and prevent overheating or thermal runaway.

5) *Power Limitations:* EVs often have power constraints due to the limitations of the electrical system or the charging infrastructure. BMS control algorithms need to optimize power utilization, considering factors such as available power, user preferences, driving conditions, and system limitations. Balancing power demands with the battery's state and performance capabilities is crucial to maximize efficiency and ensure a reliable power supply.
6) *Real-time Operation:* BMS control algorithms operate in real-time, requiring fast and efficient computation to respond to dynamic changes in the battery's state and operating conditions. Real-time constraints pose challenges in algorithm design, implementation, and computational efficiency. Efficient algorithms, hardware acceleration, and optimization techniques are necessary to meet real-time requirements while maintaining accuracy and reliability.
7) *Safety and Fault Detection:* Ensuring the safety of the battery system is a primary concern. BMS control algorithms need to incorporate robust fault detection mechanisms to identify and respond to abnormal conditions promptly. Reliable fault detection and appropriate fault mitigation strategies are vital to prevent hazardous situations and protect the battery from damage.

Addressing these control challenges requires a multidisciplinary approach that combines expertise in battery electrochemistry, control theory, signal processing, and system integration. Continuous research and development efforts are essential to improve the performance, reliability, and efficiency of BMS control algorithms in EVs. By overcoming these challenges, BMS control algorithms can enable optimal battery management, enhance the driving experience, and contribute to the widespread adoption of electric vehicles.

2.6. BMS Testing and Validation Procedure

The testing and validation procedure of the BMS is a critical step in ensuring the proper functioning, reliability, and safety of the BMS in EVs. BMS testing involves a series of tests that evaluate the performance, accuracy, and functionality of the BMS under various operating conditions. Through rigorous testing and validation, manufacturers can identify any potential issues or shortcomings in the BMS and make necessary improvements before the BMS is integrated into the EVs.

Figure 2.14 represents the development life cycle of a BMS. The development life cycle of the BMS consists of several stages to ensure the system's functionality, reliability, and compliance. It starts with the proof of concept, where initial ideas are tested to validate their feasibility. This is followed by the development of a hardware prototype, where physical components are built and tested using appropriate testing equipment. The next stage involves firmware development, where the software that controls the BMS is created and tested. Validation and verification then take place to ensure the BMS meets the specified requirements and performs as expected. Production testing is conducted to ensure consistent quality and reliability of the BMS during mass production. Finally, Return Merchandise Authorization (RMA) testing is performed to assess and address any issues or failures that occur during the use of the BMS in the field. Each stage is crucial in the overall development and lifecycle of a BMS, ensuring its functionality, reliability, and customer satisfaction.

BMS testing and validation are crucial to ensure that the BMS functions as intended, safeguarding the battery cells, optimizing their performance, and ensuring the overall efficiency of the EV. Comprehensive testing is conducted to verify that the BMS meets specified standards, regulations, and safety requirements. The testing process includes laboratory-based and real-world tests, simulating various operating conditions an EV might encounter. These tests assess the BMS's functionality, performance, accuracy, safety features, and compatibility with other vehicle systems. Validation involves confirming that the BMS meets predetermined specifications, accurately monitors battery parameters, communicates with other systems, detects faults, and maintains system reliability. By conducting thorough testing and validation, the BMS's proper functioning and safety in EVs can be assured.

Furthermore, BMS testing and validation must adhere to industry standards and regulations specific to EVs (Sørensen, et al., 2022). These standards provide guidelines and requirements for BMS performance, safety, and interoperability. Compliance with these standards ensures that the BMS meets the industry's best practices and delivers the expected level of functionality and safety. By conducting thorough testing and validation, manufacturers can identify and rectify any shortcomings or issues with the BMS, leading to improved performance, efficiency, and reliability of the EV. This section will delve into the various types of tests conducted on the BMS, the validation process, and the relevant standards that govern BMS testing and validation in the context of EVs.

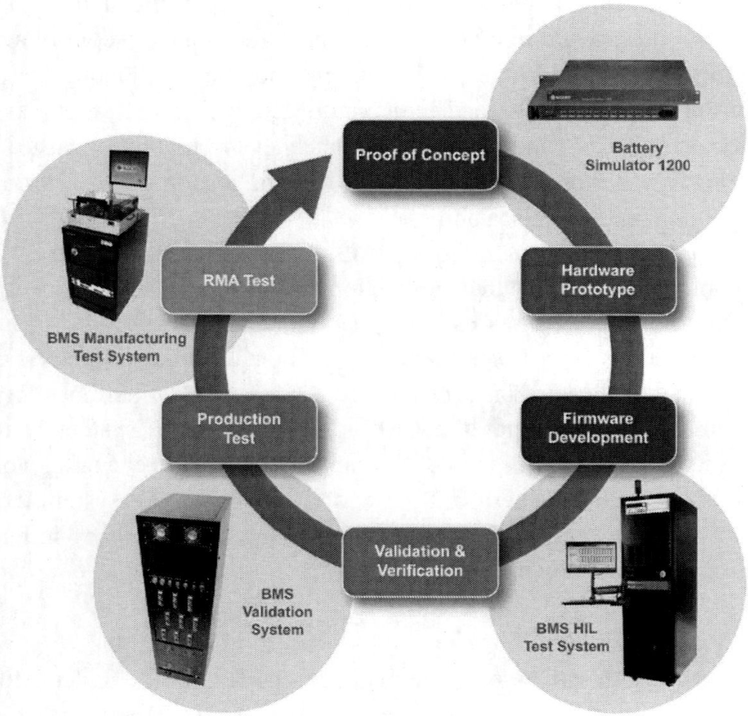

Figure 2.14. The life cycle of BMS development and the types of testing equipment used at each stage (Lombardo, 2020).

2.6.1. Types of Tests Conducted on BMS

BMS testing involves various types of tests to ensure the reliability, functionality, and safety of the system. Various types of tests are conducted on the BMS to assess its performance under different conditions and validate its compliance with industry standards. Some common types of tests conducted on BMS include performance testing, environmental testing, safety testing, endurance testing, communication testing, reliability testing, EMC testing, and software testing.

2.6.1.1. Performance Testing

This test evaluates the BMS's performance in terms of accuracy, responsiveness, and stability. It involves verifying the BMS's ability to accurately measure battery parameters, control charging and discharging processes, and provide reliable information to the vehicle's control systems. Performance tests assess the BMS's effectiveness in maintaining the battery within the desired voltage and temperature ranges, as well as its ability to handle varying load conditions. Performance testing involves measuring and analyzing key performance parameters, such as response time, throughput, resource utilization, and scalability.

Performance testing of the BMS includes response time testing, throughput testing, resource utilization testing, and scalability testing. Response time testing assesses how quickly the BMS responds to changes in battery conditions or user commands. Throughput testing measures the BMS's ability to handle data and tasks efficiently within a given timeframe. Resource utilization testing evaluates the BMS's efficient use of system resources. Scalability testing assesses the BMS's capacity to adapt to changing workload demands and maintain performance and reliability as the system expands. These tests ensure the BMS's proper functioning and efficiency in handling various scenarios and workloads in the EV.

2.6.1.2. Environmental Testing

Environmental tests assess the BMS's performance under different environmental conditions. It involves conducting tests such as temperature cycling, thermal shock, humidity testing, vibration testing, and EMI/EMC testing. These tests simulate the demanding conditions that an electric vehicle may encounter during its lifetime and enable BMS developers to identify any performance issues, failures, or malfunctions that may occur under extreme or adverse environmental conditions.

Environmental testing for BMS involves temperature testing, humidity testing, vibration testing, ingress protection (IP) testing, and salt spray testing. Temperature testing assesses the BMS's performance and reliability under extreme temperature conditions, ensuring it can operate within specified ranges without degradation or failure. Humidity testing evaluates the BMS's resistance to high humidity levels and moisture ingress, preventing corrosion and performance issues in humid environments. Vibration testing ensures the BMS can withstand vibrations during vehicle operation without compromising functionality. IP testing assesses its resistance to dust and water ingress, protecting the BMS from contaminants. Salt spray testing evaluates corrosion resistance, crucial for BMS units in marine or coastal applications. Comprehensive environmental testing enhances the BMS's robustness and reliability, contributing to EV safety and longevity.

By performing comprehensive environmental testing, manufacturers can enhance the robustness and reliability of the BMS, ensuring its effective operation in various climates and challenging environments. This testing also contributes to the overall safety and longevity of the electric vehicle, as a reliable and durable BMS is vital for the efficient and secure management of the vehicle's battery system.

2.6.1.3. Safety Testing

Safety testing is crucial to ensure the BMS's compliance with safety regulations and standards. It involves various tests and assessments to identify potential safety risks, vulnerabilities, and compliance with safety standards and regulations. The primary goal of safety testing is to prevent hazards, mitigate risks, and ensure the protection of users, the vehicle, and the surrounding environment. Safety tests include electrical safety testing (overcurrent protection, overvoltage protection), thermal safety testing (temperature monitoring), and fault detection capabilities.

Safety testing for BMS involves electrical safety testing and thermal safety testing. Electrical safety testing verifies the BMS's insulation, grounding, and protection against electrical faults, ensuring it can handle high voltage conditions and prevent short circuits. It also ensures effective isolation of the battery system to prevent electrical hazards. Thermal safety testing assesses the BMS's ability to monitor and regulate battery temperature, preventing overheating and managing thermal runaway situations. This testing ensures that the BMS has proper thermal management mechanisms to maintain safe operating temperatures and avoid thermal hazards like fire or explosions.

Another essential aspect of safety testing is functional safety assessments, which evaluate the BMS's ability to perform its functions safely. This involves verifying the BMS's prompt detection and response to faults or abnormal conditions, ensuring the safe operation of the EV, and preventing potential hazards. Safety testing also includes compliance with relevant safety standards and regulations, such as ISO 26262 for functional safety in the automotive industry. Thorough safety testing helps identify and address safety concerns, validate safety features, and ensure the BMS's compliance with industry safety standards. By prioritizing safety, manufacturers can enhance the overall safety of electric vehicles, minimize risks, and instill confidence in the reliable and secure operation of the BMS and battery system.

2.6.1.4. Endurance Testing

Endurance testing is a critical evaluation of the BMS's long-term reliability and durability. It involves subjecting the system to prolonged operation under various load conditions to simulate real-world usage and identify potential issues related to component fatigue, performance degradation, or system failures over time. This testing validates the BMS's ability to withstand continuous operation without compromising performance or safety. It evaluates system stability, reliability of individual components, energy efficiency, and battery management algorithms. It also provides insights into the BMS's expected lifespan. Endurance testing ensures that the BMS can deliver stable, reliable, efficient, and long-lasting performance throughout the electric vehicle's lifespan.

The testing is conducted under controlled laboratory conditions, continuously monitoring various performance parameters such as battery voltage, current, temperature, and cell balancing. The duration of the testing varies based on desired confidence levels and the vehicle's expected lifespan. By rigorously conducting endurance testing, manufacturers can identify and address potential weaknesses, ensuring the BMS's consistent and reliable management of the battery system over the vehicle's life. This enhances overall performance, longevity, and customer satisfaction, ultimately ensuring a robust and efficient battery management system for electric vehicles.

2.6.1.5. Communication Testing

Communication testing is a vital aspect of BMS testing and validation, ensuring the reliable exchange of data and commands within the BMS and with other vehicle systems. The objective is to verify seamless communication between different components, such as the powertrain and charging infrastructure, and synchronize the BMS's operation with the overall vehicle architecture. Tests simulate various scenarios to assess data transmission, protocol compatibility, error handling, and synchronization between components. The BMS must effectively communicate with sensors, actuators, powertrain, and charging systems for optimal performance.

Various tests are conducted during communication testing, including message validation, latency testing, protocol compatibility, and stress testing. Message validation checks data accuracy and integrity, latency testing measures transmission time, protocol compatibility ensures communication protocol alignment and stress testing evaluates performance under high data loads. Thorough communication testing identifies and resolves issues, enhancing data exchange reliability, battery management, safety, and overall vehicle performance. It ensures seamless integration and interoperability with other vehicle systems, promoting efficient electric vehicle operation.

2.6.1.6. Reliability Testing

Reliability testing is a critical part of BMS testing and validation, assessing the system's ability to consistently function under normal operating conditions over an extended period. The testing involves subjecting the BMS to stress factors like temperature variations, humidity, vibration, and electrical load to simulate real-world usage scenarios. By doing so, manufacturers can identify weaknesses and potential failure modes. It estimates the system's overall reliability and mean time between failures (MTBF). Reliability testing provides valuable insights into the BMS's performance and ability to maintain battery health and accuracy throughout its intended lifespan.

During the testing process, the BMS is monitored and analyzed under controlled conditions for an extended duration. This helps identify intermittent failures, performance degradation, and unexpected behavior over time. The testing enables manufacturers to enhance the BMS's design, components, and quality, improving its reliability and durability to withstand the demands of real-world usage. By instilling customer confidence in the BMS's performance and longevity, reliability testing contributes to the successful adoption and acceptance of electric vehicles.

2.6.1.7. Electromagnetic Compatibility (EMC) Testing

EMC testing is a crucial aspect of BMS validation, ensuring that the system does not emit excessive electromagnetic interference that could disrupt other electronic systems in the vehicle. It also verifies that the BMS can withstand external electromagnetic disturbances without compromising its operation. The testing assesses the BMS's compliance with regulatory standards for electromagnetic compatibility, ensuring it can harmoniously coexist with other onboard electronic systems without causing interference or performance issues. EMC testing involves subjecting the BMS to various radiated and conducted electromagnetic fields that simulate real-world environments. It measures the system's emissions, the electromagnetic energy it generates, and its immunity to external electromagnetic disturbances. By evaluating emissions and immunity, manufacturers can ensure the BMS meets the required electromagnetic compatibility standards and functions reliably within electric vehicles.

During EMC testing, the BMS is exposed to electromagnetic fields of varying frequencies and intensities, assessing parameters such as radiated emissions, conducted emissions, radiated immunity, conducted immunity, and susceptibility to electrostatic discharge (ESD). The results of EMC testing help identify any potential design issues or vulnerabilities in the BMS that could lead to electromagnetic interference or compromise its functionality. Manufacturers can make necessary modifications or improvements to ensure compliance with EMC standards. Successfully passing EMC testing demonstrates the BMS's ability to operate reliably without causing interference or safety hazards, making it an integral part of electric vehicle integration and operation. Complying with EMC standards ensures that the BMS does not disrupt the functioning of other vehicle systems or external communication systems, contributing to the overall reliability and safety of electric vehicles.

2.6.1.8. Software Testing

Software testing is a critical aspect of BMS development and ensures the reliability, functionality, and performance of the system's software components. It involves a systematic process of evaluating the software algorithms, functions, and interfaces to identify any defects or errors that may impact the BMS's performance or compromise its safety.

Software testing for BMS serves two primary objectives: verifying the correctness and accuracy of algorithms and calculations, and ensuring the safety and security of the system. The testing process involves evaluating the BMS software's ability to accurately measure battery parameters, estimate SoC and SoH, detect faults and make optimal decisions for battery performance. By conducting systematic testing at various stages of development, including unit, integration, and system-level testing, potential defects are identified and addressed to deliver reliable and high-performing BMS software. Different testing techniques such as black-box and white-box testing, regression testing, and stress testing are utilized to maximize test coverage and validate the software's functionality.

Furthermore, software testing plays a crucial role in bolstering the BMS's security infrastructure. Vulnerability assessments and penetration testing are conducted to identify and rectify potential weaknesses in the software, safeguarding the BMS from unauthorized access, data breaches, or malicious attacks. Ensuring the integrity and confidentiality of sensitive battery and vehicle information is a critical aspect of software testing. The iterative nature of software testing allows for continuous improvement as defects or issues are documented, reported, and addressed throughout the testing process. The ultimate goal is to deliver a BMS software that meets functional and safety requirements, providing accurate battery management and contributing to the overall reliability and security of electric vehicles.

2.6.2. BMS Validation Process

The BMS validation process is a critical step in ensuring the reliability, functionality, and safety of EVs. It involves a series of tests and evaluations to validate the BMS design, functionality, and performance against predetermined specifications and standards. The validation process typically consists of the following steps:

1) *Requirements Review:* The first step in the validation process is to review the requirements and specifications of the BMS. This involves a thorough examination of the functional and non-functional requirements to ensure clarity and completeness.
2) *Test Planning:* A detailed test plan is developed, outlining the objectives, scope, test cases, and test methods for each aspect of the BMS. The test plan serves as a roadmap for the validation process, ensuring comprehensive coverage of all BMS functionalities and features.
3) *Test Execution:* The BMS is subjected to a range of tests and evaluations as defined in the test plan. This includes functional testing, performance testing, safety testing, and other specific tests based on the BMS requirements. The tests are performed using various tools, equipment, and simulation techniques to simulate real-world conditions.
4) *Data Collection and Analysis:* During the test execution phase, data related to BMS performance, behaviour, and parameters are collected and analyzed. This includes voltage and current measurements, temperature readings, communication signals, and other relevant data. The collected data is then analyzed to assess the BMS's compliance with the defined specifications.
5) *Defect Identification and Resolution:* Any deviations, defects, or issues identified during the validation process are documented and

reported. The BMS development team then works to resolve these issues, either through software updates, hardware modifications, or design improvements. The defects are tracked and managed using a defect tracking system, ensuring that they are appropriately addressed.

6) *Documentation and Reporting:* The validation process requires comprehensive documentation of the test procedures, test results, and any modifications made to the BMS design or software. A validation report is prepared, summarizing the findings, test results, and any actions taken to address identified issues. The report serves as evidence of compliance with the required standards and specifications.
7) *Compliance and Certification:* Once the BMS has successfully undergone the validation process, it may be subjected to compliance testing based on industry standards and regulations. This may include certifications such as ISO 26262 for functional safety or other relevant standards specific to the automotive industry. Compliance testing ensures that the BMS meets the necessary safety and performance requirements.

The BMS validation process is iterative, with multiple rounds of testing and refinement to ensure the BMS's robustness and reliability. It is a collaborative effort involving various stakeholders, including BMS developers, system integrators, and regulatory bodies. Through thorough validation, the BMS can be deemed suitable for integration into EVs, ensuring optimal battery performance, safety, and longevity.

2.6.3. Standards for BMS Testing and Validation

Several standards and regulations provide guidelines for BMS testing and validation. These standards provide guidelines and requirements that BMS manufacturers and developers must adhere to during the testing and validation process. Some of the key standards for BMS testing and validation include:

- ISO 26262: This is an international standard for functional safety in automotive systems. It provides guidelines for the development and validation of safety-related systems, including BMS. ISO 26262 defines the safety lifecycle and requirements for hazard analysis, risk assessment, and validation of safety functions.

- UN ECE R100: This United Nations regulation sets out requirements for the approval of BMS and high-voltage electrical systems used in EVs. It covers safety aspects such as insulation, electrical protection, and functional performance of the BMS.
- IEC 61851: This standard addresses the electric vehicle conductive charging system. It defines the requirements and test procedures for charging equipment, including BMS functionality related to charging control, monitoring, and safety features.
- IEC 62660: This standard provides guidelines for testing lithium-ion battery packs used in electric vehicles. It covers various aspects of battery testing, including electrical, mechanical, and environmental performance.
- SAE J1772: This standard focuses on the charging interface between the EV and the charging station. It includes specifications for the BMS communication with the charging infrastructure, ensuring compatibility and interoperability during charging.
- SAE J1798: This recommended practice outlines test procedures for evaluating the performance and reliability of battery management systems for electric vehicles.
- UL 2580: This standard is specific to BMS used in EVs. It covers the safety and performance requirements for lithium-ion battery systems, including BMS components. UL 2580 provides guidelines for testing battery systems, thermal management, electrical isolation, and fault detection.
- IEEE 1679: This standard focuses on the characterization and testing of lithium-ion batteries used in electric vehicle applications. It provides guidelines for testing battery performance, capacity, cycle life, and safety aspects related to the BMS.

These standards ensure that BMS undergoes comprehensive testing and validation to meet specific safety, performance, and interoperability requirements. They help establish a common framework for BMS development, ensuring consistency across different manufacturers and facilitating the integration of BMS into EVs. Adhering to these standards not only ensures compliance with regulatory requirements but also enhances consumer confidence in the safety and reliability of EVs. BMS manufacturers and developers must follow these standards to meet industry best practices and deliver BMS solutions that meet the stringent demands of the automotive sector. Compliance with these standards also facilitates market acceptance and enables the smooth integration of BMS with other vehicle systems and charging infrastructure.

2.7. Challenges and Opportunities

The development and deployment of BMS in EVs present both challenges and opportunities. While BMS technology has made significant advancements, there are still hurdles to overcome to ensure the optimal performance, safety, and reliability of EV batteries. At the same time, there are exciting opportunities on the horizon as emerging technologies and market trends continue to shape the future of BMS in EVs.

This section explores the challenges and opportunities that lie ahead in the field of BMS for EVs. It delves into the emerging technologies that can revolutionize BMS functionality, highlights the key challenges faced in BMS development and deployment, and presents a glimpse into the future prospects of BMS in the ever-evolving landscape of EVs. By understanding and addressing the challenges, such as increasing battery capacity, improving thermal management, optimizing energy efficiency, and ensuring robust safety measures, the industry can unlock the full potential of EV batteries. Simultaneously, capitalizing on the opportunities presented by emerging technologies, such as solid-state batteries, advanced sensors, artificial intelligence, and connectivity, can pave the way for more efficient, intelligent, and sustainable BMS solutions.

As the global transition towards electrified transportation gains momentum, it is crucial to overcome these challenges and seize the opportunities to maximize the performance, range, and longevity of EV batteries. The effective integration and control of BMS with other vehicle systems, adherence to standards and regulations, and continuous research and development efforts will shape the future of BMS in EVs, driving innovation and propelling the growth of the electric mobility ecosystem. In the following subsections, we delve into the specific challenges faced in BMS development and deployment, explore the emerging technologies that hold great promise for the future of BMS in EVs, and present an overview of the opportunities that lie ahead in this dynamic and transformative field.

2.7.1. Emerging Technologies in BMS

Emerging technologies have the power to reshape the landscape of BMS, unlocking new possibilities and pushing the boundaries of performance, safety, and efficiency. As the demand for EVs continues to rise and the need for more advanced energy storage solutions becomes paramount, researchers, engineers, and innovators are exploring cutting-edge technologies that have the potential to revolutionize BMS functionality. From solid-state batteries to advanced sensor technologies, from artificial intelligence to connectivity solutions, these emerging technologies are poised to redefine the capabilities of BMS and drive the next wave of innovation in the EV industry. Lithium metal batteries have special features and abilities that present exciting opportunities to overcome challenges in traditional BMS. They can improve performance, safety, efficiency, range, charging speed, reliability, and the overall user experience of EVs. Here are some notable emerging technologies in BMS that are discussed in this section.

1) *Solid-State Batteries:* Solid-state batteries are a promising alternative to conventional lithium-ion batteries. They replace the liquid electrolyte with a solid-state electrolyte, offering higher energy density, faster charging times, improved safety, and longer lifespan. Solid-state batteries have the potential to significantly increase EV range and reduce charging times, making them a highly sought-after technology in the EV industry.
2) *Advanced Sensor Technologies:* Advanced sensors, such as impedance spectroscopy, allow for more accurate and real-time monitoring of battery health. These sensors measure various parameters, including voltage, current, temperature, and impedance, providing valuable data for precise SoC and SoH estimation. Advanced sensor technologies enable better battery management, optimization of charging and discharging cycles, and early detection of potential issues.
3) *Artificial Intelligence (AI) and Machine Learning (ML):* AI and ML techniques are being employed in BMS to analyze large amounts of data and make intelligent decisions. These technologies enable predictive battery management, optimize energy usage, and enhance overall performance. AI algorithms can learn from driving patterns, environmental conditions, and battery characteristics to adapt

charging and discharging strategies in real-time, resulting in improved efficiency and extended battery life.

4) *Connectivity and Internet of Things (IoT):* Connectivity solutions enable seamless communication between the BMS, EV, and external systems, such as charging infrastructure and cloud platforms. IoT refers to the network of interconnected devices that can communicate and share data through the internet. When applied to BMS, IoT enables seamless communication between the BMS, the EV, and external systems, unlocking a range of benefits and opportunities. IoT integration allows for remote monitoring, diagnostics, and software updates, ensuring optimal performance and enabling preventive maintenance. IoT connectivity also facilitates over-the-air (OTA) software updates for the BMS. This means that manufacturers can remotely update the BMS software to introduce new features, enhance performance, and address potential vulnerabilities or bugs. OTA updates enhance BMS functionality, keeping EVs up to date without physical service visits. Real-time connectivity enables V2G integration and empowers EV owners with remote access and control over BMS functions via mobile apps or web interfaces. However, data security is crucial, requiring industry standards and robust measures to safeguard sensitive information from unauthorized access or cyber threats.
5) *Thermal Management Systems:* Effective thermal management is crucial for maintaining optimal battery performance and lifespan. Advanced cooling and heating systems are being developed to regulate the temperature of battery cells, ensuring they operate within the desired range. There are various thermal management technologies employed in BMS, including active cooling and heating systems, passive cooling methods, and phase change materials. Thermal management in BMS includes active cooling with fans or liquid cooling, passive cooling with natural convection, and phase change materials. These systems maintain optimal battery temperature, minimizing capacity loss and ensuring battery safety and reliability by preventing overheating and thermal runaway.
6) *Energy Harvesting:* Energy harvesting technologies aim to capture and utilize wasted energy from various sources, such as regenerative braking and solar panels, to supplement the power requirements of the BMS. This approach enhances energy efficiency and reduces reliance on external charging sources. Various energy harvesting

technologies can be employed in BMS, such as solar panels, regenerative braking systems, piezoelectric materials, thermoelectric generators, and electromagnetic induction. These technologies convert ambient energy into usable electrical energy, which can be stored in batteries or capacitors for later use by the BMS. One of the key benefits of energy harvesting in BMS is the potential for increased energy efficiency. Energy harvesting from ambient sources like solar, kinetic, thermal, or electromagnetic energy can supplement or replace external power inputs in BMS, leading to extended battery life and increased EV range. It enhances reliability by reducing dependency on external power connections, making the BMS more robust and resilient.

These emerging technologies offer great potential for the future of BMS in EVs, promising improved range, faster charging, enhanced safety, and optimized energy management. Further research, development, and collaboration are crucial for successful integration and commercialization. With ongoing advancements and innovation, these technologies will continue to shape the evolution of BMS, making EVs more efficient, reliable, and sustainable.

2.7.2. Challenges in BMS Development and Deployment

The development and deployment of the BMS come with several challenges that need to be addressed to ensure their effective implementation. These challenges span various aspects, including technical, operational, and regulatory considerations (Li, Zhou, & Han, A Review on End-of-Life Battery Management: Challenges, Modeling, and Solution Methods, 2017). Here are some of the key challenges in BMS development and deployment:

1) *Battery Complexity:* EV batteries are complex systems with interconnected cells, presenting challenges in BMS development to accurately measure and balance individual cell voltages due to manufacturing variations and aging. BMS requires voltage monitoring circuits to keep cells within specified ranges. Temperature monitoring is critical to address performance and safety issues caused by heat generation. Figure 2.15 highlights the challenges and opportunities of lithium metal batteries in future EVs, with

advantages in energy density and cost. Meeting automotive industry requirements for cycle life and safety is crucial. Robust power electronics and advanced BMS algorithms are needed to optimize performance and safety with increasing energy density. Collaboration among manufacturers, BMS developers, and automotive OEMs is essential for effective BMS integration into EV battery systems.

Figure 2.15. Illustration of challenges and their solutions for Lithium metal batteries (Chen, Dai, & Cai, 2020).

2) *Safety and Reliability:* The BMS plays a critical role in ensuring the safety and reliability of EV batteries by monitoring various parameters and implementing protection mechanisms. Thermal runaway is a primary safety concern, and BMS developers use advanced fault detection algorithms to detect abnormal behavior. If a fault is detected, appropriate actions are taken, such as isolating affected cells or reducing charging/discharging rates. Extensive testing and validation are conducted to enhance safety and reliability. However, achieving high levels of safety and reliability is challenging due to the dynamic nature of battery behavior and the need for continuous monitoring and robust protection mechanisms. BMS developers must implement effective safety features and fault detection algorithms to mitigate risks.
3) *Scalability and Flexibility:* BMS should be scalable and adaptable to accommodate diverse battery chemistries, capacities, and configurations, promoting widespread EV adoption. Modular and configurable systems with adjustable algorithms and software are

essential for accommodating different battery types and pack designs. Parameterization and flexibility enable BMS to adapt to specific battery characteristics, allowing for easy upgrades and replacements without major modifications. Anticipating evolving EV market needs, BMS developers must design solutions that cater to a wide range of battery configurations, supporting various vehicle types and applications.

4) *Communication and Integration:* BMS needs seamless communication with the powertrain, charging infrastructure, and telematics systems, posing challenges in standardization, interoperability, and coordination with stakeholders. Implementing standardized communication protocols like CAN, LIN, or Ethernet is crucial for data exchange. Interoperability is essential when integrating with various vehicle platforms. Collaboration with stakeholders enables advanced functionalities and consistent data exchange. BMS developers must ensure compatibility with industry standards, design flexible communication architectures, and collaborate for seamless integration across the EV ecosystem.
5) *Cost and Affordability:* BMS development should prioritize cost-effective solutions to promote widespread EV adoption. Minimizing BMS component costs without compromising performance and safety is a challenge. Cost reduction can be achieved through efficient design, optimized manufacturing, and leveraging economies of scale. Advancements in sensor technologies and modular design can also contribute to cost savings. Balancing cost and functionality require navigating trade-offs while meeting performance and safety requirements. Continuous research and EV market growth are expected to drive down BMS component costs, making EVs more affordable for mass adoption.
6) *Standardization and Regulations:* Standardization and regulations are crucial for BMS development and deployment in EVs, considering the diverse stakeholders with varying requirements. Harmonizing BMS standards across regions requires collaboration among automakers, battery manufacturers, and regulatory bodies. Comprehensive and consistent standards ensure interoperability, compatibility, and safety, covering communication protocols, data formats, performance, and testing. Adherence to standards allows seamless integration with vehicle systems, aftermarket servicing, and battery reliability. Regulatory bodies play a vital role in defining and

enforcing these standards, which evolve with technology. Ongoing development of BMS standards fosters innovation and supports the global growth of the EV industry.

7) *Durability and Longevity:* BMS should withstand harsh operating conditions throughout the EV's lifespan, including temperature variations, vibration, and humidity. Ensuring durability and resistance to wear and aging is challenging. Optimizing algorithms to reduce computational load and incorporating diagnostics are crucial. Thorough testing under realistic conditions is necessary to ensure longevity. Striking the right balance between robustness, cost, and weight is vital. Addressing durability challenges enhances EV reliability and performance, boosting customer satisfaction and confidence.
8) *Over-the-Air Updates and Cybersecurity:* As BMS systems rely more on software, secure over-the-air updates and robust cybersecurity measures are crucial. Over-the-air updates enable remote software installation for functionality improvements and vulnerability fixes, but implementing them securely is challenging. Secure communication, source authentication, and software integrity are necessary. Robust cybersecurity measures, like encryption and intrusion detection, protect against threats. Regular updates are essential to address evolving risks. Collaboration between stakeholders is vital to establish industry standards for secure over-the-air updates and cybersecurity in BMS. Addressing these challenges ensures BMS safety and contributes to EV security.

Addressing these challenges requires continuous innovation, collaboration between industry stakeholders, and advancements in technology and regulations. Overcoming these hurdles will drive the development and deployment of robust, efficient, and safe BMS for the widespread adoption of EVs.

2.7.3. Future Prospects of BMS in EVs

Despite the challenges, BMS holds promising opportunities for the future of EVs. The future prospects of BMS are promising, driven by ongoing advancements in technology, increasing demand for EVs, and the need for improved battery performance and efficiency. Here are some key areas where BMS is expected to play a significant role in shaping the future of EVs:

1) *Enhanced Battery Performance:* BMS will continue to evolve to optimize battery performance and extend the range of EVs. Future BMS will employ advanced algorithms and predictive analytics to accurately monitor and manage battery parameters, allowing for more efficient charging and discharging, improved energy storage, and enhanced overall battery performance.
2) *Extended Range and Faster Charging:* Advancements in BMS technology, such as higher energy density batteries and improved charging algorithms, will enable EVs to achieve longer driving ranges and faster charging times, addressing the limitations of range anxiety and charging infrastructure.
3) *Integration with Vehicle Systems:* BMS will further integrate with other vehicle systems, enabling better coordination and control. This integration will allow BMS to interact with powertrain systems, charging infrastructure, navigation systems, and telematics, facilitating smart charging, optimized energy consumption, and seamless communication between various vehicle components.
4) *Advanced Safety Features:* Safety will remain a top priority in EVs, and BMS will continue to play a critical role in ensuring safe battery operation. Future BMS will incorporate advanced safety features such as improved thermal management, enhanced fault detection and isolation, and rapid response to abnormal battery conditions, minimizing the risk of battery-related incidents and maximizing overall vehicle safety.
5) *Extended Battery Lifespan:* BMS will focus on increasing the longevity of EV batteries. Advanced algorithms and state-of-the-art monitoring techniques will enable BMS to optimize battery usage, reduce wear, and mitigate degradation, resulting in longer battery lifespan and reduced maintenance costs for EV owners.
6) *Intelligent Energy Management:* BMS will become more sophisticated in optimizing energy usage, considering factors such as driving conditions, traffic patterns, and user preferences. AI and ML algorithms will enable adaptive energy management strategies to maximize efficiency and extend battery life. The integration of AI and ML techniques in BMS will enable intelligent decision-making and adaptive control strategies. AI-powered BMS will continuously learn

from battery behaviour, environmental conditions, and user patterns to optimize battery management and provide personalized energy management solutions.

7) *Vehicle-to-Everything (V2X) Integration:* BMS will play a crucial role in enabling V2X communication, allowing EVs to interact with other vehicles, infrastructure, and the grid. BMS will facilitate dynamic energy management, grid integration, and grid services, enabling EVs to contribute to grid stability and support various applications such as vehicle-to-home (V2H) and vehicle-to-grid systems. BMS will facilitate the integration of EVs into the electricity grid through V2G technology. BMS will enable bidirectional energy flow, allowing EVs to not only consume energy but also feed excess energy back into the grid during peak demand periods. This integration will promote renewable energy utilization, grid stability, and energy cost optimization.
8) *Standardization and Interoperability:* With the growing adoption of EVs worldwide, standardization of BMS will become crucial. Establishing industry-wide standards for BMS will ensure interoperability, compatibility, and seamless integration between different EV models, charging stations, and energy management systems.
9) *Sustainability and Environmental Impact:* BMS will continue to contribute to the sustainability and environmental impact of EVs. Advanced BMS will optimize energy usage, reduce battery waste, and facilitate recycling and second-life applications for retired EV batteries.
10) *Advanced Diagnostics and Prognostics:* BMS will continue to evolve in its ability to diagnose and predict battery health and performance. Advanced diagnostic algorithms and prognostic models will enable early detection of potential issues, allowing for proactive maintenance and minimizing the risk of battery failure.
11) *Energy Storage Systems Integration:* BMS technology can extend beyond the realm of EVs and be applied to other energy storage applications. BMS can be utilized in stationary energy storage systems, such as residential and commercial battery storage, to optimize energy usage, manage power flow, and enhance the overall performance and longevity of the energy storage system.
12) *Enhanced User Experience:* BMS will contribute to improving the overall user experience of EV owners. Advanced user interfaces, real-

time data visualization, and smartphone applications will provide users with valuable information about battery status, charging options, and energy consumption, enhancing convenience and control over their EVs.

The future of BMS in EVs holds immense potential for enhancing battery performance, safety, and overall vehicle efficiency. The future prospects of BMS in EVs are intertwined with the overall growth and development of the electric mobility ecosystem. As the demand for EVs continues to rise, BMS will play a pivotal role in enhancing battery performance, extending range, ensuring safety, optimizing energy usage, and enabling seamless integration with smart grids and other energy systems. The ongoing advancements and innovation in BMS technology will contribute to the widespread adoption of EVs and the transition to a sustainable and electrified transportation future. Continued research, technological advancements, and collaboration between industry stakeholders will drive the evolution of BMS.

Conclusion

In conclusion, this chapter has explored various aspects of BMS in EVs. It has provided an in-depth understanding of the importance of BMS in ensuring the optimal performance, safety, and longevity of EV batteries. The chapter discussed the fundamental components and functionalities of BMS, including cell balancing, SoC and SoH estimation, fault detection and diagnosis, integration with other vehicle systems, control algorithms, testing and validation procedures, as well as the challenges and future prospects of BMS in EVs. The importance of BMS in the success of EVs has been highlighted, emphasizing its crucial role in managing and optimizing the performance of battery systems. The future directions for BMS research and development have been discussed, including advanced monitoring and control, integration with smart grids, adaptation to new battery technologies, standardization and interoperability, and enhanced safety and cybersecurity measures. Overall, BMS represents a critical technology in the evolution of EVs, and continued advancements in BMS will pave the way for more efficient, reliable, and sustainable electric transportation.

Summary of the Chapter

In this chapter, we embarked on a journey into the fascinating realm of BMS and its vital role in the electrifying world of EVs. The intricate web of components and functionalities that constitute a BMS was unraveled, with a focus on the watchful eye of cell monitoring, the delicate art of balancing, and the temperature-taming sensors. With each step, we delved deeper into the heart of BMS testing and validation, unveiling the rigorous processes that ensure their reliability and performance in the face of demanding environments. However, this chapter is not just a mere exploration of technology; it is a testament to the indispensable role of BMS in the triumph of EVs. BMS stands as the guardian of the battery pack, orchestrating the symphony of power and endurance that fuels these vehicles. With its watchful gaze, it optimizes battery performance, safeguards against overcharging and over-discharging, and paves the way for a longer and more fruitful battery lifespan. The safety of EVs rests firmly in the hands of BMS, as it deftly navigates the treacherous waters of thermal runaway risks and stands as a stalwart protector against faults and vulnerabilities.

As we cast our gaze towards the future, the road ahead for BMS research and development gleams with possibilities. The horizon is adorned with dreams of enhanced energy management, where advanced algorithms and control strategies dance to the tune of optimized efficiency and extended driving ranges. The integration of BMS with smart grids beckons, offering a gateway to bidirectional energy flow and the tantalizing concept of V2G capabilities. AI and ML, those wondrous companions, stand poised to unlock new realms of BMS performance, precise state estimation, and predictive maintenance. But it is not just about the technologies; it is about the unyielding pursuit of standardization and interoperability. The quest to establish comprehensive protocols and harmonize the symphony of BMS across different EV models and charging infrastructure sings a song of seamless integration and compatibility. And as we traverse this path, the challenge of cost reduction whispers in our ears, reminding us of the need to make EVs more accessible, and more affordable to all who yearn to embrace the future of sustainable transportation.

In this grand narrative, the story of BMS reaches its climax with a resounding call for cybersecurity. As BMS evolves into a connected entity, its vulnerabilities must be fortified, and its defences fortified against the ever-present threats that loom in the digital realm. With each turn of the page, this chapter unveils the pivotal role of BMS, an enigmatic force that propels EV towards a brighter and greener future. As we bid farewell, we embrace the exhilarating prospects that lie ahead, beckoning us to continue pushing the boundaries of BMS research and development. For it is in these endeavours that we illuminate the path towards safer, more efficient, and more reliable EVs, ensuring that the wheels of progress keep turning towards a sustainable tomorrow.

Importance of BMS in the Success of EVs

The importance of BMS in the success of electric vehicles cannot be overstated. BMS serves as the backbone, the guardian, and the orchestrator of the EV's power source, the battery pack. It is the invisible force that ensures the optimal performance, safety, and longevity of the battery, thus directly impacting the overall success and acceptance of EVs in the market.

First and foremost, BMS plays a critical role in maximizing the efficiency and range of EVs. By continuously monitoring and balancing individual cell voltages, BMS ensures that the battery operates at its optimal state of charge, minimizing energy wastage and maximizing the driving range. Through advanced algorithms and control strategies, BMS optimizes charging and discharging processes, ensuring that the battery delivers its full potential, thus enhancing the overall driving experience and convenience for EV owners. Safety is another paramount aspect in the success of EVs, and BMS serves as the guardian of the battery pack. It monitors critical parameters such as cell temperature, voltage, and current, and triggers protective measures to prevent overcharging, over-discharging, and thermal runaway. BMS acts as a shield, ensuring the safe operation of the battery pack and minimizing the risk of hazards such as fires or explosions. The assurance of safety provided by BMS instils confidence in consumers, promoting trust and acceptance of EVs as a reliable mode of transportation. Furthermore, BMS contributes to the longevity and durability of EV batteries. By implementing sophisticated algorithms for SoH estimation, BMS can accurately assess the battery's capacity and aging, enabling proactive maintenance and extending the battery's lifespan. This not only reduces the cost of ownership for EV owners but also supports the sustainability of the entire EV ecosystem by reducing the need for frequent battery replacements and minimizing environmental impact.

The success of EVs also relies on their integration into the broader energy ecosystem, including smart grids and vehicle-to-grid capabilities. BMS plays a crucial role in enabling bidirectional energy flow, facilitating energy storage, and providing grid stabilization services. By coordinating with charging infrastructure and smart grid systems, BMS contributes to the overall efficiency and stability of the energy grid, supporting the integration of renewable energy sources and promoting the transition to a sustainable energy future. Moreover, the importance of BMS in the success of EVs cannot be overstated. Its role in maximizing efficiency, ensuring safety, extending battery life, and enabling integration with energy systems is pivotal. As the demand for EVs continues to grow and technology advances, BMS will continue to evolve, unlocking new possibilities and further propelling the success of EVs as a sustainable and mainstream mode of transportation.

Future Directions for BMS Research and Development

The field of BMS is constantly evolving, driven by the rapid advancements in EV technology and the growing demand for more efficient and reliable energy storage solutions. As we look to the future, several key directions emerge for BMS research and development:

- *Advanced Monitoring and Control:* One of the major focuses of future BMS development is to enhance monitoring and control capabilities. This includes improving the accuracy and speed of sensor technologies, developing advanced algorithms for real-time data analysis, and integrating artificial intelligence and machine learning techniques for predictive maintenance and fault detection. By leveraging these technologies, BMS can provide more precise and proactive management of battery systems, optimizing their performance and extending their lifespan.
- *Integration with Smart Grids and Energy Management Systems:* The future of BMS lies in its integration with smart grids and energy management systems. BMS can play a crucial role in facilitating bidirectional energy flow, enabling V2G capabilities, and supporting the integration of renewable energy sources. By effectively managing the charging and discharging processes of EV batteries, BMS can contribute to grid stability, demand response programs, and peak load management. This integration opens up new opportunities for optimizing energy consumption, reducing costs, and promoting a sustainable energy ecosystem.

- *Solid-State Batteries and Beyond:* The emergence of solid-state battery technology holds great promise for the future of EVs, and BMS will play a vital role in their successful implementation. Solid-state batteries offer higher energy density, faster-charging rates, and improved safety compared to conventional lithium-ion batteries. BMS will need to adapt to the unique characteristics of solid-state batteries, developing new algorithms and control strategies to ensure their optimal performance and longevity. Furthermore, BMS research will extend beyond solid-state batteries to explore new and innovative energy storage technologies, such as lithium-sulfur or lithium-air batteries, opening up new possibilities for EVs.
- *Standardization and Interoperability:* As the EV industry continues to grow, standardization and interoperability of BMS across different vehicle platforms and charging infrastructure will become increasingly important. Efforts should be made to establish common communication protocols, data formats, and safety standards to ensure seamless integration and compatibility. This will not only promote interoperability between different BMS components but also enable efficient data exchange and collaboration among different stakeholders, including automakers, battery manufacturers, and charging station operators.
- *Enhanced Safety and Cybersecurity:* Safety and cybersecurity will remain critical areas of focus in BMS research and development. BMS must continue to evolve to address emerging safety challenges, such as thermal management in high-performance EVs and the safe handling of high-voltage systems. Additionally, robust cybersecurity measures will be essential to protect BMS from potential cyber threats, ensuring the integrity and reliability of EV battery systems. Continual advancements in safety and cybersecurity technologies will be necessary to maintain consumer trust and confidence in EVs.

Overall, the future of BMS research and development is driven by the goal of enhancing monitoring and control capabilities, integrating with smart grids and energy management systems, adapting to new battery technologies, promoting standardization and interoperability, and ensuring enhanced safety and cybersecurity. These advancements will contribute to the continued success and widespread adoption of EVs, enabling a sustainable and efficient transportation future.

References

Abdelaal, A. S., Mukhopadhyay, S., & Rehman, H. (2022). Battery Energy Management Techniques for an Electric Vehicle Traction System. *IEEE Access, 10*, 84015-84037.

Bae, J. H., Zhiguo, B., Koo, B. J., Kim, H. R., & Jang, D. G. (2013). Battery management system by using CAN communication based on DSP platform. *Fifth International Conference on Ubiquitous and Future Networks (ICUFN).* Da Nang, Vietnam.

Chen, S., Dai, F., & Cai, M. (2020). Opportunities and Challenges of High-Energy Lithium Metal Batteries for Electric Vehicle Applications. *ACS Energy Letters, 5*(10), 3140–3151.

Chen, S., Xiong, J., Qiu, Y., Zhao, Y., & Chen, S. (2023). A bibliometric analysis of lithium-ion batteries in electric vehicles. *Journal of Energy Storage, 63*, 107109.

Cheng, K. W., Divakar, B. P., Wu, H., Ding, K., & Ho, H. F. (2011). Battery-Management System (BMS) and SOC Development for Electrical Vehicles. *IEEE Transactions on Vehicular Technology, 60*(1), 76-88.

Discharge charateristics of Lithium-ion batteries. (2020). (MakerMax) Récupéré sur https://makermax.ca/articles/discharge-characteristics-of-lithium-ion-batteries/.

Dorian, D. (2023). *2023 Tesla Roadster*. (Car and Driver) Récupéré sur https://www.caranddriver.com/tesla/roadster.

Ebbs Picken, T., Silva, C. M., & Amon, C. H. (2023). Design optimization methodologies applied to battery thermal management systems: A review. *Journal of Energy Storage, 67*, 107460.

Gaton, B. (2022, December). *Tesla launches new EV charging battle, but the Plug War is already over*. (The Driven) Récupéré sur https://thedriven.io/2022/12/21/tesla-launches-new-ev-charging-battle-but-the-plug-war-is-already-over/.

General Motors EV1. (2021). (Wikipedia) Récupéré sur https://en.wikipedia.org/wiki/General_Motors_EV1.

Global EV sales by scenario, 2020-2030. (2022, Oct). (International Energy Agency) Récupéré sur https://www.iea.org/data-and-statistics/charts/global-ev-sales-by-scenario-2020-2030.

Green Car Reports. (2022). Récupéré sur https://www.greencarreports.com/news/1095924_imagine-electric-car-charging-stations-every-25-miles-france-will-have-them.

Greenwood, D., Marco, J., McMahon, R., & Sattar, A. (2020). *Electric Vehicles: The now, the near future and the never again*. (WMG, International Manufacturing Centre, University of Warwick) Récupéré sur https://warwick.ac.uk/fac/sci/wmg/people/wmginsight/electric-vehicles/.

Hossain Lipu M. S. a, M. H., Hussain, A., Saad, M. H., Ayob, A., Miah, M. S., & Mahlia, T. I. (2021). Intelligent algorithms and control strategies for battery management system in electric vehicles: Progress, challenges and future outlook. *Journal of Cleaner Production, 292*, 126044.

Hu, X., Zhang, K., Liu, K., Lin, X., Dey, S., & Onori, S. (2020). Advanced Fault Diagnosis for Lithium-Ion Battery Systems: A Review of Fault Mechanisms, Fault Features, and Diagnosis Procedures. *IEEE Industrial Electronics Magazine, 14*(3), 65-91.

Johnson, J., & Ekler, M. (2020). *Li-ion battery cell management ensures safety- and range-optimized EVs*. (AspenCore) Récupéré sur https://www.edn.com/lithium-ion-battery-cell-management-ensures-safety-and-range-optimized-evs/.

Kumar, B., Khare, N., & Chaturvedi, P. (2018). FPGA-based design of advanced BMS implementing SoC/SoH estimators. *Microelectronics Reliability, 84*, 66-74.

Lambert, F. (2017, May). *Electrek*. Récupéré sur https://electrek.co/2017/05/11/tesla-roadster-tzero-electric-car-lost-fire/.

Li, J., Zhou, S., & Han, Y. (2017). A Review on End-of-Life Battery Management: Challenges, Modeling, and Solution Methods. Dans *Advances in Battery Manufacturing, Service, and Management Systems* (pp. 79-98). Wiley-IEEE Press.

Li, J., Zhou, S., & Han, Y. (2017). *Advances in Battery Manufacturing, Service, and Management Systems*. Wiley-IEEE Press.

Liu, W., Placke, T., & Chau, K. (2022). Overview of batteries and battery management for electric vehicles. *Energy Reports, 8*, 4058-4084.

Lombardo, T. (2020, February). *A look at BMS validation testing*. (Charged EV (Magazine)) Récupéré sur https://chargedevs.com/features/a-look-at-bms-validation-testing/.

Recharging India's electric vehicle ambition by electrifying public transport . (2019). (Deloitte) Récupéré sur https://www2.deloitte.com/in/en/pages/public-sector/articles/EV.html.

Saiteja, P., & Ashok, B. (2022). Critical review on structural architecture, energy control strategies and development process towards optimal energy management in hybrid vehicles. *Renewable and Sustainable Energy Reviews, 157*, 112038.

Solovev, A., & Petrova, A. (2021). *Designing a BMS (Battery Management System) for a Stationary Battery Storage Solution*. (Integra Sources) Récupéré sur https://www.integrasources.com/blog/bms-battery-management-system-battery-storage-solution/.

Sørensen, A., Wang, H., Remy, M. J., Kjettrup, N., Sørensen, R. B., Nielsen, J. J., . . . Madueño, G. C. (2022). Modeling and Experimental Validation for Battery Lifetime Estimation in NB-IoT and LTE-M. *IEEE Internet of Things Journal, 9*(12), 9804-9819.

Tesla's Roadster. (2021). (Tesla) Récupéré sur https://www.tesla.com/en_eu/roadster.

The history and development of batteries: by Jose Alarco And Peter Talbot, The Conversation. (2015). Récupéré sur https://phys.org/news/2015-04-history-batteries.html.

Thomas, S. (2021, May). *Understanding the Role of BMS in Electric Vehicles*. (eInforchips: An Arrow Company) Récupéré sur https://www.einfochips.com/blog/understanding-the-role-of-bms-in-electric-vehicles/.

Turksoy, A., Teke, A., & Alkaya, A. (2020). A comprehensive overview of the dc-dc converter-based battery charge balancing methods in electric vehicles. *Renewable and Sustainable Energy Reviews, 133*, 110274.

Waseem, M., Ahmad, M., Parveen, A., & Suhaib, M. (2023). Battery technologies and functionality of battery management system for EVs: Current status, key challenges, and future prospectives. *Journal of Power Sources, 580*, 233349.

Weng, A., Mohtat, P., Attia, P. M., Sulzer, V., Lee, S., Less, G., & Stefanopoulou, A. (2021). Predicting the impact of formation protocols on battery lifetime immediately after manufacturing. *Joule, 5*(11), 2971-2992.

Xiong, R., Zhang, Y., Wang, J., He, H., Peng, S., & Pecht, M. (2019). Lithium-Ion Battery Health Prognosis Based on a Real Battery Management System Used in Electric Vehicles. *IEEE Transactions on Vehicular Technology, 68*(5), 4110-4121.

Zhang, Z., Cai, Y. Y., Zhang, Y., Gu, D. J., & Liu, Y. F. (2016). A Distributed Architecture Based on Microbank Modules With Self-Reconfiguration Control to Improve the Energy Efficiency in the Battery Energy Storage System. *IEEE Transactions on Power Electronics, 31*(1), 304-317.

Zhao, J., & Burke, A. F. (2022). Electric Vehicle Batteries: Status and Perspectives of Data-Driven Diagnosis and Prognosis. *Batteries, 8*(10), 142.

Chapter 3

Probabilistic Load Demand Modeling of HEVs in the Distribution System

Jaber Pouladi*
Taher Abedinzadeh
and Ali Daghigh
Department of Electrical Engineering, Shabestar Branch, Islamic Azad University, Shabestar, Iran

Abstract

During the past few years, the Hybrid Electric Vehicle (HEV) has gained momentum as a capable solution in overcoming energy and environmental issues. Designing suitable residential distribution systems using a wide array of HEVs requires prediction of the load in existing networks. This chapter offers a comprehensive and exact model in regard to the actual statistics for HEVs aggregation. For this reason, Cumulative Distribution Functions (CDF) is obtained from analysing existing . Also, calculating the initial State of Charge (SOC) in batteries of HEVs are discussed in different models. The chapter explores the influence of Heating, Ventilation, and Air Conditioning (HVAC) power usage on the charging demand of HEVs. It introduces a novel model for estimating the SOC that takes into account vehicle fuel usage and various driving factors, including speed, road slope, and more. Subsequently, based on the point estimate method, the probabilistic load flow simulates the stochastic aspects of HEV charging and examines HEV impact on the residential distribution system. In contrast to existing models that provide fixed outputs for the impacts of HEVs, this chapter examines the

* Corresponding Author's Email: pouladi.jaber@gmail.com.

In: The Future of Hybrid-Electric Vehicles
Editors: Pawan Kumar Pathak and Anil Kumar Yadav
ISBN: 979-8-89113-515-4

probability of HEV effects. Additionally, the proposed models are employed to assess the performance and accuracy by applying them to the IEEE-37 node test feeder.

Keywords: hybrid electric vehicles (HEVs), initial state of charge, charging profile, probabilistic load flow, HVAC usage, fuel usage

3.1. Introduction

The transportation industry with its high usage of fossil fuels is the cause of many environmental problems around the world. To remedy this situation, HEVs play a significant role as they have found increasing applications in eco-friendly and efficient vehicles that utilize electrical energy as an alternative to petroleum-based fuels. They offer desirable benefits by reducing greenhouse gas emissions [1]. However, from a distribution system operator's perspective, these HEVs consume extra load power, which might negatively impact the various components of the distribution system [2-4]. An important issue is that required data are inaccessible because of the low penetration of HEVs. Therefore, gathering them requires expensive devices such as Global Positioning System (GPS) units [5, 6]. Hence, because of the unavailability of such data, researchers have offered various models for defining the charging demand of HEVs via statistics. Also, they have investigated the effectrs of HEV charging on distribution systems. The SOC parameter is a key parameter that directly influences the charging demand of HEVs. The initial SOC of a HEV is determined by factors such as driving style, driving conditions, and the use of accessories.

Previous studies have made certain assumptions to simplify their analysis. For instance, in [2], a multi-objective optimization solution for optimized charging of Electric Vehicles (EVs) in unbalanced three-phase distribution systems is presented. It assumes that all existing batteries have an initial SOC of 20%. Conversely, [7] assumes an initial SOC of zero at the arrival time, which can have an impact on the output results. The initial SOC of vehicles in [5, 6] varies between 0.3 and 0.9. Additionally, [3, 7] assume that every HEV starts the morning with a full charge (SOC = 100%). Moreover, there are different perspectives on battery capacities in HEVs among various studies. For example, [1, 2] consider the same battery capacities for all vehicles. However, it's important to note that the battery capacity in a HEV can vary depending on the vehicle type and the All Electric Range (AER), which refers to the mileage achievable in electrical mode. Furthermore, models presented in [7, 8] have considered the initial SOC as a function of daily mileage, vehicle type, and battery capacity. These models take into account factors such as mileage in electrical mode and energy usage of the HVAC system, which also affect the rate of battery usage. Additionally, [9-11] introduce a new parameter to estimate the operational status of HEVs, representing the fraction of the total input power supplied by the battery. Besides, the impact of HVAC systems on the overall energy usage of the vehicle has been addressed in [12, 13]. However, the specific impact of HVAC energy usage on the initial SOC of the battery is an area that still requires further study.

The previous study [14] utilized a Spatial-Temporal model and employed a Monte Carlo simulation approach to determine the charging demand of EVs at different times. However, the model overlooked influential parameters and assumed a linear decrease in SOC based solely on distance driven, which could lead to inaccurate results when assessing the network impact. Another study [15] developed an agent-based framework to model the spatial distribution of HEV adoption in a residential network, but it failed to consider the initial SOC and internal power management of HEVs, crucial aspects for accurately representing charging behavior and energy usage patterns. In a separate study [16], a real-time energy management algorithm for HEV smart charging in a parking lot was proposed, incorporating charging priority levels. However, the study assumed random initial SOC values, which do not align with practical scenarios where the initial SOC depends on driving behaviors and previous charging patterns. Similarly, a study [17] investigating HEVs' contribution to a fuel shifting demand response program used the EV Scenario Simulator (EVeSSI) tool but assumed a uniform initial battery SOC (30%), potentially impacting result accuracy. Another study [4] presented a stochastic method based on queuing theory to estimate the battery curve of a Plug-in Electric Vehicle (PEV) fleet but did not consider the internal power management of the vehicles or the percentage of electric mode usage, affecting the reliability of the estimates. These studies collectively emphasize the need for comprehensive models that account for various factors like driving behaviors, initial SOC, and internal power management of HEVs and EVs to ensure accurate analysis of network impact and charging behavior.

Although most studies have evaluated the impact of HEV charging on distribution systems, they have generally provided constant outputs, neglecting the stochastic nature of HEVs. For example, [18] examined the effects of optimal HEV charging on total load usage and HEV charging demands to meet the expectations of HEV owners. Similarly, researchers [14] investigated the impact of HEV charging on load voltages and powers, while others [19] introduced two online methods for coordinating PEV charging and assessed their impact on power generation costs, overall system power usage, total system loss, and voltage drop. In a recent study [20], an optimal design method was proposed to satisfy PEV charging demand in the secondary distribution system.

To account for the inherent uncertainty in HEV charging behavior, it is crucial to incorporate probabilistic analysis when studying distribution systems. Probabilistic Load Flow (PLF) analysis offers a suitable approach for understanding the steady-state characteristics of the network under various uncertainties. Different methods can be employed for PLF analysis, including analytical, simulation, and approximation approaches. Among these, Monte Carlo (MC) simulation [21] is commonly used but can be computationally intensive and time-consuming compared to analytical methods that simplify the problem using mathematical assumptions. On the other hand, approximation methods like the Point Estimate Method (PEM) strike a balance between computational efficiency and accurate results [22, 23]. In this chapter, the PEM method is utilized to investigate the probabilistic effects of HEVs on the distribution system, aiming to achieve the following main goals:

- Develop an inclusive and precise model that integrates realistic assumptions and statistical data to examine the collective behavior of HEVs and their effects on the distribution system.
- Introduce a novel approach for calculating the initial SOC in EVs, considering the fuel usage of the vehicles as a determining factor.
- Present a new hourly model that accounts for various uncertainties, including the influence of HVAC systems on the Probabilistic Load Flow (PLF) problem.
- Precisely estimate the HEV load demand by considering factors such as the percentage of vehicle usage in electric mode and internal power management strategies implemented in the vehicles.
- Accurately identify the characteristics of HEVs and distribute them appropriately among the distribution system and households.
- Estimate the increase in load demand over the specified timeframe by utilizing pertinent reports such as the Transportation Information of America report [24], the America Energy Department report [25], and the Electric Power Research Institute report [26].

To accomplish these goals, the research relies on reliable sources of information to comprehend the behavior of vehicle owners, the characteristics of HEVs, and other crucial parameters such as load growth. By leveraging these reports, the study aims to provide a robust and reliable analysis of HEV integration into the distribution system, taking into account real-world factors and ensuring accurate modeling of the system.

3.2. HEVs Aggregation

The energy storage capacity of individual EVs is relatively insignificant when compared to the overall distribution system. As a result, the concept of aggregating HEVs as a controllable energy source has been proposed to enhance the reliability of power systems. Although it is challenging to determine the exact parking or driving status of a specific vehicle at a given time, the behavior of aggregated vehicles tends to be more predictable. Therefore, the theory of HEV aggregation has been explored in industrial projects such as MOBILE [27] and PLACE BETTER [28]. These projects aim to leverage the collective behavior of HEVs to optimize energy management and improve the overall performance of the power system. In this study, to determine for the daily energy requirements of HEVs, we assume that at all vehicles are charged at home as described in a report by the United States Department of Energy. Results indicate that roughly 80% of customers tend to charge their vehicle at home while less than 6% of customers attempt to charge their vehicles at public parking lots [29]. To conduct a more precise study, extensive research has been carried out on the charging model of HEVs. Taking into account all the influential factors on aggregated vehicles, a comprehensive model has been developed, emphasizing the concept of excessive aggregation of these vehicles.

3.2.1. Charging Model of a HEV

The charger of a HEV consists of components such as a rectifier, low-pass filter, and DC-DC converter, as depicted in Figure 3.1 [30]. Traditionally, the rectifier in a charging system is based on diode circuits. However, it is possible to enhance the charging power factor by replacing it with a power factor correction circuit, resulting in power factors between 0.97 and 0.99. This modification significantly reduces the reactive power of the charger, making it substantially smaller compared to the active power. Furthermore, the combined efficiency of the DC-DC converter and low-pass filter typically exceeds 96% [23]. To simplify the modeling of HEVs, it is common to neglect the charger reactive power and the losses associated with the DC-DC converter. Instead, the charging behavior of a HEV can be represented by an active power curve, which describes the power drawn by the vehicle's battery during the charging process. HEVs commonly utilize Lithium-ion (Li-ion) batteries due to their extended lifespan, ability to be fully charged and discharged, and high efficiency [1]. Therefore, this chapter focuses specifically on Li-ion batteries for HEV applications. According to [31], the characteristics of Li-ion batteries differ significantly from other battery types. During the charging process, Li-ion batteries draw nearly constant power until they reach their maximum voltage level. Subsequently, the power usage of the battery decreases rapidly until it reaches zero. Considering the widespread adoption of Li-ion batteries in HEVs, it is possible to calculate the Charging profile if the nominal voltage, battery capacity, and initial SOC of the lithium battery are known. Consequently, for the determination of the Charging profile of aggregated vehicles, it is assumed that all HEVs in the aggregation are equipped with Li-ion batteries and receive a fixed power during the charging process. This simplification enables the estimation of the collective charging behavior of the aggregated HEVs.

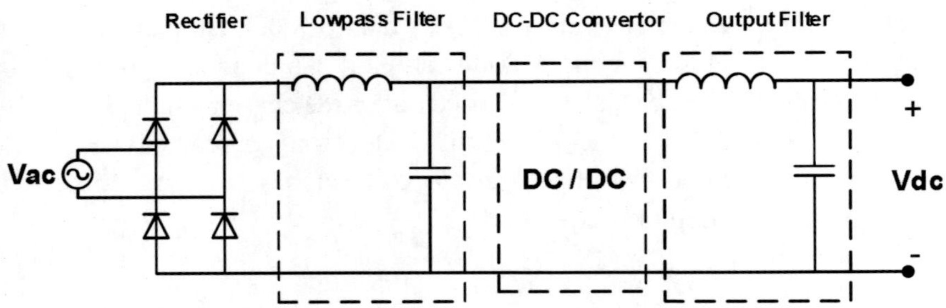

Figure 3.1. The circuit Diagram of HEV charging.

3.2.2. Charging Model of Aggregated HEVs

To determine the Charging profile of the aggregated HEVs, specific data related to the vehicles and their charging patterns is required. However, acquiring accurate information on individual vehicles is practically impossible, and existing reports often lack the necessary parameters as their focus is primarily on transportation rather than electrical energy. Therefore, it is necessary to analyze and evaluate raw data to extract the relevant information. In this chapter, the necessary data for accurately modeling the Charging profile of the aggregated HEVs is extracted from transportation reports, ensuring a satisfactory level of accuracy. The primary source of comprehensive transportation information is the National Household Travel Survey-2009 (NHTS-2009), which serves as the main reference for this study. By leveraging this data, the Charging profile of the aggregated HEVs can be calculated with a reasonable degree of precision. To achieve this, it is essential to determine the effective parameters, which can be categorized into the following three groups:

Individual driving behaviors or patterns that influence the charging process: This group encompasses factors such as the daily milage traveled by each vehicle, the duration of their connection to the charging network, and the daily fuel usage, which is influenced by driver preferences and habits. Vehicle characteristics and specifications: This group includes parameters such as the AER, the type of HEV, the penetration rate of HEVs in the fleet, and the daily load profile. These factors are crucial for investigating the broader impacts of HEVs on distribution systems. Distribution system characteristics and other influential factors: This group encompasses parameters related to the distribution system, such as load growth, which is essential for understanding the additional impacts of HEVs on the system. By considering and incorporating these characteristics and parameters from the three mentioned groups, a comprehensive and accurate model of the Charging profile for the aggregated HEVs can be developed. These parameters provide crucial insights into the driving patterns, vehicle specifications, and distribution system characteristics, enabling a more realistic representation of HEV charging behavior and its impact on the power grid.

3.3. Different Uncertainties Related to HEVs

In order to determine the impact of these vehicles on distribution system, it is needed to investigate of the behavior of an EV driver. Hence, it is crucial to carefully consider the probability function of the relevant factors to accurately account for their impact.

3.3.1. Modeling of Personal Driving Habits

3.3.1.1. Daily Mileage

The daily mileage of a vehicle, or in other words, the daily milage a driver tends to drive their car is a random parameter dependent on the behaviour of every driver. NHTS-2009 proposed the BESTMILES parameter that indicates the distance travelled by each vehicle during the past 12 months [32]. Therefore, dividing the parameter value by 365 gives the Daily mileage by each vehicle. The daily milage traveled by a specific vehicle can be described by a Probability Density Function (PDF), denoted as equation (3.1). This PDF aligns with the lognormal distribution function, with parameter values μ = 3.2 and σ = 0.9. Figure 3.2 provides a visualization of both the Cumulative Distribution Function (CDF) and a histogram depicting the distribution of the daily milage traveled by the vehicle.

$$f\left(d^i\right)=\frac{1}{d^i.\sqrt{2\pi\sigma}}e^{\frac{\left(\ln d^i-\mu\right)}{2\sigma^2}} \qquad d^i>0 \tag{3.1}$$

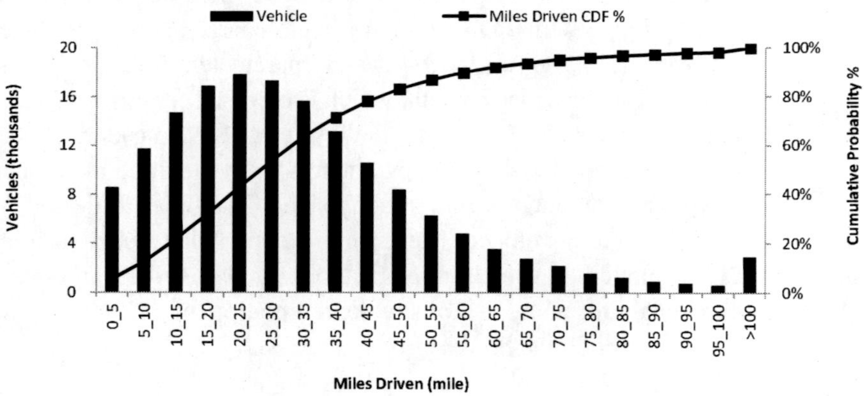

Figure 3.2. Histogram & CDF related to Daily Mileage.

3.3.1.2. Start and End Times for Vehicle Charging

A similar method is used to determine the probability function of the departure and arrival times. Assuming that a HEV is charged once a day, the arrival time from the last trip could be taken as the starting time of charging. This assumption has been considered in majority of the literature [2, 15, 18, 19, 33-35]. Therefore, according to the NHTS-2009 report, the end time of a day trip (signified by the ENDTIME parameter) could be considered as the starting time of charging. Correspondingly, considering many people start their daily commute in the morning, a random time between 7 AM to 10 AM could be considered for the end time of charging the vehicle,. In this article, the PDF of the starting time for each vehicle charging session is based on a predictable value as noted in Equation (3.2), where the shape parameter equals k = -0.5, the scale parameter is σ = 267.3, and the location parameter possesses a value of μ = 922.7. In addition, the rectangular distributed PDF is tailored to the home departure time as shown in Equation (3.3). Figure 3.3 depicts the CDF and histogram of a vehicle's arrival time.

$$f(t_s) = \frac{1}{\sigma}\left(1 + k\frac{(t_s - \mu)}{\sigma}\right)^{-(1+\frac{1}{k})} .e^{-\left(1+k\frac{(t_s-\mu)}{\sigma}\right)^{-\frac{1}{k}}} \qquad t_s > 0 \tag{3.2}$$

$$f(t_e) = \frac{1}{b-a} \qquad a \le t_e \le b \quad a = 7,\ b = 10 \tag{3.3}$$

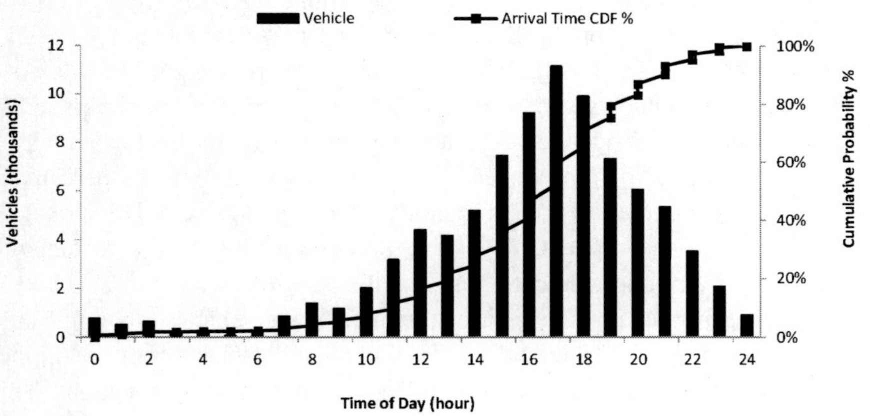

Figure 3.3. Histogram & CDF related to Arrival Times.

3.3.1.3. Daily Fuel Usage Quantity

Based on the NHTS-2009 report, the parameter GSYRGAL represents the annual fuel usage for a vehicle, which is comparable to BESTMILE. To determine the daily fuel usage, this parameter can be divided by 365 to obtain an average value. Figure 3.4 illustrates both the CDF and a corresponding histogram, depicting the distribution of daily fuel usage for the vehicle. The graph highlights that the majority of vehicles consume approximately one U.S. gallon of fuel per day on average.

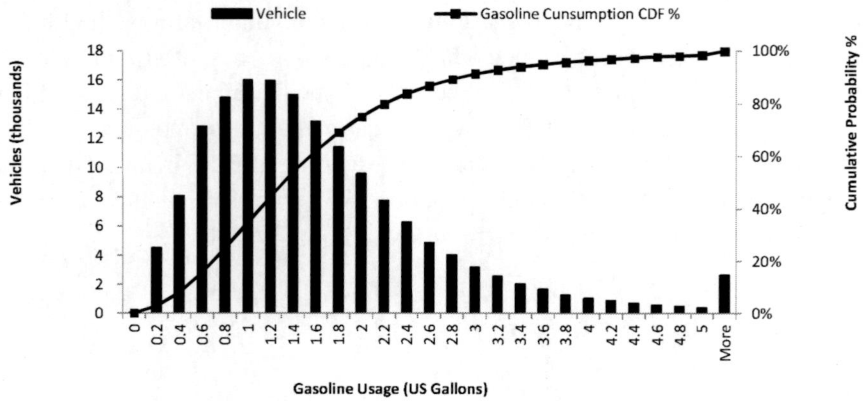

Figure 3.4. Histogram & CDF related to Daily Gasoline Usage.

3.3.1.4. The Rate of Operating the Vehicle in Electric Mode

Because of inadequate data, it is hard to determine whether a hybrid vehicle has worked on battery or gasoline during a day trip. According to [36], HEVs use roughly 40%-60% less fuel than internal combustion engine vehiclesAs a result, it can be concluded that over 50% of the daily trips are completed in electric mode. To model the Driving Status (DS) of the hybrid vehicle, a parameter called DS is introduced. This parameter represents the percentage of the vehicle's operation in electric mode. To assign a value to DS, a random number within the range of [0.5-1] is assumed, following a rectangular probability distribution function (PDF) with parameters a = 0.5 and b = 1. Multiplying the value by the distance travelled by the vehicles in a single day, the Daily mileage in electrical mode is obtained, which results in higher accuracy of the results, or simply put, leads to output error reduction . This outcome is also established in [10, 11, 37].

3.3.2. HEV Characteristic Modeling

3.3.2.1. Vehicle Type, AER, and Battery Capacity Rating

The VEHTYPE parameter proposed by the NHTS-2009 report offers four types of vehicles [32], which consume a certain amount of electrical energy per mile displayed by eTraction [38]. Moreover, since the NHTS-2009 report did not address matters associated to electric energy and HEVs, about it does not provide any information regarding the AER of the vehicles [32]. Therefore, in this study, the percentage of HEVs with different AERs are chosen based on [39]. It should be noted that the product of eTraction and AER determines the battery capacity. Hence, if the probability related to the HEV characteristics is considered based on Table 1, we could say that the PDF associated to the eTraction of each HEV is fitted to the following equation:

$$f_{e^i_{Traction}}(x)=\begin{cases}0.30 & 0\le x\le 0.09\\ 0.46 & 0.09\le x\le 0.25\\ 0.38 & 0.25\le x\le 0.47\\ 0.26 & 0.47\le x\le 1\end{cases} \tag{3.4}$$

where x is the uniform random variable. The CDF for the AER parameter and battery capacity of HEVs can be determined based on the data provided in Table 3.1.

Table 3.1. HEVs characteristics used in this chapter [25, 30]

		Capacity of Battery (kWh)			
vehicle Type (percentage)		Automobile (53%)	Van (9%)	Sports utility (22%)	Pickup truck (16%)
$e^i_{Traction}$ (kWh/mile)		0.26	0.30	0.38	0.46
aer (percentage)	30 mile (21%)	7.8	9	11.4	13.8
	40 mile (59%)	10.4	12	15.2	18.4
	60 mile (20%)	15.6	18	22.8	27.6

3.3.2.2. Number of Vehicles in Each House

Table 3.2 from the National Household Travel Survey (NHTS) 2009 provides the distribution of households based on the number of vehicles they possess. The data indicates that the average number of vehicles per household is approximately 2.11.

Table 3.2. Distribution of households by vehicle count

Household percentage	Vehicles per household
3.4%	0
25.3%	1
42.6%	2
18.7%	3
6.6%	4
2.2%	5
1.2%	More than 5
Mean vehicles per household = 2.11	

3.3.3. Other Effective Factors Modeling

3.3.2.3. Charge Level

In this study, EVs are connected to the low-voltage level of the distribution system, thereby receiving a slow charging mode. Thus, as shown in Table 3.3, the charge level-1 will be employed in accordance to the SAE J1772 standard [40]. Correspondingly, in this study, to be able to determine the daily energy requirement for HEVs, we assume that is the vehicles are charged exclusively at home as reported in a research by the US Department of Energy. The results indicate that while almost 80% of customers tend to charge their vehicle at home, only 6% are willing to use public parking spaces as the charging location [29].

Table 3.3. Charge levels based on SAE J1772 standard [23]

Level	Single Phase Voltage (V)	Current (A)	Power (kW)
Level-1 (Normal)	120	12	1.44
Level-2 (Rapid)	240	30	7.2

3.3.2.4. HEV Penetration

The quantity of EVs in the distribution system can be determined by considering their penetration level. Here, the penetration level was predicted according to [26] and the results revealed three levels of low, average, and high HEV penetration levels of 26%, 50%, and 60%, until 2025, respectively. Yet, considering the penetration levels may vary in different sections of the distribution system, thus, in this study, apart from the average penetration value of 50%, the HEV penetration level is simulated and analysed between 0-100% in 20% increments.

3.3.2.5. Calculation of HEV Counts in the Distribution System

To determine the Charging profile of the aggregated HEVs, it is crucial to specify the total number of HEVs in the system. The calculation for different penetrations is as follows:

$$N_{PHEV} = X_{Penetration} \times N_{Household} \times A \quad (3.5)$$

where NHEV is number of HEVs in a specific penetration, Xpenetration is penetration of HEVs, Nhousehold is the number of houses in the studied system and the average number of vehicles per household, denoted as A, has been calculated as 2.11 vehicles based on the NHTS-2009 data, as discussed in previous sections.

3.3.2.6. HEV Distribution in the Network

It is important to note that the distribution system operator has no control over the charging location of the HEVs. Thus, a precise estimation of the location and distribution of the HEVs across the distribution system is impossible. Indeed, the distribution systems do not have the joint PDF of the charging demand power of HEVs related to different houses. Consequently, HEVs are distributed randomly between the load-points depending on the number of houses available in every load-point.

3.3.2.7. Daily Load Profile

In this chapter, [10] was referenced for generating the average weekday profile for the daily residential loads in both the summer and winter seasons, which are depicted in Figure 3.5. The uncertainty around load demand is usually modelled as a normal PDF [23, 41]. Therefore, based on the current average load profile, a normal PDF with a explicit power factor is generated for the active load demand to enable modelling of such uncertainties at each time interval. The equivalent PDF can be characterized as follows:

$$f\left(P_{d_t}\right) = \frac{1}{\sqrt{2\pi\sigma_t}} e^{-\left(\frac{(P_{d_t} - \mu_t)}{2\sigma_t^2}\right)} \quad (3.6)$$

The load demand, denoted as P_{d_t}, can be represented as a function of time. In this equation, μt represents the mean value of the load demand at the t-th time interval, and σt represents the standard deviation of the load demand at the same time interval.

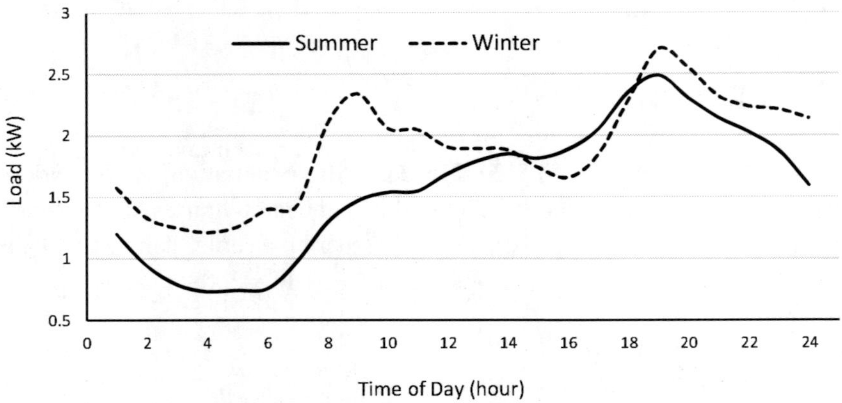

Figure 3.5. Average Daily Residential Load Profile.

3.3.2.8. Load Growth

To analyze the influence of HEVs on the load demand, it is necessary to determine the level of load growth during the specified period. According to reports [25], the average load growth in residential buildings is estimated to be 1.3%, primarily attributed to the utilization of new electrical facilities.

3.4. Probability-Based Modeling of Aggregated HEV Charging Demand

In order to estimate the daily energy needs of a HEV, it is crucial to determine the initial charge or SOC of the vehicle upon arrival. This parameter plays a significant role in determining the amount of energy required for charging the vehicle.

3.4.1. SOC Calculation

3.4.1.1. Method 1: The Traditional and the Customary Method (Old Model)

The approach described above has been widely employed in various research papers [10, 42-44]. The initial SOC is influenced by factors such as the daily milage traveled, the battery capacity of the vehicle's, and the vehicle type. The uncertainties associated with these parameters have been accurately determined. To calculate the SOC, the aforementioned parameters are first determined using Cumulative Distribution Functions (CDFs). Subsequently, the initial SOC upon the vehicle's arrival at home can be calculated using the following equation:

$$SOC_0^i = (1 - \frac{W_{Traction}^i}{C^i}) \tag{3.7}$$
$$SOC_0^i = (1 - \frac{OS^i \times d^i \times e_{Traction}^i}{C^i}) = (1 - \frac{OS^i \times d^i}{AER^i})$$

In the given equation, i represents the number of HEVs, SOC0i denotes the initial SOC, Eti represents the total electric energy usage during the day trip for the vehicle's traction, OSi indicates the operation status, di represents the daily milage traveled in miles, Eci represents the energy usage per mile for the vehicle's traction in kilowatt-hours per mile (kWh/mile), and Ci denotes the battery capacity in kilowatt-hours (kWh). It is important to note that (OSi × di) represents the distance traveled in the electric mode.

3.4.1.2. Method 2: Calculating the SOC by Considering the HVAC Usage Effect (HVAC Model)

The existing equations found in references [8, 18, 20] for calculating the initial SOC of EVs consider factors such as daily milage traveled, vehicle type, and battery capacity. However, they do not account for other important factors that directly affect the battery, such as the percentage of mileage traveled in electric mode and the energy usage of HVAC systems.

Modern vehicles are equipped with various amenities that provide comfort and convenience to passengers, including air conditioning systems. It is crucial to thoroughly study the effect of HVAC energy usage on EVs since all the energy required for traction and HVAC systems is supplied solely by the vehicle's battery. HVAC energy usage refers to the electrical energy needed for heating and cooling the vehicle's interior. Studies have shown that the energy usage for heating and cooling can significantly impact the fuel economy of PEVs, reducing it by approximately 20% to 50% under specific climate conditions [45]. In another study [40], the power usage of HVAC systems in EVs and Internal Combustion Engine (ICE) vehicles was compared. The study concluded that HVAC systems contribute to a 20% increase in total energy usage for EVs and a 9% increase for ICE vehicles.

HVAC systems can consume substantial amounts of energy depending on the climate conditions. For example, a study [19] analyzed the energy usage of HVAC systems in different PEVs, considering the HVAC energy usage model, driving behaviors, and local weather. It was found that, under equal driving conditions, a PEV consumes nearly 1000 kWh more electrical energy per year in warm weather conditions compared to the same vehicle in cooler climates [19]. Therefore, the impact of HVAC systems on EVs is highly dependent on driving behavior and weather conditions. Another study investigated the performance of EVs in terms of driving behavior and weather conditions [41]. It examined the impacts of three different HVAC settings on PEV energy usage: no HVAC, traditional cooling with Positive Thermal Coefficient (PTC) heating systems, and traditional cooling with more efficient Heat Pump (HP) heating systems. The study was conducted in Los Angeles, California (mild climate), Phoenix, Arizona (hot climate), and Minneapolis, Minnesota (cold climate). According to [46], using traditional cooling and PTC heating systems in mild, hot, and cold weather conditions increases the Energy Usage Per Mile (ECPM) of EVs by 9%, 12%, and 24%, respectively. These values demonstrate the rate of ECPM increase in EVs when accounting for the effect of HVAC energy usage, denoted by the variable R_{HVAC}.

In the present study, an assumption of RHVAC = 12% is considered for a specific hot climate condition like Phoenix, Arizona. Since HVAC energy usage includes both cooling and heating the interior space of an EV, it is necessary to determine the ratio of cooling to heating loads. According to [19], the ratio for Arizona is 6.3, indicating that 86.3% of the HVAC energy usage is for cooling purposes, while 13.7% is consumed for heating purposes in this region. Therefore, based on the PDF associated with the energy usage of EVs without HVAC (as denoted by Eq. 4), the PDF of daily energy usage for HVAC systems in EVs can be calculated using the following equation:

$$f_{e^i_{HVAC}}(x) = R_{HVAC} \times mode \times f_{e^i_{Traction}}(x) \tag{3.8}$$

In the given equation, x represents a uniform random variable, f(x) represents the PDF associated with the energy usage of EVs without considering HVAC, RHVAC represents the rate of increase in energy usage when HVAC is taken into account, and mode represents the probability of using HVAC during day trips.

It is important to consider that the probability of utilizing HVAC systems during daily trips is a stochastic parameter influenced by factors such as the start time of travel, daily milage traveled, and weather conditions. In this study, we can assume a random number between [0.7-1] for the HVAC usage mode, following a rectangular probability distribution function (PDF) with parameters a = 0.7 and b = 1. Figure 3.6 illustrates the PDF of daily energy usage per mile by HVAC (including cooling and heating) for a single PEV.

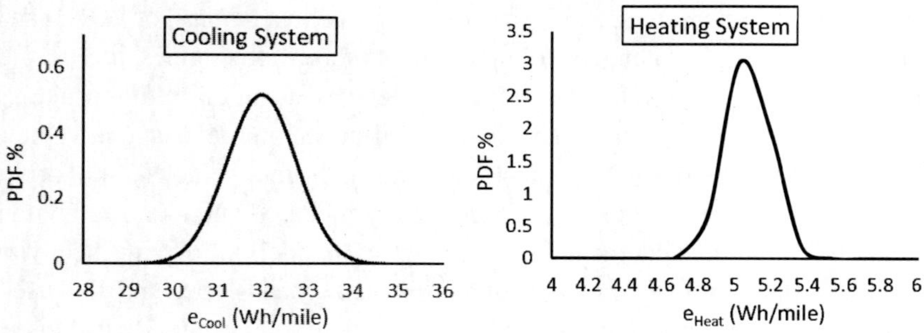

Figure 3.6. CDF for the daily energy usage related to HVAC (Cooling and Heating) in PEVs.

Hence, the calculation of the initial SOC for EVs, taking into account the energy usage from HVAC, can be expressed as follows:

$$SOC_0^i = \left(1 - \frac{W_{Traction}^i + W_{HVAC}^i}{C^i}\right) \tag{3.9}$$

$$SOC_0^i = \left(1 - \frac{OS^i \times d^i \times e_{Traction}^i + OS^i \times d^i \times e_{HVAC}^i}{C^i}\right)$$

where W_{HVAC}^i is the total HVAC energy usage during the day trip, $e_{Traction}^i$ is the traction energy usage per mile (kWh/mile), and e_{HVAC}^i is the energy usage per mile (kWh/mile) by the HVAC system in a specific area.

Contrarily, equation (3.9) shows the likelihood of the complete discharge of the battery, which is practically incorrect, It is noteworthy that the vehicle's energy is originally supplied by its battery, and the moment the SOC reduces to 30%, the system automatically switches to the ICE mode to maintain a longer battery life [47]. Hence, equation (3.9) is improved as follows:

$$SOC_0^i = Max\left(30\%, \left(1 - \frac{OS^i \times d^i}{C^i} \times (\mathrm{e}_{\mathrm{Traction}}^{\mathrm{i}} + \mathrm{e}_{\mathrm{HVAC}}^{\mathrm{i}})\right)\right) \tag{3.10}$$

3.4.1.3. Method 3: SOC Calculation through Fuel Usage (Fuel Model)

It is important to acknowledge that existing equations in previous research papers for calculating the initial SOC of vehicle batteries consider factors such as daily milage traveled, vehicle type, and battery capacity. However, these equations do not account for other crucial variables that directly impact the vehicle battery, such as speed, road slope, usage of cooling or heating systems, and other relevant factors. Since the energy stored in a battery, or SOC, can be analogized to the fuel gauge level in internal combustion engines, it is possible to estimate the initial SOC of a vehicle battery by using daily fuel usage as a reference. Driving behaviors have a direct influence on fuel usage in vehicles. By analyzing the daily fuel usage data of gasoline vehicles, the driving behavior of vehicle owners can be statistically modeled and identified with an acceptable level of accuracy. In the developed model, all driving behaviors, including daily milage traveled, driving speed, road slope, and the use of cooling or heating systems, are represented as daily fuel usage. Real statistical data on fuel usage for individual vehicles is readily accessible, and its CDF has been computed in the preceding section. The percentage of remaining fuel (PRF) in a vehicle, influenced by driving behavior and the fuel tank capacity, can be determined using Equation (3.11).

$$PRF^{i} = \left(1\text{-}PFC^{i}\right)\times 100\%$$
$$PFC^{i} = \frac{\text{Daily } \mathit{Fuel} \text{ Consumption}}{\text{Fuel Capacity}}\times 100\% \quad (3.11)$$

In the equation, PRFi represents the percentage of remaining fuel in the vehicle, while PFCi denotes the percentage of fuel usage in the HEVs.

Therefore, taking into account various factors that impact fuel usage, such as daily milage traveled, driving speed, road slope, and the use of cooling or heating systems, the remaining percentage of fuel in the tank of internal combustion engine vehicles is analogous to the remaining charge in the battery of EVs. This remaining charge represents the initial SOC. Nevertheless, it is crucial to acknowledge that the percentage of electricity and fuel usage may not be equivalent for a given distance, and this distinction should be considered. While obtaining accurate statistical data for electrical energy usage in vehicles can be challenging due to low penetration rates, real statistical data for fuel usage is readily available. This availability enables the precise calculation of the initial SOC for vehicle batteries. The proposed method incorporates all the significant factors affecting the daily usage of vehicle batteries. By conducting a comparative analysis of different HEVs and comparing the fuel usage in electric and fuel modes over a fixed distance of 100km, it has been observed that the average energy usage of a vehicle in electric mode surpasses that in fuel mode for the same distance [36, 48]. Table 3.4 presents an overview of the usage patterns of commonly available vehicles in the market. To conduct a more detailed analysis, the CDF of the electricity to fuel usage ratio in HEVs can be computed by considering a significant number of HEVs, as documented in [36, 48]. This ratio is visually represented in Figure 3.7.

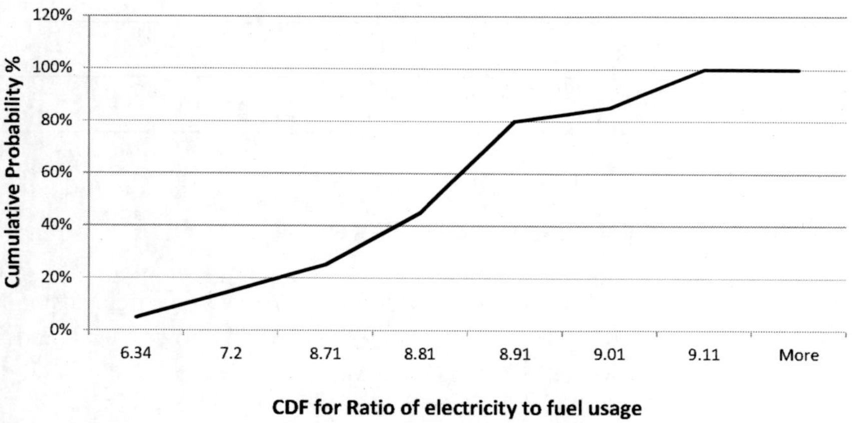

Figure 3.7. Electric-to-fuel consumption ratio in HEVs.

Now, the initial SOC equation can be calculated as follows:

$$SOC_0^i = Max\left(30\%,\ \left(1\text{-}PEC^i\right)\times 100\%\right) \tag{3.12}$$

In the given equation, PEC^i represents the percentage of electricity usage. The calculation of the PEC^i parameter is determined using the following equations:

$$\begin{aligned} PEC^i &= CDF^{-1} \times PFC^i \\ PEC^i &= \frac{\text{Daily Electric Consumption}}{\text{Battery Capacity}} \times 100\% \end{aligned} \tag{3.13}$$

where CDF^{1} represents the ratio of electricity to fuel usage in HEVs and it was determined according to Figure 3.7.

Table 3.4. Comparison of fuel usage in some HEVs

Manufacturer Model	Usage		
	Electric (kWh/100 km)	Fuel (liter /100 km)	Electric/Fuel (kWh/liter)
FORD FUSION ENERGY	20.9	2.4	8.7
TOYOTA PRIUS PLUG-IN	18	2.5	7.2
CHEVROLET VOLT	21.4	2.4	8.9
NISSAN LEAF	18.4	2.1	8.7

3.4.2. Energy Needed to Charge the Battery

The energy needed to charge a HEV battery upon arrival home depends on its initial SOC. The required energy is supplied by the power grid and is calculated according to the following equation:

$$E^{i} = \frac{C^{i}}{\eta^{i}_{Charger}} \times (SOC^{i}_{sch} - SOC^{i}_{0}) \qquad (3.14)$$

In the given equation, E^{i} represents the total energy needed to charge vehicle i in kilowatt-hours (kWh), SOC_0^{i} denotes the initial SOC of vehicle i, $SOCS_{ch}^{i}$ represents the scheduled SOC of vehicle i, C^{i} represents the battery capacity of vehicle i, and η Chargei represents the efficiency of the battery charger of vehicle i, which is assumed to be 90% according to reference [49].

3.4.3. Modelling the Charging Profile of a HEV

The Charging profile of a HEV is calculated based on the following parameters:

- The charging Start time
- Available charge level

- The time required for charging the battery.

As stated previously, we assume that vehicle owners tend to charge their vehicles upon their arrival home. Hence, arrival time could be taken as the initial recharging time. The charge level, thus, is dependent on the distribution system's structure and the charger's control system.

The time required to charge the battery is determined by two crucial parameters: Ei (energy required to charge) and the available charge level. Consequently, by specifying the start time of the charge, it becomes possible to calculate the Charging profile for each HEV with a reasonable level of accuracy. This process is repeated for all vehicles under consideration. For every HEV, denoted as HEV^i, the charging program is defined as a matrix of Pi, which represents the charging schedule, in the following manner:

$$P^i = [P_1^i, P_2^i, ..., P_t^i ..., P_T^i] \tag{3.15}$$

In the equation, T represents the charge period, which is 24 hours. $P_t^i \geq 0$ denotes the power usage of vehicle i at time t. If t_s^i and t_e^i are considered as the start time (plugging or arrival of the vehicle) and end time of the charge (vehicle exit) for HEV^i, respectively, the time interval between tsi and tei represents the charging period for the vehicle. Outside this period, $P_t^i = 0$. To calculate the aggregated values for all HEVs, a diagram similar to Figure 3.8 must be computed for each HEV. Then, the values from all HEVs are summed up to form a fixed time axis.

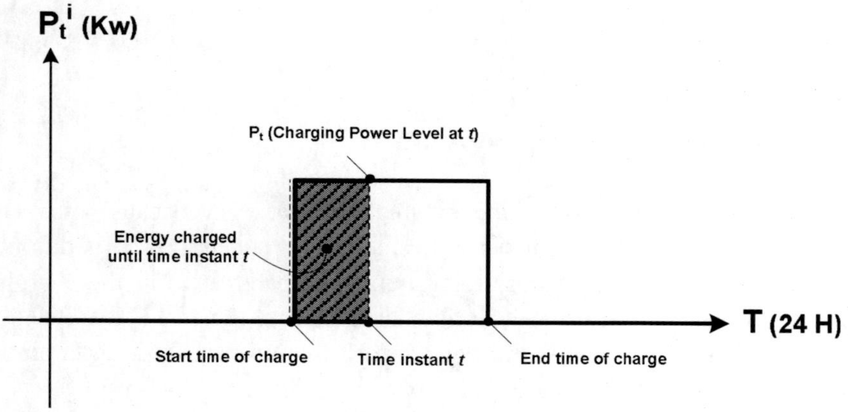

Figure 3.8. Daily Charging profile of the HEVi.

The energy required to charge HEV^i, denoted as Ei, can be determined by calculating the Integrated area of the daily charging profile of HEVi. Hence, by taking into account the start and end times of charging, along with the power usage of the vehicle, Ei can be determined using the following formula:

$$E^i = \sum_{t=t_s}^{t_e} P_t^i \tag{3.16}$$

The duration (t_e^i - t_s^i) should be equal to or greater than the required time for a complete charge of the vehicle's battery to ensure a full charge is achieved. If there is additional time available beyond (t_e^i - t_s^i), various charging periods can be selected, but without control. In such cases, when the vehicle is plugged in, the battery is charged using the maximum power available from the distribution system.

3.4.4. Modelling the Charging Profile of the Aggregated HEVs

Based on the information provided in the previous sections, the Charging profile for the aggregated HEVs can be obtained by summing up the Charging profiles calculated for each individual vehicle over a 24-hour period. The total load of the vehicles at a specific time instant, denoted as t, is derived through the following formula:

$$P_t^{PHEVs} = \sum_{i=1}^{N_{PHEV}} P_t^i \qquad 0<t \leq 24h \tag{3.17}$$

3.4.5. Probabilistic Load Flow Analysis

Conducting a deterministic analysis might not be easy because of varied driving behaviours, HEV characteristics, and HEV penetration, in addition to other factors. Therefore, considering numerous parameters in the charging demand of HEVs possess stochastic characteristics, thus, PLF is utilized to study and predict the characteristics of the distribution system for uncertainties pertaining to the HEVs and related network load points.

3.4.5.1. Monte Carlo Method

In the Monte Carlo method, the values of parameters such as daily mileage, arrival time, departure time, etc., can be randomly defined based on the specified PDFs for HEVs. Subsequently, values related to the initial SOC and the energy required for charging each HEV are calculated. The charging demand for each subscriber is then determined considering the vehicle's arrival time and the available charge level. Subsequently, the load profiles of the load points are established, and deterministic load flow calculations are carried out using the OpenDSS software. OpenDSS is a widely used for electric utility distribution systems. It provides multiple built-in solution modes, such as duty cycle, peak day, daily, and yearly power flow simulations [50]. The Monte Carlo simulation results can be used as inputs to generate PDFs for the desired outputs.

3.4.5.2. Point Estimate Method (PEM: 2m+1 Scheme) for PLF

In this chapter, the (PEM: 2m+1) is utilized as a solution for the Power Load Forecasting (PLF) problem, as described in [22]. The PDF of each random variable is expressed within each time interval, incorporating both the typical daily load power and the HEV charging demand demand, as defined by equations (3.5) and (3.14) respectively. Next, three points are calculated for each random input variable based on its PDF. The output variables in the PLF model are then evaluated a total of 2m+1 times, considering m random input variables. These output variables are determined in accordance with equation (3.18):

$$Z = F(X) = F(x_1, x_2, \ldots, x_m, \ldots, x_n) \tag{3.18}$$

In the equation, n represents the total number of variables, and m denotes the number of input random variables.

For each time interval, three locations and corresponding weights are determined for every random variable, as shown below:

$$x_{l,k} = \mu_{x_l} + \xi_{l,k}\sigma_{x_l} \quad (l = 1,2,\ldots,m \quad and \quad k = 1,2,3) \tag{3.19}$$

In the given equation, μ^i represents the mean of the i-th random variable, σ^i represents the standard deviation of the i-th random variable, θ_K^i is the coefficient proportional to the k-th location of concentration, and m represents the number of input variables. The coefficient of θ_K^i is determined as follows:

$$\xi_{l,k} = \frac{\lambda_{l3}}{2} + (-1)^{(3-k)} \cdot \sqrt{\left(\lambda_{l4} - \frac{3}{4}\lambda_{l3}^2\right)} \quad (l = 1,2,...,m \quad k = 1,2) \tag{3.20}$$
$$\xi_{l,3} = 0$$

In the equation, λ_{l3} and λ_{l4} represent the skewness and kurtosis of the input random variable x_l, respectively. Skewness and kurtosis correspond to the third and fourth standard central moments of x_l, respectively. Furthermore, the weight of each location ($w_{l,k}$) associated with $x_{l,k}$ is calculated using equation (3.21).

$$w_{l,k} = \frac{(-1)^{(3-k)}}{\xi_{l,k}(\xi_{l1} - \xi_{l2})} \quad (l = 1,2,...,m \quad k = 1,2)$$
$$w_{l,3} = \frac{1}{m} - \frac{1}{\lambda_{l4} - \lambda_{l3}^2} \tag{3.21}$$

To solve the load flow problem with desirable certainty, every random variable is substituted by three points based on equation (3.19), and the remaining random variables maintain their constant mean value. Besides, the input random variables are initially transformed to (2m+1) vectors of X according to equation (3.22), and subsequently, the deterministic load flow problem is solved (2m+1) times.

$$X_i = [\mu_1, \mu_2, ..., x_{l,k}, ..., \mu_m] \quad (l = 1,2,...,m \quad k = 1,2)$$
$$X_{2m+1} = [\mu_1, \mu_1, ..., \mu_m] \tag{3.22}$$

Consequently, the j-th order moment of the Z output variables can be determined using the provided expressions. Subsequently, the PDF of the Z output variables can be reconstructed using the calculated output statistical moments as described in references [51].

$$E\left(Z^j\right) = \sum_{l=1}^{m}\sum_{k=1}^{3} w_{l,k} \times \left(Z_{l,k}\right)^j \tag{3.23}$$

Figure 3.9 illustrates the general flowchart of the PLF method. The following procedures can be performed for each time interval, denoted as t, once the PDFs related to the HEVs are defined. These steps include information such as daily milage traveled, arrival time, departure time, expected charging power of the HEVs, and load point demands, among others.

- Choosing the anticipated charging power of the HEVs and load point powers for the tth time interval.
- Defining three load points and their corresponding weights for each random variable using equations (3.19) to (3.21).
- Randomly allocating the aggregated charging demand power of the HEVs to the load points, taking into account the total of dwellings in each load point.

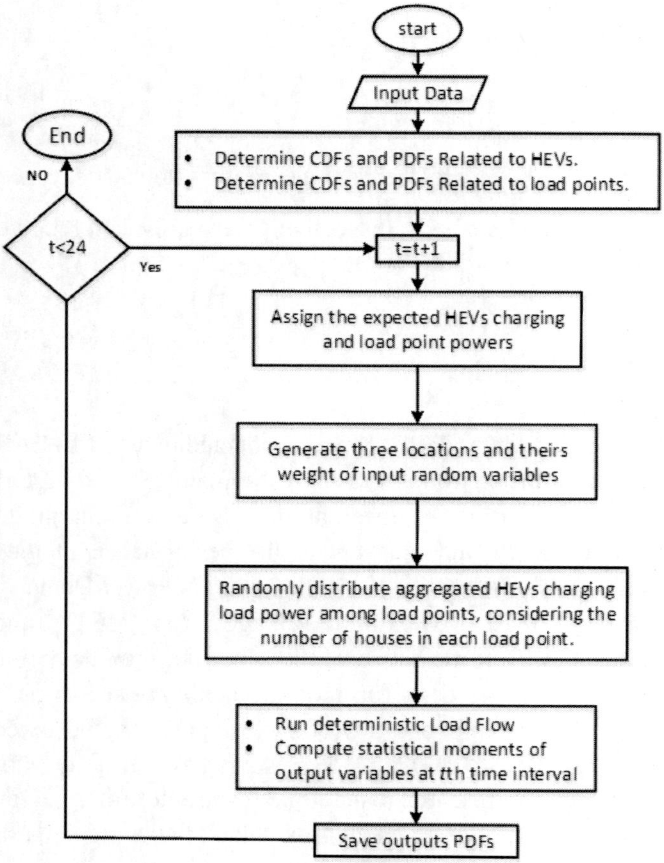

Figure 3.9. Monte Carlo simulation flowchart.

- Evaluating the charging demand demand of the HEVs by performing deterministic load flow calculations and determining the statistical moments of the output variables.
- Iterating the aforementioned steps for HEVs in various locations and across all time intervals.

Ultimately, to evaluate the precision of the suggested approach, the Monte Carlo method is used as a reference with 2500 iterations. The results obtained from the proposed method are then compared with those from the Monte Carlo method. To achieve this, the relative error for the output variables is computed using the following equation.

$$err_{\mu_{v_i}} = \left| \frac{\left(\mu_{MC_{v_i}} - \mu_{PEM_{v_i}}\right)}{\mu_{PEM_{v_i}}} \right| \times 100\% \tag{3.24}$$

where $\mu_{MC_{v_i}}$ and $\mu_{PEM_{v_i}}$ are the average of the stochastic variable V_i for Monte Carlo and PEM methods, respectively. An equivalent relative error can be determined for standard deviation.

3.5. Simulation Studies

The Point Estimated Method (PEM) was conducted in the MATLAB software and the power flow problem was solved by the OpenDSS software. Therefore, this section aims to portray the different impacts of HEV charging profile on the distribution system and to evaluate the performance of the proposed models. These impacts comprise Network Apparent Power Demand (NAPD), Feeder Real Loss (FRL), Maximum Voltage Drop (MVD), and Voltage Imbalance Index (VII) of the network. PLF does not provide definite results, and thus, should be stated in form of probability distribution functions. Consequently, only the probabilities of such impacts should be considered. Yet, some studies [10, 34, 39, 42, 52, 53] resolved to only descriptions of such impacts. The present research used this feature to improve one of the significant disadvantages of the models explained in the previous studies.

3.5.1. Test System

As shown in Figure 3.10, the present study employs a modified IEEE-37 node test feeder. Assuming a 5kVA feed for every house, the count of residential customers of the load points and the total subscribers can be computed according to the rated capacity of the load points, which results in 544 residential subscribers. Considering the average number of vehicles in each household is 2.11, the total number of vehicles in the system under test amounts to 544 × 2.11 = 1148. Also, the total number of HEVs in the system will be calculated based on the penetration level of HEVs. In this study, the charging demand of HEVs and its impact on the importante parameters of the distribution system are computed and assessed by considering the effects of HVAC systems and fuel usage models. The base case, thus, constitutes the load of the system in the absence of HEVs. The following sections elucidate the various specifications of the system under test in the presence of HEVs.

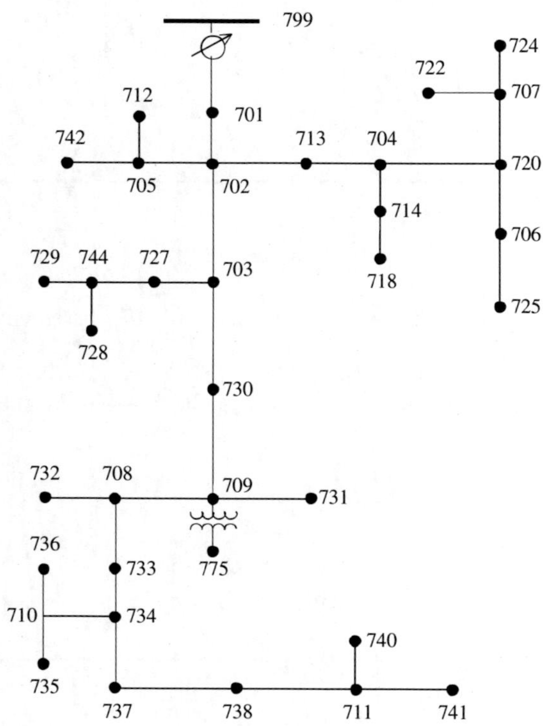

Figure 3.10. IEEE-37 node test feeder [48].

3.5.2. Required Energy to Recharge HEVs

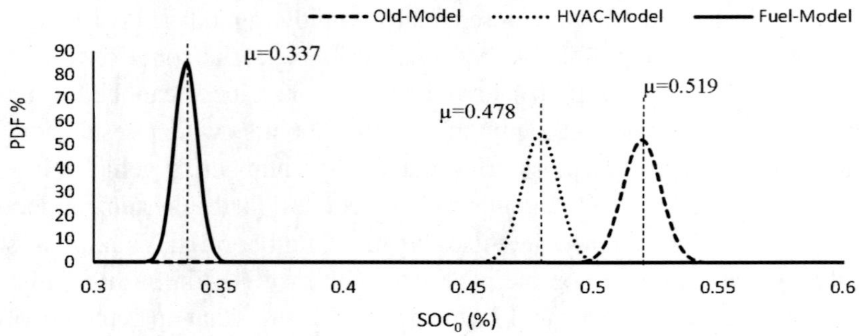

Figure 3.11. PDF of initial SOC based on different models and 50% penetration level of HEV.

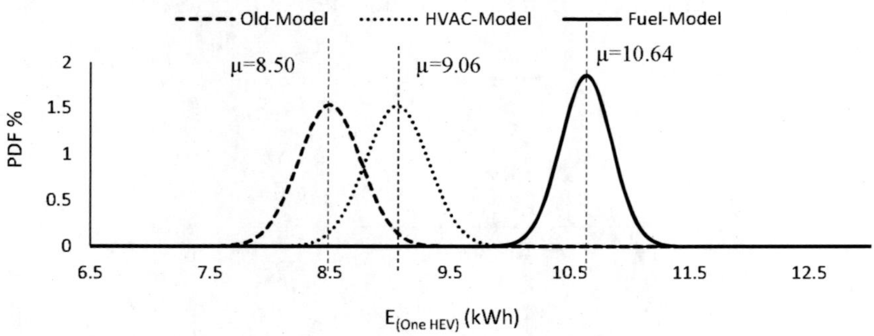

Figure 3.12. PDF of energy required to recharge one HEV based on different models (HEV penetration level: 50%).

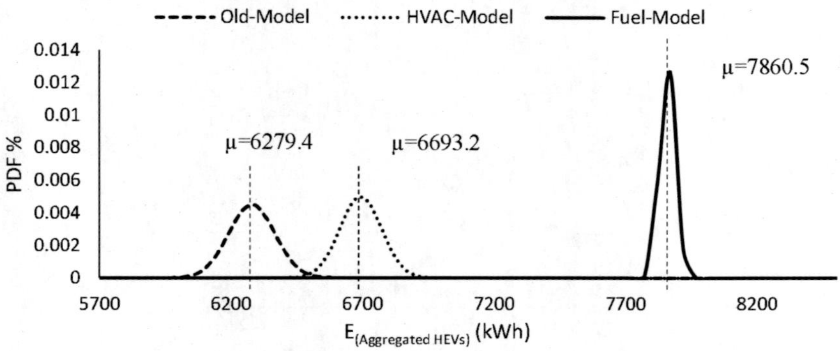

Figure 3.13. PDF of energy required to recharge aggregated HEVs based on different models (HEV penetration level: 50%).

As previously mentioned, this study takes into account all factors that influence vehicle energy usage, including traction and HVAC loads, based on available statistics regarding HVAC and fuel usage in HEVs. These loads result in a decrease in the initial SOC of a HEV and a significant increase in the amount of electrical energy required for charging the vehicle. The PDF related to the reduction in the initial SOC and the PDF related to the increased energy requirement for recharging, based on different models, are depicted in Figures 11-13, respectively. According to these figures, the average value of the initial SOC has decreased from 0.519 to 0.478 and 0.337, respectively, for each vehicle, when considering the HVAC and fuel models. This decline has led to an increase in the mean value of the energy required to recharge the aggregated HEVs based on different models, from 6279.4 kWh to 6693.2 kWh and 7860.5 kWh, respectively.

3.5.3. Network Apparent Power Demand (NAPD)

Figure 3.14 displays the 24-hour average value of Network Average Power Demand (NAPD) using different models and the base case. It demonstrates that NAPD increases in both the typical and suggested models with the presence of HEVs. Moreover, the average NAPD value is higher in the proposed model compared to the old model. Notably, the proposed model shifts the peak-load time to the early evening, indicating a convergence between vehicle plug-in time and the grid's peak-load period. Furthermore, Figure 3.14 shows that the mean NAPD remains consistent across different models and the base case between 5 AM and 10 AM but varies significantly during peak-load hours. Thus, despite the overall increase in network load due to HEVs, the increase is more pronounced from 6 PM to 9 PM. Consequently, a detailed assessment of NAPD is conducted as follows.

To investigate uncertainties in the network, a stochastic analysis is performed based on the description provided earlier. Table 3.5 compares the mean NAPD values obtained from the Monte Carlo (MC) and 2m+1 PEM methods at different time periods to evaluate the proposed method. The table shows that the relative error for NAPD is less than 0.37%. Additionally, since power engineers are primarily concerned about the maximum load capacity of the substation transformer, Figure 3.15 presents the PDF of NAPD at 7 PM supplied by the substation transformer for a 50% penetration level of HEVs. In this figure, the mean values are μ = 1.94 MVA for the base case, μ = 2.52 MVA for the old model, μ = 2.56 MVA for the HVAC model, and μ = 2.63

MVA for the Fuel model. Furthermore, for distribution system planning studies, estimating the probability of the substation load exceeding a specific MVA value (e.g., 0.9×Smax) is crucial. Hence, the area under the curve to the right of Srated-Limit = 0.9×Smax = 2.5 MVA in Figure 3.15 represents the frequency of cases in the shaded area, known as the Unallowable Index (UI %). According to the same figure, the UI index is zero for the base case, old model, and proposed models with a 50% penetration level of HEVs. However, a higher penetration level of HEVs is likely to cause substation transformer overload. Therefore, the UI index is introduced in Section 3.5.7 for different HEV penetration levels.

3.5.4. Feeder Real Loss (FRN)

Figure 3.16 displays the average value of feeder real power losses in each hour using different models. It shows that the power usage of HEVs leads to increased system losses compared to the base case throughout all time intervals. Similar to NAPD, the different models shift the peak total loss of the network to the early evening period, coinciding with the peak NAPD. The findings indicate that network losses are influenced by the method employed to calculate the initial SOC for HEV charging, with the proposed methods resulting in higher power losses than the old method. Figure 3.16 also reveals that the minimum feeder power losses for the base case occur at 6 AM, with a value of 5.6 kW, and increase for the different models with a 50% HEV penetration level. Likewise, the maximum feeder power losses for the base case, old model, HVAC model, and Fuel model at 7 PM increase to 63.9 kW, 107.7 kW, 110.9 kW, and 117.2 kW, respectively.

Table 3.6 presents the results of the stochastic analysis of Feeder Resistance Loss (FRL) at different times using the Monte Carlo (MC) and 2m+1 PEM methods based on different models. It demonstrates that the proposed method performs more accurately than the MC method. Additionally, Figure 3.17 illustrates the PDF of FRL at 7 PM for a 50% penetration level, indicating μ = 107.7 kW for the base case, μ = 107.7 kW for the old model, μ = 110.9 kW for the HVAC model, and μ = 117.2 kW for the Fuel model. Furthermore, at 7 PM, FRL increases by approximately 68.5%, 73.5%, and 83.4% compared to the base case for the old, HVAC, and Fuel models, respectively. These extra losses make the integration of HEVs into existing residential distribution systems a costly endeavor. Equivalent outcomes were obtained for the reactive losses of the power system.

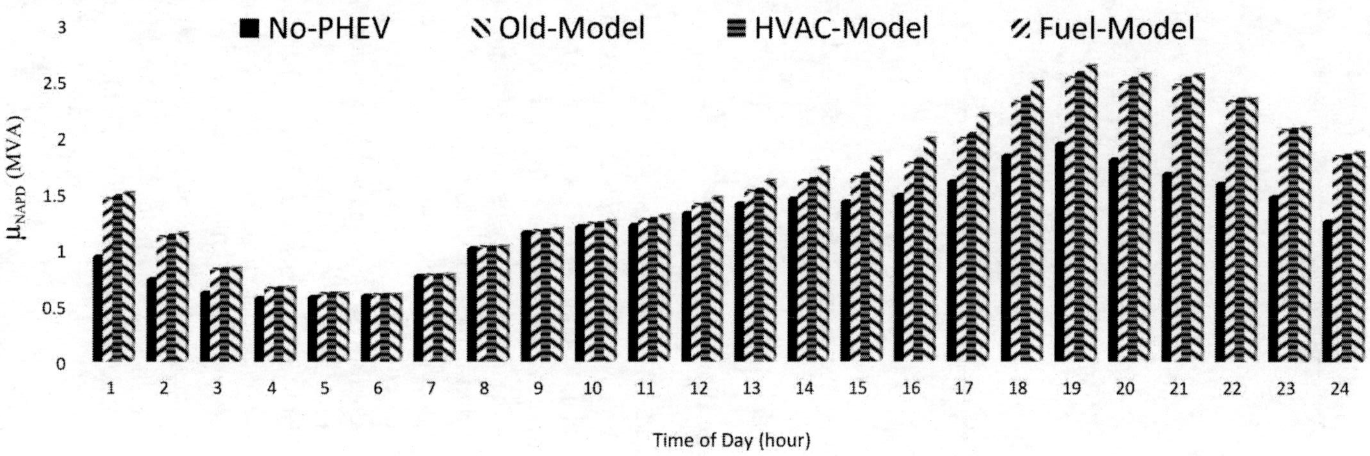

Figure 3.14. The impact of various models on the average NAPD parameter at a 50% HEV penetration level.

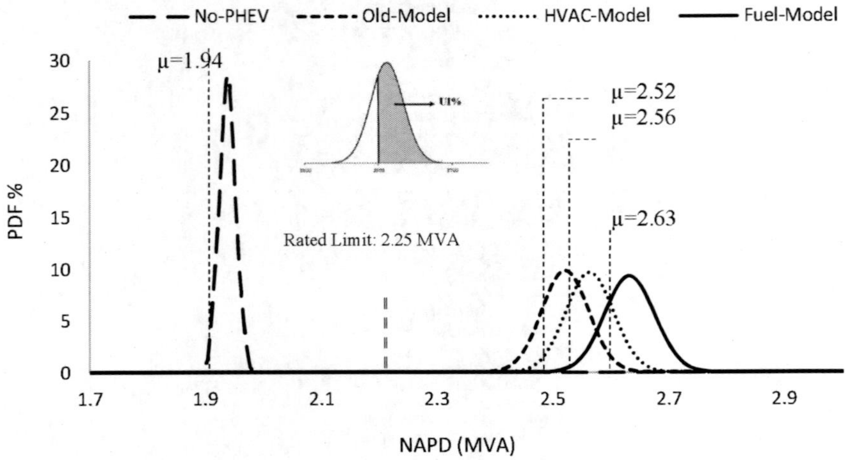

Figure 3.15. PDF of NAPD at 7 PM based on different models and 50% penetration level of HEV.

3.5.5. Maximum Voltage Drop (MVD)

One important concern with HEVs is the potential voltage drops at the final load points during the peak charging period. Figure 3.18 illustrates the average voltage at the worst load point (load point: 741) at each time based on different models and a 50% penetration level. Similar voltage trends are observed at other load points as well. The figure indicates that while minimum voltage disturbances (MVD) occur during the peak-load period, the charging demands have minimal impact on system voltages, even during the peak-load period. This outcome can be attributed to the presence of voltage regulators in the network that effectively maintain system voltages within acceptable limits. To accurately calculate the end load point voltages, a PEM analysis was conducted in this study. Table 3.7 reveals that, unlike other parameters, the mean value of the worst load point experiences negligible changes at different time intervals in the presence of HEVs. The probabilistic analysis of this parameter confirms the accuracy of the PEM method. Therefore, Figure 3.19 presents the PDF of MVD at the end load point at 7 PM for different models. The figure highlights the probability of voltage limit violations at the end load point when using different SOC calculation methods. The probability of violating voltage limits is zero for the base case, but for a 50% penetration level of HEVs, the probability reaches 100% for the different models.

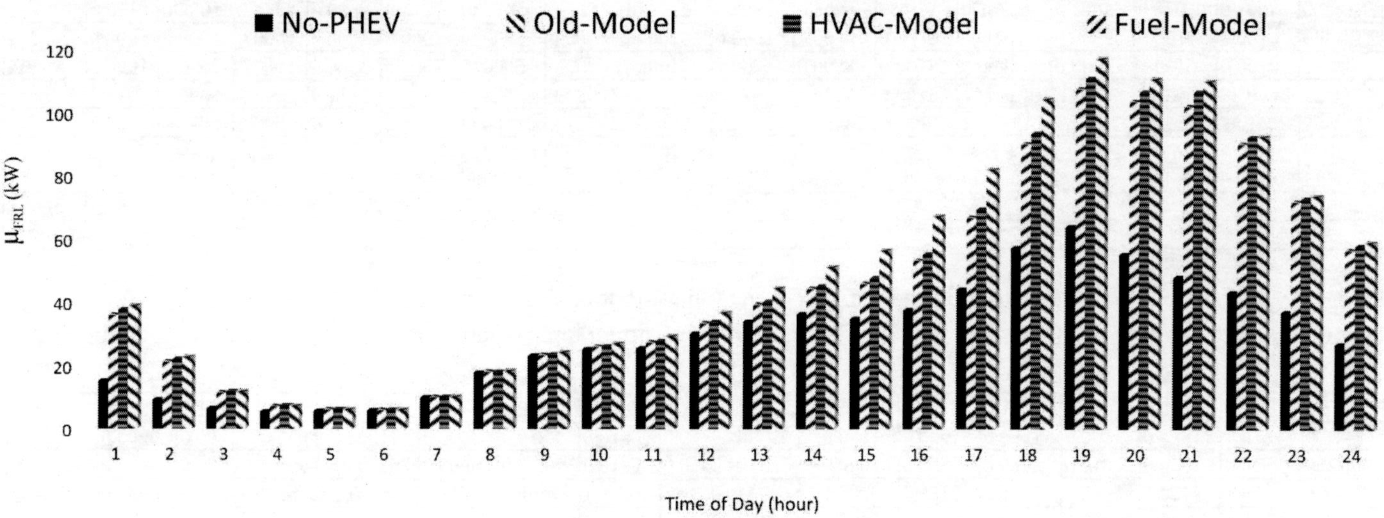

Figure 3.16. The impact of various models on the average FRL parameter at a 50% HEV penetration level.

Table 3.5. Comparing the average and standard deviation values of NAPD at different time between MC and PEM methods, considering various models and a 50% HEV penetration level

NAPD (MVA)																
Times	6:00				12:00				18:00				24:00			
Methods	No-ev	Old model	HVAC model	Fuel model	No-ev	Old model	HVAC model	Fuel model	No-ev	Old model	HVAC model	Fuel model	No-ev	Old model	HVAC model	Fuel model
MC: μ	0.5809	0.5887	0.5890	0.5948	1.3247	1.3926	1.4033	1.4568	1.8319	2.2983	2.3599	2.4932	1.2508	1.8206	1.8448	1.8674
PEM: μ	0.5807	0.5875	0.5878	0.5947	1.3225	1.3934	1.3985	1.4620	1.8329	2.3046	2.3513	2.4883	1.2502	1.8255	1.8378	1.8652
MC: Σ	0.0048	0.0062	0.0062	0.0070	0.0098	0.0178	0.0182	0.0252	0.0126	0.0401	0.0430	0.0467	0.0101	0.0291	0.0264	0.0226
PEM: Σ	0.0046	0.0047	0.0057	0.0071	0.0098	0.0188	0.0190	0.0266	0.0105	0.0431	0.0440	0.0355	0.0075	0.0281	0.0294	0.0198
Err: μ %	0.0388	0.1989	0.2030	0.0109	0.1659	0.0618	0.3469	0.3518	0.0507	0.2761	0.3659	0.1997	0.0501	0.2669	0.3772	0.1155

Table 3.6. Comparing the average and standard deviation values of FRL at different time between MC and PEM methods, considering various models and a 50% HEV penetration level

FRL (kw)																
Times	6:00				12:00				18:00				24:00			
Methods	No-ev	Old model	hvac model	Fuel model	No-ev	Old model	hvac model	Fuel model	No-ev	Old model	hvac model	Fuel model	No-ev	Old model	hvac model	Fuel model
MC: μ	5.8258	5.9834	5.9897	6.1083	30.3047	33.4497	33.9611	36.5457	57.2522	89.4685	94.2468	104.9132	27.0540	56.6027	58.1228	59.6349
PEM: μ	5.8123	5.9607	5.9642	6.1136	30.1780	33.4958	33.8002	36.8585	57.3030	90.2257	93.4332	104.5913	27.0133	56.8099	58.0642	59.3126
MC: σ	0.0001	0.0001	0.0001	0.0002	0.0005	0.0009	0.0009	0.0013	0.0009	0.0033	0.0036	0.0042	0.0005	0.0019	0.0018	0.0015
PEM: σ	0.0001	0.0001	0.0001	0.0002	0.0005	0.0009	0.0009	0.0015	0.0007	0.0034	0.0038	0.0035	0.0004	0.0016	0.0019	0.0015
Err: μ %	0.2310	0.3801	0.4254	0.0876	0.4181	0.1377	0.4739	0.8559	0.0887	0.8464	0.8633	0.3069	0.1505	0.3661	0.1008	0.5405

Table 3.7. Comparing the average and standard deviation values of 741 load point at different time between MC and PEM methods, considering various models and a 50% HEV penetration level

V-741 (p.u)																
Times	6:00				12:00				18:00				24:00			
Methods	No-ev	Old model	hvac model	Fuel model	No-ev	Old model	hvac model	Fuel model	No-ev	Old model	hvac model	Fuel model	No-ev	Old model	hvac model	Fuel model
MC: μ	0.9723	0.9720	0.9720	0.9717	0.9364	0.9332	0.9329	0.9305	0.9132	0.8916	0.8890	0.8827	0.9398	0.9138	0.9128	0.9116
PEM: μ	0.9726	0.9721	0.9722	0.9714	0.9369	0.9338	0.9322	0.9303	0.9130	0.8907	0.8902	0.8823	0.9403	0.9140	0.9128	0.9127
MC: σ	0.0010	0.0010	0.0010	0.0011	0.0020	0.0024	0.0024	0.0029	0.0026	0.0045	0.0047	0.0052	0.0020	0.0041	0.0043	0.0042
PEM: σ	0.0009	0.0009	0.0010	0.0012	0.0024	0.0026	0.0025	0.0037	0.0026	0.0039	0.0053	0.0055	0.0017	0.0047	0.0042	0.0039
Err: μ %	0.0232	0.0099	0.0253	0.0225	0.0561	0.0615	0.0731	0.0252	0.0241	0.0962	0.1315	0.0495	0.0498	0.0176	0.0032	0.1157

Table 3.8. Comparing the average and standard deviation values of VII at different time between MC and PEM methods, considering various models and a 50% HEV penetration level

VII (%)																
Times	6:00				12:00				18:00				24:00			
Methods	No-ev	Old model	hvac model	Fuel model	No-ev	Old model	hvac model	Fuel model	No-ev	Old model	hvac model	Fuel model	No-ev	Old model	hvac model	Fuel model
MC: μ	0.3550	0.3606	0.3625	0.3634	0.8283	0.8709	0.8708	0.9062	1.0983	1.3503	1.3768	1.4623	0.7867	1.0842	1.0951	1.1116
PEM: μ	0.3501	0.3589	0.3515	0.3693	0.8290	0.8548	0.8941	0.9079	1.1105	1.3859	1.3352	1.4594	0.7775	1.0744	1.0956	1.0931
MC: σ	0.0250	0.0297	0.0282	0.0310	0.0594	0.0695	0.0746	0.0847	0.0757	0.1478	0.1568	0.1682	0.0542	0.1460	0.1452	0.1516
PEM: σ	0.0261	0.0300	0.0318	0.0314	0.0472	0.0881	0.0773	0.0947	0.0615	0.1412	0.1735	0.2083	0.0419	0.1360	0.1719	0.1241
Err: μ %	1.3806	0.4669	3.0357	1.6376	0.0744	1.8504	2.6789	0.1909	1.1110	2.6397	3.0234	0.1982	1.1613	0.9003	0.0449	1.6677

Table 3.9. Characteristics of distribution system based on different models at 7 PM for different HEV penetration levels and average weekday summer load profile

HEV Penet.	Characteristics of the network (μ)																	
	μNAPD (mva)			μfrl (%)			μmvd (%)			μVII (%)			UI_{NAPD} (%)			UI_{mvd} (%)		
	Old model	hvac model	Fuel model	Old model	hvac model	Fuel model	Old model	hvac model	Fuel model	Old model	hvac model	Fuel model	Old model	hvac model	Fuel model	Old model	hvac model	Fuel model
No-HEV	1.94			63.89			9.15			1.14			0.00			0.23		
30%	2.30	2.32	2.37	89.00	90.97	94.86	10.74	10.90	11.09	1.32	1.33	1.36	0.94	0.97	1.00	0.97	0.98	0.99
50%	2.51	2.57	2.63	106.57	111.58	117.16	11.78	12.14	12.45	1.44	1.50	1.55	1.00	1.00	1.00	1.00	1.00	1.00
70%	2.75	2.81	2.87	127.21	132.13	138.22	12.95	13.21	13.38	1.59	1.65	1.66	1.00	1.00	1.00	1.00	1.00	1.00
90%	2.95	3.04	3.13	146.13	155.21	165.37	13.93	14.30	14.81	1.73	1.80	1.83	1.00	1.00	1.00	1.00	1.00	1.00
100%	3.05	3.14	3.27	156.95	165.34	179.51	14.34	14.70	15.28	1.78	1.80	1.89	1.00	1.00	1.00	1.00	1.00	1.00

Table 3.10. Characteristics of distribution system based on different models at 7 PM for different HEV penetration levels and average weekday winter load profile

HEV Penet.	Characteristics of the network (μ)																	
	μNAPD (mva)			μfrl (%)			μmvd (%)			μVII (%)			UINAPD (%)			UImvd (%)		
	Old model	hvac model	Fuel model	Old model	hvac model	Fuel model	Old model	hvac model	Fuel model	Old model	hvac model	Fuel model	Old model	hvac model	Fuel model	Old model	hvac model	Fuel model
No-HEV	2.09			74.47			9.87			1.23			0.00			29.24		
30%	2.46	2.48	2.53	102.46	103.40	108.33	11.59	11.63	12.01	1.43	1.43	1.48	1.00	1.00	1.00	1.00	1.00	1.00
50%	2.70	2.72	2.80	123.01	125.22	131.91	12.68	12.84	13.08	1.55	1.60	1.64	1.00	1.00	1.00	1.00	1.00	1.00
70%	2.93	2.96	3.06	145.23	147.18	157.48	13.80	13.96	14.31	1.73	1.74	1.80	1.00	1.00	1.00	1.00	1.00	1.00
90%	3.15	3.19	3.33	166.77	171.19	185.53	14.80	14.89	15.67	1.88	1.84	1.93	1.00	1.00	1.00	1.00	1.00	1.00
100%	3.27	3.30	3.45	178.88	182.21	199.33	15.30	15.55	16.12	1.85	1.94	1.99	1.00	1.00	1.00	1.00	1.00	1.00

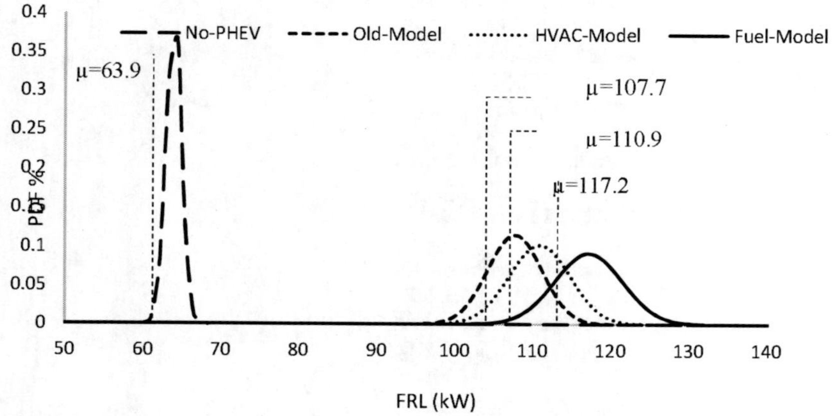

Figure 3.17. PDF of FRL at 7 PM based on different models and 50% penetration level of HEV.

3.5.6. Voltage Imbalance Index (VII)

The load demand of subscribers varies with time. Voltage imbalance in a three-phase system occurs when there is a disparity in the magnitude of phase or line voltages, deviation in phase angles from balanced conditions, or a combination of both factors. This event can be defined using the following equation [41, 54, 55]:

$$VII^{t} = \frac{1}{N}\sum_{i=1}^{N}\left(\left|\frac{V_{-}^{h}(i)}{V_{+}^{h}(i)}\right|\right) \tag{3.25}$$

In the equation, N represents the total nodes of the network, V+ denotes the positive sequence voltage component, V- represents the negative sequence voltage component, and VII represents the average voltage unbalancing of the network. Figure 3.20 demonstrates an increase in VII when HEVs are present across all time intervals. The results of the system analysis are presented in Table 3.8 and Figure 3.21. Based on Table 3.8, we can observe that the average value of VII at different times shows minimal error when calculated using the PEM method compared to the MC method. Additionally, Figure 3.21 displays the PDF of VII at 19:00 for various models with a 50% penetration level. It indicates that the choice of different models for estimating HEV charging demand demand has a negligible impact on the VII index, with approximately 1.41% for the base case and approximately 1.65% for the old and different models.

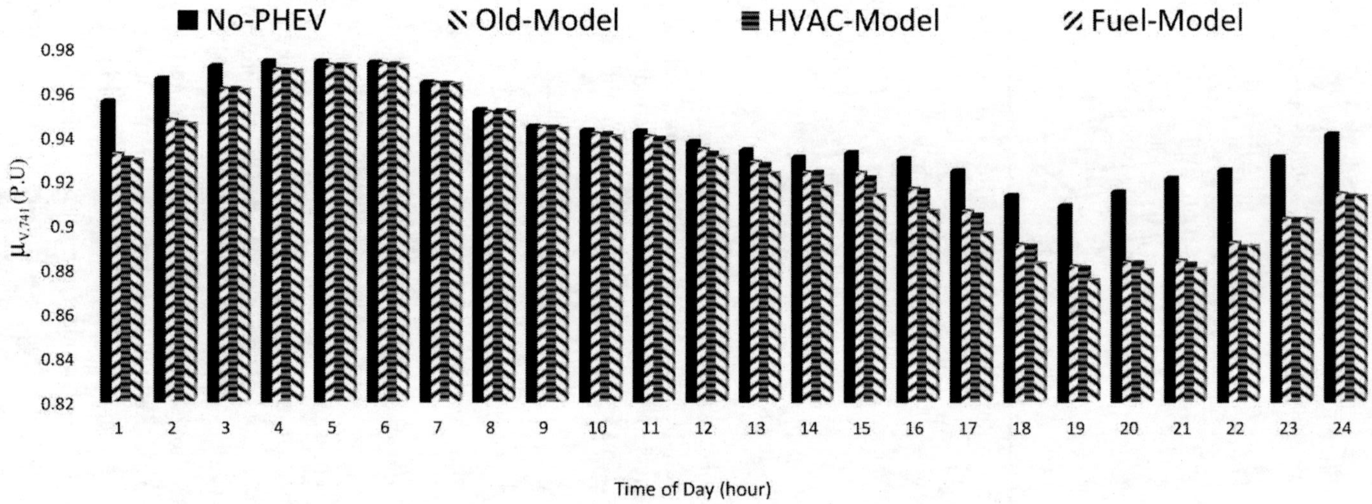

Figure 3.18. The impact of various models on the mean voltage of load point 741 at a 50% HEV penetration level.

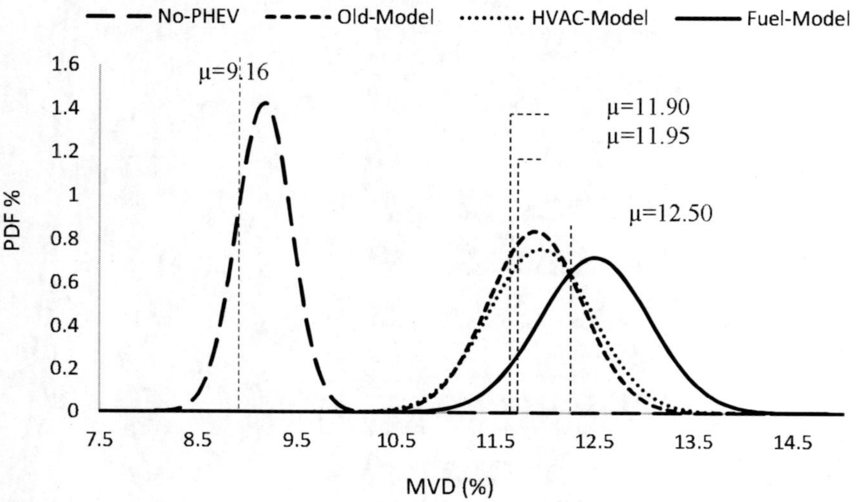

Figure 3.19. PDF of MVD at 7 PM based on different models and 50% penetration level of HEV.

3.5.7. Sensitivity Analysis

To investigate the effects of different penetration levels of HEVs on distribution system characteristics, a sensitivity analysis is conducted. It is recognized that HEV penetration can vary across different areas of the distribution system, thus influencing simulation outcomes. The analysis aims to explore the impacts of various penetration levels on the distribution system. At 7 PM, which represents peak load time, the distribution system's response to HEVs is analyzed for penetration levels ranging from 0% to 100% in 20% increments. The focus is on the effects of different penetration levels on distribution system characteristics. The results of this analysis, considering HVAC energy usage and different fuel models, are presented in Table 3.9 using the average weekday summer load profile.

Clearly, all parameters of the distribution system experience a decline as the HEV penetration level increases. Notably, the feeder power usage significantly increases with higher penetration levels, raising the risk of feeder transformer overload. The probability of such an overload, referred to as UINAPD, is demonstrated in Table 3.9. A similar situation is observed for FRL, indicating that feeder transformer overload is likely to occur for HEV

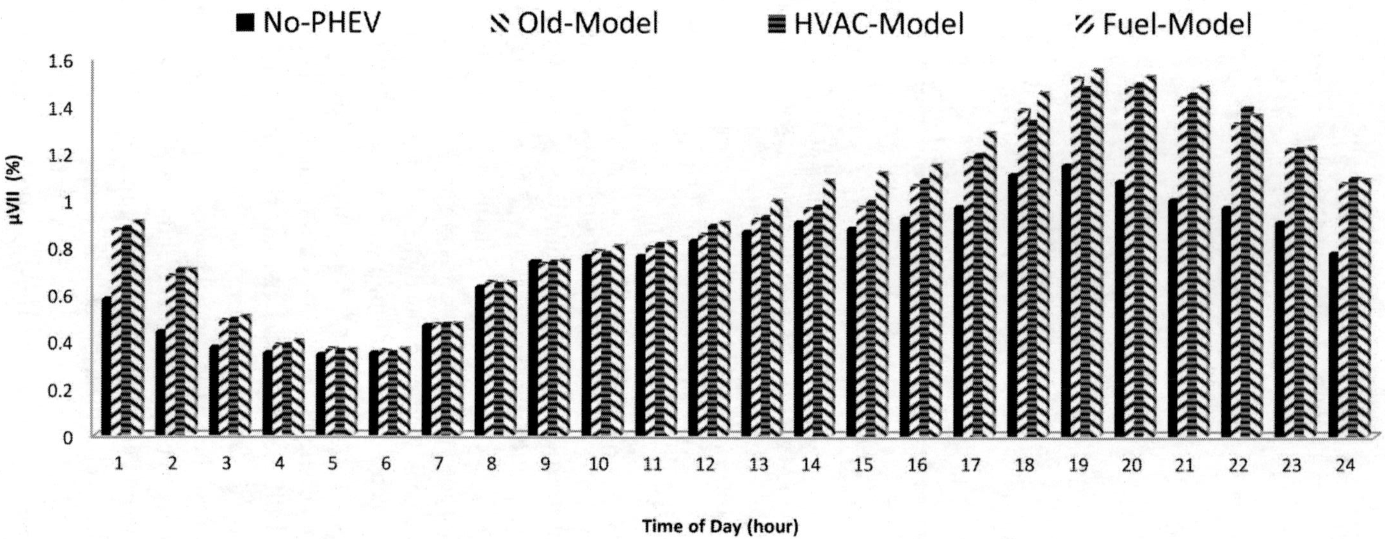

Figure 3.20. The impact of various models on the average VII parameter at a 50% HEV penetration level.

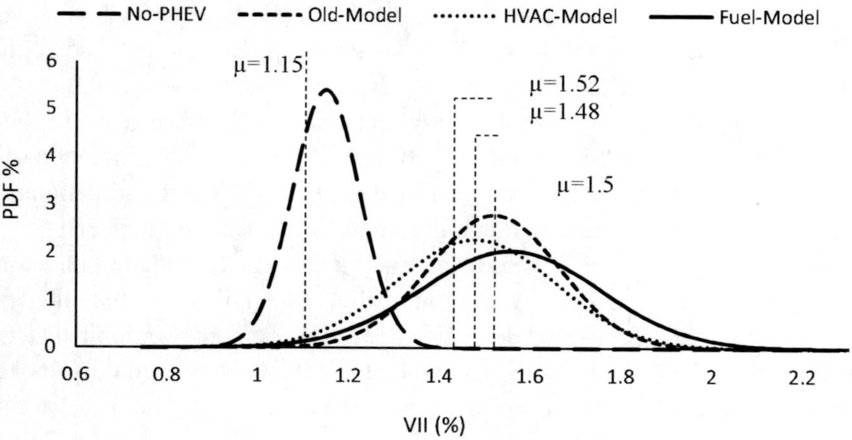

Figure 3.21. PDF of VII at 7 PM based on different models and 50% penetration level of HEV.

penetration levels above 50%. This excess penetration can further exacerbate voltage drops in the system, potentially disrupting the operation of home electrical systems. Additionally, the increased penetration level of HEVs also leads to an increase in VII (average voltage unbalancing) of the network. This is expected since the increased HEV usage load is distributed unequally across the three phases of the distribution system.

Similarly, a similar analysis can be conducted for the winter load profile, with comparable results to those presented in Table 3.9. The only difference would be the substitution of daily energy usage values for the cooling system with those for the heating system. It is worth mentioning that operating the heating system in EVs reduces the initial SOC of the vehicle battery in winter. In contrast, the fuel usage due to operating the heating system in conventional internal combustion engine (ICE) vehicles is insignificant. Table 3.10 illustrates the probabilistic analysis of the network based on PEM method at 19:00 for the mean weekday winter load profile. The analysis considers different models and penetration levels of HEVs.

Conclusion

This chapter presents an extensive and reliable model for aggregating Hybrid Electric Vehicles (HEVs) based on actual statistics and reports. Its primary focus is to investigate the probabilistic effects of HEVs on residential power distribution systems. Furthermore, a statistical model is introduced that utilizes data on HVAC system energy usage in EVs to calculate the initial state of charge (SOC) and estimate the charging demand of HEV fleet.

Given the limited penetration of HEVs and due to the absence of easily accessible registered data, researchers have suggested different estimations for the HEV charging profile utilizing statistical models. In this chapter, a stochastic approach is adopted; leveraging data logged for traditional fossil fuel vehicles, for determining the initial SOC of HEVs and predicting the charging demand profile of the HEV fleet. The outcomes exhibit that the proposed method takes into account crucial factors that have been overlooked in previous studies, leading to an enhanced understanding of the charging demand profile in the presence of HEVs.

To address the uncertainties associated with HEVs, a probabilistic method based on the Probabilistic Energy Management (PEM) approach is employed. It is important to note that the PEM approach yields probabilistic results, which are expressed using probability distribution functions. The findings reveal that feeder power transformer overloads occur when HEV penetration levels exceed 50%. Moreover, it can be concluded that HEVs, considering hourly HVAC energy usage, contribute to increased charging demand, real power loss, voltage drop, and voltage unbalancing in the power system.

A sensitivity analysis is conducted to examine the effects of different HEV penetration levels. Additionally, a winter load profile is generated to investigate the influences of varying penetration levels and average weekday load profiles on power network parameters. The results confirm that all parameters related to HEV energy usage increase during winter, albeit to a lesser extent compared to summer due to reduced HVAC system energy usage in winter.

References

[1] Maulik, A.: 'Probabilistic power management of a grid-connected microgrid considering electric vehicles, demand response, smart transformers, and soft open points,' *Sustainable Energy, Grids and Networks*, 2022, 30, pp. 100636.

[2] Esmaili, M., and Goldoust, A.: 'Multi-objective optimal charging of plug-in electric vehicles in unbalanced distribution networks,' *International Journal of Electrical Power & Energy Systems*, 2015, 73, pp. 644-652.

[3] Neagoe-Stefana, A., Eremia, M., Toma, L., and Neagoe, A.: 'Impact of charging Electric Vehicles in residential network on the voltage profile using Matlab,' in

Editor (Ed.)^(Eds.): '*Book Impact of charging Electric Vehicles in residential network on the voltage profile using Matlab'* (IEEE, 2015, edn.), pp. 787-791.

[4] Angelim, J. H., and Affonso, C. M.: 'Effects of long-range electric vehicles on distribution system using probabilistic analysis,' *International Journal of Electrical Power & Energy Systems*, 2023, 147, pp. 108868.

[5] Speidel, S., and Bräunl, T.: 'Driving and charging patterns of electric vehicles for energy usage,' *Renewable and Sustainable Energy Reviews*, 2014, 40, pp. 97-110.

[6] Li, X., Wang, Z., Zhang, L., Sun, F., Cui, D., Hecht, C., Figgener, J., and Sauer, D. U.: 'Electric vehicle behavior modeling and applications in vehicle-grid integration: An overview,' *Energy*, 2023, pp. 126647.

[7] Tovilović, D. M., and Rajaković, N. L.: 'The simultaneous impact of photovoltaic systems and plug-in electric vehicles on the daily load and voltage profiles and the harmonic voltage distortions in urban distribution systems,' *Renewable Energy*, 2015, 76, pp. 454-464.

[8] Xiang, Y., Liu, J., and Liu, Y.: 'Optimal active distribution system management considering aggregated plug-in electric vehicles,' *Electric Power Systems Research*, 2016, 131, pp. 105-115.

[9] Jahromi, A. J., Mohammadi, M., Afrasiabi, S., Afrasiabi, M., and Aghaei, J.: 'Probability density function forecasting of residential electric vehicles charging profile,' *Applied Energy*, 2022, 323, pp. 119616.

[10] Shafiee, S., Fotuhi-Firuzabad, M., and Rastegar, M.: 'Investigating the Impacts of Plug-in Hybrid Electric Vehicles on Power Distribution Systems,' *IEEE Transactions on Smart Grid*, 2013, 4, (3), pp. 1351-1360.

[11] Tan, J., and Wang, L.: 'Integration of plug-in hybrid electric vehicles into residential distribution grid based on two-layer intelligent optimization,' *IEEE Transactions on Smart Grid*, 2014, 5, (4), pp. 1774-1784.

[12] Kambly, K. R., and Bradley, T. H.: 'Estimating the HVAC energy consumption of plug-in electric vehicles,' *Journal of Power Sources*, 2014, 259, pp. 117-124.

[13] ElNozahy, M. S., and Salama, M. M. A.: 'A Comprehensive Study of the Impacts of PHEVs on Residential Distribution Networks,' *IEEE Transactions on Sustainable Energy*, 2014, 5, (1), pp. 332-342.

[14] Mu, Y., Wu, J., Jenkins, N., Jia, H., and Wang, C.: 'A Spatial–Temporal model for grid impact analysis of plug-in electric vehicles,' *Applied Energy*, 2014, 114, pp. 456-465.

[15] Cui, X., Kim, H. K., Liu, C., Kao, S. C., and Bhaduri, B. L.: 'Simulating the household plug-in hybrid electric vehicle distribution and its electric distribution network impacts,' *Transportation Research Part D: Transport and Environment*, 2012, 17, (7), pp. 548-554.

[16] Mohamed, A., Salehi, V., Ma, T., and Mohammed, O.: 'Real-time energy management algorithm for plug-in hybrid electric vehicle charging parks involving sustainable energy,' *IEEE Transactions on Sustainable Energy*, 2014, 5, (2), pp. 577-586.

[17] Morais, H., Sousa, T., Soares, J., Faria, P., and Vale, Z.: 'Distributed energy resources management using plug-in hybrid electric vehicles as a fuel-shifting

demand response resource,' *Energy Conversion and Management*, 2015, 97, pp. 78-93.

[18] Shaaban, M., Eajal, A., and El-Saadany, E.: 'Coordinated charging of plug-in hybrid electric vehicles in smart hybrid AC/DC distribution systems,' *Renewable Energy*, 2015, 82, pp. 92-99.

[19] Hajforoosh, S., Masoum, M. A., and Islam, S. M.: 'Real-time charging coordination of plug-in electric vehicles based on hybrid fuzzy discrete particle swarm optimization,' *Electric Power Systems Research*, 2015, 128, pp. 19-29.

[20] Abdelsamad, S., Morsi, W., and Sidhu, T.: 'Optimal secondary distribution system design considering plug-in electric vehicles,' *Electric Power Systems Research*, 2016, 130, pp. 266-276.

[21] Carpinelli, G., Caramia, P., and Varilone, P.: 'Multi-linear Monte Carlo simulation method for probabilistic load flow of distribution systems with wind and photovoltaic generation systems,' *Renewable Energy*, 2015, 76, pp. 283-295.

[22] Morales, J. M., and Perez-Ruiz, J.: 'Point estimate schemes to solve the probabilistic power flow,' *IEEE Transactions on Power Systems*, 2007, 22, (4), pp. 1594-1601.

[23] Shargh, S., Mohammadi-ivatloo, B., Seyedi, H., and Abapour, M.: 'Probabilistic multi-objective optimal power flow considering correlated wind power and load uncertainties,' *Renewable Energy*, 2016, 94, pp. 10-21.

[24] Santos, A., McGuckin, N., Nakamoto, H. Y., Gray, D., and Liss, S.: 'Summary of travel trends: 2009 national household travel survey,' in Editor (Ed.)^(Eds.): '*Book Summary of travel trends: 2009 national household travel survey*' (2011, edn.), pp.

[25] Duvall, M., Knipping, E., Alexander, M., Tonachel, L., and Clark, C.: 'Environmental assessment of plug-in hybrid electric vehicles,' *EPRI*, July, 2007

[26] Andersen, P. H., Mathews, J. A., and Rask, M.: 'Integrating private transport into renewable energy policy: The strategy of creating intelligent recharging grids for electric vehicles,' *Energy Policy*, 2009, 37, (7), pp. 2481-2486.

[27] America, E. N.: 'The EV Project Q4 2012 Report,' in Editor (Ed.)^(Eds.): '*Book The EV Project Q4 2012 Report*' (February, 2013, edn.), pp.

[28] Etezadi-Amoli, M., Choma, K., and Stefani, J.: 'Rapid-Charge Electric-Vehicle Stations,' *IEEE Transactions on Power Delivery*, 2010, 25, (3), pp. 1883-1887.

[29] Jinghong, Z., Xiaoyu, W., Kun, M., Chun, Z., and Shouzhen, Z.: 'Aggregation Model-Based Optimization for Electric Vehicle Charging Strategy,' *IEEE Transactions on Smart Grid*, 2013, 4, (2), pp. 1058-1066.

[30] Chenrui, J., Jian, T., and Ghosh, P.: 'Optimizing Electric Vehicle Charging: A Customer's Perspective,' *IEEE Transactions on Vehicular Technology*, 2013, 62, (7), pp. 2919-2927.

[31] Darabi, Z., and Ferdowsi, M.: 'Aggregated Impact of Plug-in Hybrid Electric Vehicles on Electricity Demand Profile,' *IEEE Transactions on Sustainable Energy*, 2011, 2, (4), pp. 501-508.

[32] Glueck, J., and Le, H. T.: 'Impacts of Plug-in Electric Vehicles on local distribution feeders,' in Editor (Ed.)^(Eds.): '*Book Impacts of Plug-in Electric Vehicles on local distribution feeders*' (IEEE, 2015, edn.), pp. 1-5.

[33] Li, G., and Zhang, X. P.: 'Modeling of plug-in hybrid electric vehicle charging demand in probabilistic power flow calculations,' *Smart Grid, IEEE Transactions on*, 2012, 3, (1), pp. 492-499.

[34] Kintner-Meyer, M., Schneider, K., and Pratt, R.: 'Impacts assessment of plug-in hybrid vehicles on electric utilities and regional US power grids, Part 1: Technical analysis,' *Pacific Northwest National Laboratory* (a), 2007, pp. 1-20.

[35] Pieltain Fernandez, L., T. Gomez San Roman, R. Cossent, and Frias, C. M. D. a. P.: 'Assessment of the Impact of Plug-in Electric Vehicles on Distribution Networks,' *IEEE Transactions on Power Systems*, 2011, 26, (1), pp. 206-213.

[36] Yilmaz, M., and Krein, P. T.: 'Review of battery charger topologies, charging power levels, and infrastructure for plug-in electric and hybrid vehicles,' *IEEE Transactions on Power Electronics*, 2013, 28, (5), pp. 2151-2169.

[37] Ruiz-Rodriguez, F., Hernández, J., and Jurado, F.: 'Voltage unbalance assessment in secondary radial distribution networks with single-phase photovoltaic systems,' *International Journal of Electrical Power & Energy Systems*, 2015, 64, pp. 646-654.

[38] Clement-Nyns, K., Haesen, E., and Driesen, J.: 'The Impact of Charging Plug-In Hybrid Electric Vehicles on a Residential Distribution Grid,' *IEEE Transactions on Power Systems*, 2010, 25, (1), pp. 371-380.

[39] Liu, Y., Huang, B., Yang, Y., Lei, Z., Zhang, Y., and Chen, Z.: 'Hierarchical speed planning and energy management for autonomous plug-in hybrid electric vehicle in vehicle-following environment,' *Energy*, 2022, 260, pp. 125212.

[40] Lian, J., Wang, X. R., LI, L. H., Zhou, Y. F., Yu, S. Z., and Liu, X. J.: 'Plug-in HEV energy management strategy based on SOC trajectory,' *International Journal of Vehicle Design*, 2020, 82, (1-4), pp. 1-17.

[41] Yu, X., Ergan, S., and Dedemen, G.: 'A data-driven approach to extract operational signatures of HVAC systems and analyze impact on electricity consumption,' *Applied Energy*, 2019, 253, pp. 113497.

[42] Neubauer, J., and Wood, E.: 'Thru-life impacts of driver aggression, climate, cabin thermal management, and battery thermal management on battery electric vehicle utility,' *Journal of Power Sources*, 2014, 259, pp. 262-275.

[43] Shengnan, S., Pipattanasomporn, M., and Rahman, S.: 'Challenges of PHEV penetration to the residential distribution network,' in Editor (Ed.)^(Eds.): '*Book Challenges of PHEV penetration to the residential distribution network*' (2009, edn.), pp. 1-8.

[44] http://www.nrcan.gc.ca/energy/efficiency/transportation/cars-light-trucks/buying/7487'.

[45] Kejun, Q., Chengke, Z., Allan, M., and Yue, Y.: 'Modeling of Load Demand Due to EV Battery Charging in Distribution Systems,' *IEEE Transactions on Power Systems*, 2011, 26, (2), pp. 802-810.

[46] Sedghi, M., Ahmadian, A., Pashajavid, E., and Aliakbar-Golkar, M.: 'Storage scheduling for optimal energy management in active distribution network considering load, wind, and plug-in electric vehicles uncertainties,' *Journal of Renewable and Sustainable Energy*, 2015, 7, (3), pp. 033120.

[47] Papadopoulos, P., Skarvelis-Kazakos, S., Grau, I., Cipcigan, L. M., and Jenkins, N.: 'Predicting Electric Vehicle impacts on residential distribution networks with

Distributed Generation,' in Editor (Ed.)^(Eds.): '*Book Predicting Electric Vehicle impacts on residential distribution networks with Distributed Generation*' (2010, edn.), pp. 1-5.

[48] Shaaban, M. F., Atwa, Y. M., and El-Saadany, E. F.: 'PEVs modeling and impacts mitigation in distribution networks,' *IEEE Transactions on Power Systems*, 2013, 28, (2), pp. 1122-1131.

[49] Pillay, P., and Manyage, M.: 'Definitions of voltage unbalance,' *IEEE Power Engineering Review*, 2001, 21, (5), pp. 50-51.

[50] Shahnia, F., Ghosh, A., Ledwich, G., and Zare, F.: 'Predicting voltage unbalance impacts of plug-in electric vehicles penetration in residential low-voltage distribution networks,' *Electric Power Components and Systems*, 2013, 41, (16), pp. 1594-1616.

Chapter 4

Electric Vehicle: Types, Charging Topologies, Energy Storage Management System and Applicability Micro-Grids – A Review

Ashish Kumar Sinha*
Prashant Kumar
and Prince Kumar
[1]Dept. of Electrical Engineering, GLA University, Mathura, Uttar Pradesh, India
[2]Dept. of Mechanical, Robotics and Energy Engineering,
Dongguk University-Seoul, Seoul, South Korea
[3]Dept. of Electrical Engineering, Meerut Institute of Technology, Meerut,
Pradesh, India

Abstract

With the looming scenario of an alarming rate of conventional energy level depletion, snowballing crude oil prices and ever-increasing environmental constraints, such as global warming and ozone layer depletion, associated with the same, energy efficiency has become a priority numero-uno. A substantial proportion of overall crude oil consumption is dedicated to road vehicle transportation systems. Therefore, battery-powered electric vehicles are becoming ever more popular all over the world in the ongoing scenario.

In recent years, electric vehicles, whose requirements have grown rapidly in accordance with green energy demand, are being recognised as a viable alternative to conventional vehicle engines, i.e., internal combustion engines (ICE).

* Corresponding Author's Email: ashish.sinha@gla.ac.in.

In: The Future of Hybrid-Electric Vehicles
Editors: Pawan Kumar Pathak and Anil Kumar Yadav
ISBN: 979-8-89113-515-4

Electric vehicles in various configurations and capacities like Pure/ Battery Electric Vehicles (PEV/ BEV), Plug-in Hybrid Electric Vehicles (PHEV), and Fuel-Cell Electric Vehicles (FCEV) are being researched and developed to meet current market demand. Moreover, emphasis is being given to battery technology for electric vehicles to obtain a faster rate of charging and increase the discharging time, i.e., increasing the range of travel in a single charging cycle of an EV battery.

Even while electric vehicles have a broad spectrum of applications and requirements, the ongoing research spectrums are also accompanied by some major roadblocks, disadvantages and glitches. These include time-constraint of charging, the longevity of battery lifetime and standardisation. Significant research in recent years has been dedicated towards accelerating the charging time of EV batteries through various mechanisms. Herein, the charging mechanism is typically seen to adopt two major techniques, namely constant current (CC) technique and constant voltage (CV) technique. Furthermore, the charging topologies include unidirectional and bi-directional, wherein vehicle-to-grid and grid-to-vehicle schemes are the present areas of concentrated research. Moreover, in accordance with the available literature, the bi-directional power flow model is clearly seen to have an advantage over the traditional unidirectional power flow topology w.r.t. the charging techniques generally employed for electric vehicles. In this regard, an effective battery management system also becomes indispensable and a comprehendible area of vital research.

Research shows that bi-directional topology with the incorporation of microgrids into the scheme forms a highly exciting and comprehendible area of future research and may potentially overcome the existing drawbacks of the schemes and methodologies that are currently being used in an attempt to accelerate the battery charging time and improve the overall working performance of the electric vehicles.

Keywords: electric vehicles (EV), charging topologies, micro-grids, energy storage management system

4.1. Introduction

With the looming scenario of conventional energy level depletion, increasing crude oil prices and ever-increasing environmental constraints associated with the same, energy efficiency has become priority numero-uno. Substantial proportion of overall oil consumption is dedicated to road vehicles run for transportation systems. As per the reports, by 2030, the transportation sector shall increase its oil consumption stake in the world market to about 55%

(Cheng 2009). In recent years, electric vehicles (EVs), whose requirements have grown rapidly in accordance with green energy demand, are being recognised as a viable alternative for conventional vehicle engine i.e., ICE (F. Nenglian et al. 2011; A. Yilmaz et al. 2012) EVs in various configurations and capacities like Pure/ Battery Electric Vehicles (PEV/ BEV), Plug-in Hybrid Electric Vehicles (PHEV) (A. Yilmaz et al. 2012), and Fuel-Cell Electric Vehicles (FCEV) are being researched and developed to meet current market demand. Moreover, emphasis is being given on the battery technology for EVs to obtain a faster rate of charging and increasing the discharging time, i.e., increasing the range of travel in a single charging cycle of EV battery. Herein, there are three vital glitches for a charging station: time-constraint of charging, longevity of battery lifetime and standardisation (A. Yilmaz et al. 2012; Cheng 2009). A charging mechanism is typically seen to adopt two major techniques, namely constant current (CC) technique and constant voltage (CV) technique. In contrast to CV, which offers reduced temperature escalation with a longer charging period, CC minimises the period for charging a battery, albeit at greater temperatures. Due to the limited quantifiable values of voltage per cell (V/Icell) and maximum charging current (I_{max}) from battery suppliers, CC and CV methods cannot satisfy the requirements of future EV customers for a fast-charging time and a low-temperature rise (A. Yilmaz et al. 2012). Pulse charging technology has off-late become a viable replacement in order to realise rapid charging at high current rate as compared to conventional charging method (Ahn et al. 2013; Tseng et al. 2002). The most critical part of EV plug-to-wheels (PTW) drive train efficiency is "charger." This is analogous to tank-to-wheels (TTW) technology of a conventional vehicle. The flow of energy from Plug to battery (PTB) is essentially the intermediate charging stage. The efficacy of PTW scheme with regards to EVs generally ranges from 45% to 50%. Therefore, analysis of the same becomes quintessential w.r.t. the efficiency quantification of PTW energy of both the EV in contention as well as the drive train system. Herein, PTB efficiency is most imperative. Commercial EVs today have on-board charging system consisting of: (a) an AC/DC rectifier, (b) a DC/DC boost power factor correction stage, and (c) a high frequency (HF) DC/DC charger converter. The above-said arrangement has three alteration steps which leads to substantial power losses. Hence, overall energy efficiency of PTW is significantly low (in the range of 15-20%) (Lee et al. 2011; Ahn et al. 2013). The main contemporaries of energy storage in EVs generally include ultracapacitors, electrochemical batteries and complete cells. However, the most used energy storage technology is still electrochemical batteries. Nevertheless, typical employment of aforementioned charging mechanisms is accomplished in conjunction with ultracapacitors. This is done with the sole motive of storing energy during specific epochs of operation like during regenerative braking in a EV (Szalai et al. 2014; Lipu et al. 2021). Furthermore, the flow of electric power from the EV charging station to the EVs is essentially a unidirectional phenomenon, with the regenerative braking phase acting as a brief period of operation wherein the flow may be considered as bidirectional. Incorporation if a bi-directional mode of electric power flow throughout the entire duration of operation still remains a nascent and under-explored category of research (Somnatha and Gupata 2019).

In this regard, the current study aims to amass and analyse the different classifications of EVs, their energy storage management techniques and the prospect of inducting bi-directional power flow into the scheme. This shall help the prospective readers gain a first-hand impression of the intended area of research which shall further assist in identifying the shortcomings in the currently used technique to optimise the EV performance.

This paper is organised in six sections. Section 4.2 encapsulates the classification of the different types of EVs. Furthermore, section 4.3 details the different existing topologies of EV battery charging. Section 4.4 attempts to provide an insight into the prevalent aspects of energy storage management system with an attempt to optimise the same. Furthermore, section 4.5 enlists the applicability and scope of integration of micro-grids and EVs in a single working steup. Finally, section 4.6 concludes the work.

4.2. Classification of Electric Vehicles

Taking the propulsion devices and drive section mechanism into account, EVs may be broadly categorised into three diverse forms: (a) pure electrical vehicle (PEV), (b) hybrid electrical vehicle (HEV) and (c) fuel cell electrical vehicle (FCEV) (Chan, Chau, and others 2001; Ramadhas 2011; Folkson 2014). Table 4.1 shows the classification of different types of EVs on the basis of the selection of drive propulsion section.

Furthermore, Figure 4.1 depicts the categorisation of EVs based on using different energy sources and deploying viable propulsion devices.

Table 4.1. Classification of various EVs

Types	PEVs	HEVs	FCEVs
Drive section	Electric motor	Electric motor and internal combustion engine (ICE)	Electric motor
Energy sources	Battery/ ultracapacitor	Battery/ultra-capacitor, ICE unit	Fuel cell

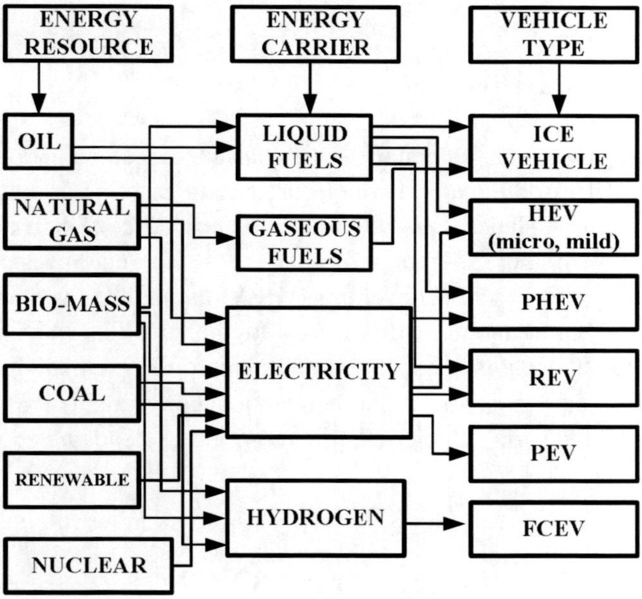

Figure 4.1. Classification in EVs on the basis on the energy resource and energy carrier.

4.2.1. Pure Electrical Vehicles (PEVs)

An electric motor is solely responsible for the process of propulsion in PEVs. In a PEV, the battery technology is entirely responsible for the energy storage capacity. This results in zero discharge emission, which significantly adheres to its advantage. PEVs are less appealing than ICE vehicles under the same economic and driving conditions due to the PEV on-board battery technology's limitations in providing real-time performance status information. Longer charging time in batteries is achieved by having higher power densities albeit lower energy densities. This holds true even for fast charging technologies. The major and key drawbacks associated with PEV's mainly engulf its limited driving range inculcated with the higher initial cost. Furthermore, there is also an inadequacy with regards to the availability of charging infrastructure (Hannan, Azidin, and Mohamed 2014). Moreover, providing the information regarding the battery size employed in EV along with the location where the installation has been done inside the PEV is indispensible for practical implementation (Chau and Li 2014).

4.2.2. Hybrid Electrical Vehicles (HEVs)

The driving propulsion system of HEV encapsulates a combination of electric motor and ICE. Additionally, both electricity and gasoline or diesel are used as power sources. Hence, HEV combines the properties of ICE and PEV. In accordance with the different refuelling or recharging measures, further classification of HEVs are: conventional HEVs and non-conventional or grid-able HEVs. Furthermore, with regards to the level of the combination, conventional HEVs are further classified as: micro HEVs, mild HEVs and full HEV. Further categorisation of non-conventional or grid-able HEVs includes: plug-in hybrid electric vehicles (PHEVs) or range-extended electric vehicles (REVs) (Chau and Li 2014).

4.2.3. Fuel Cell Electrical Vehicles (FCEVs)

The main sources of power of fuel cell (FC) in FCEVs are: hydrogen, methanol, ethanol or gasoline. They are inherently driven by an electric motor. FC is an electrochemical energy conversion device that harvests electrical energy output in DC form through a chemical process. FCEVs overcome one of the major drawbacks associated with HEVs, i.e., limited driving range. Driving range of FCEV is comparable to ICEV. However, the elevated preliminary price and scarcity of refueling stations are still observed as noteworthy hinderances on the road to success for FCEVs (Rao and Wang 2011). Furthermore, the supply electricity continuity of FCs has lesser levels of reliability in comparison to the battery used in PEVs and HEVs which essentially employs conventional mechanism of operation. The decisive benefit of PEV and FCEV is the negligibly minimal roadside emission and abridged levels of air contamination.

4.3. Charging Systems for Battery Management: Uni-Dierectional and Bi-Directional Topoloogies

The electrical power grids are not conventionally designed for this EV battery charging purpose. The task is to restructure the power grids with emphasis on time factor, keeping the circuit simplicity intact and keeping the entire scheme environmentally friendly. The major problems associated with EV charging station is the prospective occurrence of concurrent charging for a hefty number of EVs, which can potentially overload the grid. Furthermore, there is always a potential threat of grid over-loading owing to non-sinusoidal nature of current consumption in the battery charging system. This, generally "on-board," charging mechanism is traditionally AC-DC power converter circuit whose parameters ought to be controlled so as to achieve the nominal battery performance characteristics whilst preserving their lifespan. The power circuit must have the provision to constantly monitor the batteries for prospective damages, as and when the may occur, both during charging as well as the discharging processes. On the basis of the requisition of performance qualities of the system, a wide array of applicable topologies of the AC-DC power circuit may be employed (M. Yilmaz and Krein 2012). Figure 4.2 exemplifies the core categories of AC-DC power circuit topologies for charging systems of EVs.

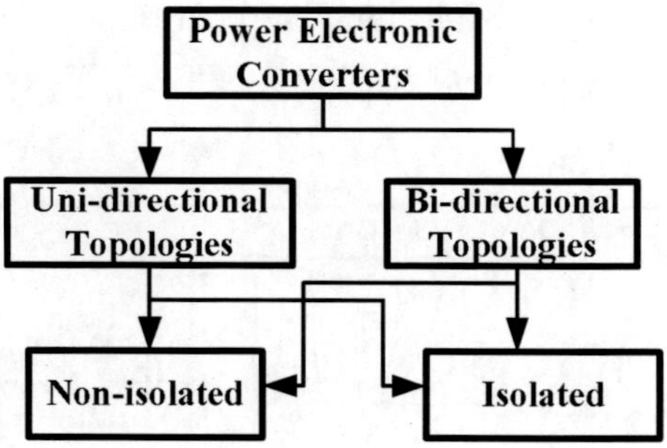

Figure 4.2. Core categories of AC-DC power circuit topologies for charging systems of EVs.

Two power electronics converters: (a) AC-DC converter and (b) DC-DC converter typically make up the power electronic circuit topology of the charging setup w.r.t. the EV batteries. Both converters can be configured in various ways, both with and without isolation, and with different topologies.

The power grid's AC voltage is rectified to a DC voltage using the AC-DC converter. For the purpose of controlling the battery charging process, the rectified voltage is adjusted using a DC-DC converter to a voltage level that is compatible with the battery rated voltage.

4.3.1. Unidirectional Topology

Diodes are the only component which possess the property of rectifying. Therefore, these are the utmostly rudimentary AC-DC power converter components. AC-DC power converters have simplistic and economic construction, making them less prone to prospective damages. However, the consumed current and the output voltage are not variables that can be controlled. As a result, the consumed current's waveform is not entirely sinusoidal, which has an impact on the value of the grid's power quality. This is shown in Figure 4.3. Moreover, the output voltage and consumed current can be categorised under controllable parameters by using power switching semiconductors. It has the advantage of pure sinusoidal current consumption, which does not affect the quality of the power grid despite forgoing the simple design and increased maintenance costs. Furthermore, addition a DC-DC boost converter to the existing circuit enhances the capability of the circuit regarding Power Factor Correction (PFC).

Although it is a viable substitute for controlling the wave form of the current consumed, it only works in one direction, from the grid to the vehicle (G2V).

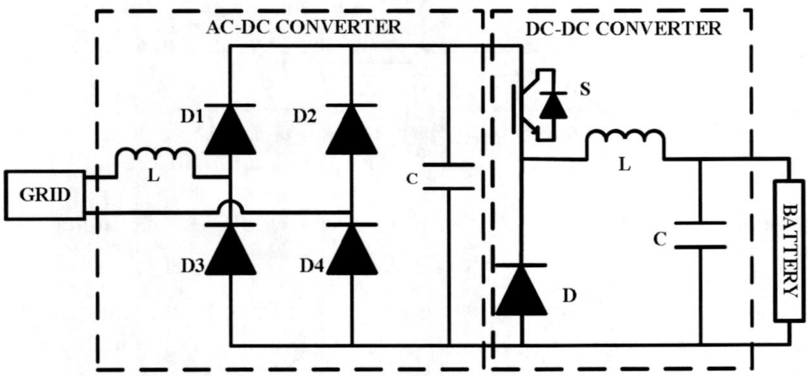

Figure 4.3. Typical scheme of unidirectional EV charging topology.

4.3.2. Bi-directional Topology

The bi-directional charging topology for EVs allows for power flow in both the vehicle-to-grid (V2G) mode and the G2V mode with the sole purpose of interchanging electrical power to and from the EV end and the grid end and vice-versa. It serves as an inverter when power is transmitted from the grid-end to the EV-end which comprises of the rectifying unit and the DC power module. Furthermore, when the tables are turned, i.e., power flow is required in the opposite direction then the same unit works as a rectifying module. The G2V charging topology is employed with the foresight of using the excess electricity, which is accumulated during off-peak hours to charge the EV's battery for the purpose of providing mobility to the same. Furthermore, on the prospective requirement of additional power during the peak hours, the V2G operating mode is employed to inoculate electricity back from the EV-end to the grid-end. Thus, making the power supply more dependable.

If, in a power surplus state of the grid, the state of charge (SOC) value of the linked EV is either superior to or equal to the $(SOC)_{max}$ then the EV power grid and the AC power grid will not interchange electrical power. However, the connected EV will begin the process of EV battery charging. This process shall continue uninterrupted until the EV battery is charged to its limits, i.e., if SOC value of the EV battery lies in a range which is in between the quantified values of $(SOC)_{max}$ and $(SOC)_{min}$. Moreover, in an event of encountering shortage of power in the power grid, the EV must be equipped to essentially fill the void in power requirement. The fulfillment of this condition preceeds the alreagy mentioned quantification of SOC value of EV battery i.e., between $(SOC)_{min}$ and $(SOC)_{max}$. Also, in an event of the SOC value of the PEV falling to a magnitude which is lesser than $(SOC)_{min}$, the power flow between the grid and the EV will cease to exist (Islam et al. 2021).

Different kinds of converters serve as intermediaries for any two-way power transfer between the AC power system and the EV grid. These mediators function both as one-way and two-way power converters. Different controllers are employed for efficient management of various mediators according to their performances. A typical scheme of bi-directional EV charging topology is depicted in Figure 4.4.

Power electronic converters control the charging and discharging of batteries to extend battery life, monitor the SOC, and assure power transfer between PEVs and the grid (Lipu et al. 2021). The combination of conventional controllers, such as the cascade topology of Static VAR Compensator (SVC) cascaded along with Fractional Order Proportional

Integral Derivative (FOPID) controller, is seen to enhance the performance of the overall power system with regards to its quantified stability (Benayad, Zabaiou, and Bouafassa 2020).

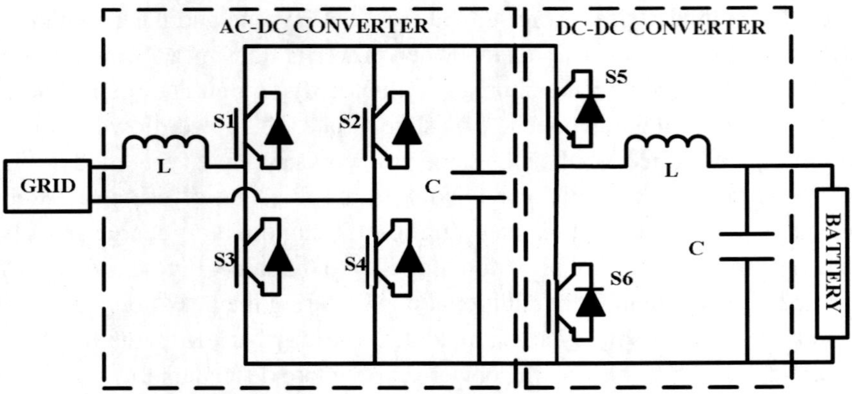

Figure 4.4. Typical scheme of bi-directional EV charging topology.

The hybrid energy storage system (HESS) is recharged using a PI controller to store energy (Elwarfalli et al. 2016). It is evident that the use of PI controller for the purpose of V2G and G2V mode of operation flattens the power output characteristics of the load without accounting for the early overshoot and distortion demonstrated by the PEV battery power performance characteristics (Khemakhem, Rekik, and Krichen 2017). The overall battery performance and lifespan may be gradually lowered by overshoot and distortion. Although PI is a simple controller, its effectiveness can be increased by hastening its operation through hybridisation based on artificial intelligence (Gasbaoui, Nasri, and Abdelkhalek 2016).

Furthermore, a different study (Arun et al. 2019) shows that fuzzy-based PI controllers outperform existing ones in terms of better energy resource usage through bi-directional exchange of system power, quick responses with reduced magnitude of error during steady-state mode of operation and improved dynamic flexibility.

The applicability of PI controller regarding sophisticated power management has been steadily declining in recent years (Arun et al. 2019). Technical limitations such as a slow or nonexistent reaction to a prospective outburst of abrupt nature, consensus regarding overshoot quantification during brief periods of time, vulnerable nature of gain magnitude in the employed controller, inadequate adaptability, prolonged time-delayed processing, accurate mathematical modelling, etc. are to blame for this.

Using various mathematical models, fuzzy and adaptive neuro-fuzzy controllers (ANFIS) methods have largely eliminated the PI controller's drawbacks. When an accurate scientific estimate is difficult and the load changes depict non-linearity, the utility of the same strategy increases (Dyanamina and Kakodia 2021). In situations when simple adjustment is necessary, with or without human involvement, fuzzy controllers are very trustworthy. Compared to a conventional PI controller, the development of a fuzzy logic controller (FLC) is simple and affordable. Moreover, FLC has greater robustness across a wider range of operations (Osinski, Leandro, and da Costa Oliveira 2019). However, FLC continues to experience reliability concerns. This is tackled by the ANFIS controller, which performs better than the conventional controller in terms of time efficiency and membership function (MF) optimisation (Krasopoulos, Beniakar, and Kladas 2018). Identifying the connection status of the system's cars is the main goal of this control logic. To achieve better outcomes with regards to all the EVs present in the scheme of things, the SOC value per capita (i.e., each EV) is then computed and used to pre-set the time needed for charging and discharging.

4.4. Energy Storage Management System

An effective energy storage management system (ESMS) in EV technology essentially counters most of the critical issues which hamper the optimised running of an EV. ESMS comprises of several sub-divisions, namely: communication unit, data acquisition unit and battery position estimation model. Furthermore, a thermal organisation unit and high voltage management unit is introduced for further improvement in battery safety. The development of battery chemistries, materials, dependable and efficient ESMSs for the purpose of achieving efficacy with regards to battery health monitoring and management, the management of charging and discharging process, regulation of system temperature, cell balancing and predictive healthmanagement is still indispensable for large-scale battery energy storage (Phanomchoeng and Chancharoen 2019). Poor electro-thermal maneuver and upkeep procedures may result in catastrophic hazards and premature and unplanned fiasco and downtime respectively in the absence of appropriate ESMSs. Therefore, in terms of performance, safety, dependability, and economics, battery management is essential for the integration of battery energy storage into the electric grid. (Haiying et al. 2014; Feng et al. 2016).

The following are some of an ESMS's key functions (Fu, Zhu, and Du 2017; Xiong et al. 2017):

a) Data acquisition: the evaluation and gathering of information on current, voltage, temperature, etc.,
b) State estimation: SOC, state of power (SOP), state of health (SOH), state of temperature, etc., are gauged with high accuracy.
c) Charge/discharge control: Regulation of the charge current and voltage, power electronics interface, etc.
d) Cell balancing: SOC and active as well as passive voltage equalisation,
e) Thermal management: effective temperature regulation between battery cells of the EV,
f) Safety protection: to prevent overcharging.

4.5. Applicabilty and Scope of Micro-Grid Integration

Electric vehicles (EVs) and micro-grids are two significant advancements in the energy sector that have the potential to complement each other (Shakeel and Malik 2019). The applicability of electric vehicles in micro-grids in dimension mentioned hereunder:

- Vehicle-to-Grid (V2G) Technology: One of the key intersections between electric vehicles and micro-grids is the concept of Vehicle-to-Grid (V2G) technology. V2G allows electric vehicles to not only draw power from the grid but also return excess electricity back to the grid when needed. In this scenario, electric vehicles become mobile energy storage units, capable of supplying power to the micro-grid during peak demand or grid outages, thereby improving grid stability and resilience.
- Energy Storage and Peak Sharing and Demand Response: EVs can act as distributed energy storage devices when connected to micro-grids. During times of low energy demand, such as at night when most vehicles are parked and charging, excess electricity can be stored in the EV batteries. Then, during periods of high demand, such as during

the day, this stored energy can be fed back into the micro-grid, reducing the need to draw power from the main grid and contributing to peak sharing. Furthermore, EVs in a micro-grid setting can be integrated into demand response programs. During times of high electricity demand or when electricity prices are at their peak, EV charging can be temporarily reduced or shifted to off-peak periods, thus balancing the load on the micro-grid and contributing to overall grid stability.

- Renewable Energy Integration: Micro-grids are often designed to incorporate renewable energy sources such as solar panels and wind turbines. However, the intermittent nature of renewable energy can be challenging for grid stability. EVs, with their storage capacity, can help mitigate this issue by storing excess renewable energy when generation exceeds demand and releasing it when demand surpasses generation.
- Islanded Micro-grids: In remote areas or places with unreliable grid connections, micro-grids operate in isolation from the main power grid. Electric vehicles can play a crucial role in these islanded micro-grids by serving as a local energy source, reducing the dependence on fossil fuel-based generators, and providing a more sustainable energy solution.
- Community Resilience: In case of natural disasters or grid failures, micro-grids equipped with EVs can ensure a continuous power supply to critical facilities and essential services. The electricity stored in EV batteries can be used to power hospitals, emergency response centers, and other vital infrastructure during emergencies.

While the applicability of electric vehicles in micro-grids shows great promise, there are some challenges that need to be addressed. These include managing the impact on EV battery life due to frequent charging and discharging, developing standardised communication protocols for V2G integration, and addressing the infrastructure requirements for widespread adoption.

Overall, the synergy between electric vehicles and micro-grids has the potential to enhance energy efficiency, grid stability, and the integration of renewable energy sources, contributing to a more sustainable and resilient energy future.

Conclusion

With the ever-increasing need of green energy, the demand for EVs as primary mode of prospective transportation is on a constant incline. Therefore, optimisation of EV performance characteristics has become a critical issue to comprehend. A comprehensive study of existing literature with regards to the classification and charging mechanisms of EVs has been presented in this paper. It is seen from the available literature that hybrid EVs gain a substantial preference over traditional EVs owing to the better performance characteristics of the former. Moreover, the bi-directional power flow model is clearly seen to have an advantage over the traditional unidirectional power flow topology *w.r.t.* the charging techniques generally employed for EVs. In this regard, an effective ESMS has also been encapsulated in the present study in accordance with recent studies. Furthermore, the authors of the present work have also identified the application of micro-grids to EV charging system as a rather un-cultivated research area. In conclusion, it can be stated that bi-directional power flow topology using micro-grid instead of traditional grids can potentially overcome the existing drawbacks of the schemes ad methodologies that are currently being used. This forms a fascinating and comprehendible area of future research.

Participants

Lorem ipsum dolor sit amet, consectetuer adipiscing elit. Maecenas porttitor congue massa. Fusce posuere, magna sed pulvinar ultricies, purus lectus malesuada libero, sit amet commodo magna eros quis urna.

Nunc viverra imperdiet enim. Fusce est. Vivamus a tellus.

Pellentesque habitant morbi tristique senectus et netus et malesuada fames ac turpis egestas. Proin pharetra nonummy pede. Mauris et orci.

References

Ahn, Suk-Ho, Ji-Woong Gong, Hong-Je Ryoo, and Sung-Roc Jang. "Implementation of 60-kW fast charging system for electric vehicle." In *IECON 2013-39th Annual Conference of the IEEE Industrial Electronics Society*, pp. 7256-7261. IEEE, 2013.

Arun, Shankar VK, Umashankar Subramaniam, Sanjeevikumar Padmanaban, Mahajan Sagar Bhaskar, and Dhafer Almakhles. "Investigation for performances comparison

PI, adaptive PI, fuzzy speed control induction motor for centrifugal pumping application." In *2019 IEEE 13th International Conference on Compatibility, Power Electronics and Power Engineering (CPE-POWERENG)*, pp. 1-6. IEEE, 2019.

Benayad, Khadidja, Tarik Zabaiou, and Amar Bouafassa. "Wide-area based SVC-fractional order PID controller for damping inter-area oscillations." In *2020 6th IEEE International Energy Conference (ENERGYCon)*, pp. 610-615. IEEE, 2020.

Chan, Chung Chow, and K. T. Chau. *Modern electric vehicle technology*. Vol. 47. Monographs in Electrical and E, 2001.

Chau, K. T., and Wenlong Li. "Overview of electric machines for electric and hybrid vehicles." *International Journal of Vehicle Design* 64, no. 1 (2014): 46-71.

Cheng, Ka Wai Eric. "Recent development on electric vehicles." In *2009 3rd International Conference on Power Electronics Systems and Applications (PESA)*, pp. 1-5. IEEE, 2009.

Dyanamina, Giribabu, and Sanjay Kumar Kakodia. "Adaptive neuro fuzzy inference system based decoupled control for neutral point clamped multi level inverter fed induction motor drive." *Chinese Journal of Electrical Engineering* 7, no. 2 (2021): 70-82.

Elwarfalli, Hamed, Akram Muntaser, Jitendra Kumar, and Guru Subramanyam. "Design and implementation of PI controller for the hybrid energy system." In *2016 IEEE National Aerospace and Electronics Conference (NAECON) and Ohio Innovation Summit (OIS)*, pp. 170-172. IEEE, 2016.

Feng, Yong, Zhenwei Cao, Weixiang Shen, Xinghuo Yu, Fengling Han, Rita Chen, and Jiangfeng Wu. "Intelligent battery management for electric and hybrid electric vehicles: A survey." In *2016 IEEE International Conference on Industrial Technology (ICIT)*, pp. 1436-1441. IEEE, 2016.

Folkson, Richard, ed. *Alternative fuels and advanced vehicle technologies for improved environmental performance: towards zero carbon transportation*. Elsevier, 2014.

Fu, Lin, Chenghua Zhu, and Tao Du. "A new type battery management system for Li-ion cell packs." *Journal of interdisciplinary mathematics* 20, no. 6-7 (2017): 1427-1430.

Gasbaoui, Brahim, Abdelfatah Nasri, and Othmane Abdelkhalek. "An efficiency PI speed controller for future electric vehicle in several topology." *Procedia Technology* 22 (2016): 501-508.

Haiying, Li, Jia Yongli, and Zhang Dan. "Application of electric vehicle battery intelligent monitoring and management system." In *2014 IEEE Conference and Expo Transportation Electrification Asia-Pacific (ITEC Asia-Pacific)*, pp. 1-5. IEEE, 2014.

Hannan, Mahammad A., F. A. Azidin, and Azah Mohamed. "Hybrid electric vehicles and their challenges: A review." *Renewable and Sustainable Energy Reviews* 29 (2014): 135-150.

Islam, Md Ariful, Jai Govind Singh, Israt Jahan, MS Hossain Lipu, Taskin Jamal, Rajvikram Madurai Elavarasan, and Lucian Mihet-Popa. "Modeling and performance evaluation of ANFIS controller-based bidirectional power management scheme in plug-in electric vehicles integrated with electric grid." *IEEE Access* 9 (2021): 166762-166780.

Khemakhem, Siwar, Mouna Rekik, and Lotfi Krichen. "A flexible control strategy of plug-in electric vehicles operating in seven modes for smoothing load power curves in smart grid." *Energy* 118 (2017): 197-208.

Krasopoulos, Christos T., Minos E. Beniakar, and Antonios G. Kladas. "Multicriteria PM motor design based on ANFIS evaluation of EV driving cycle efficiency." *IEEE transactions on transportation electrification* 4, no. 2 (2018): 525-535.

Lee, Jung-Hyo, Jung-Song Moon, Yong-Seok Lee, Young-Real Kim, and Chung-Yuen Won. "Fast charging technique for EV battery charger using three-phase AC-DC boost converter." In *IECON 2011-37th Annual Conference of the IEEE Industrial Electronics Society*, pp. 4577-4582. IEEE, 2011.

Lipu, Molla S. Hossain, Mohammad Faisal, Shaheer Ansari, Mahammad A. Hannan, Tahia F. Karim, Afida Ayob, Aini Hussain, Md Sazal Miah, and Mohamad Hanif Md Saad. "Review of electric vehicle converter configurations, control schemes and optimizations: Challenges and suggestions." *Electronics* 10, no. 4 (2021): 477.

Nenglian, Feng, Yong Jiawang, Bin Yang, Peng Jiankun, and Tang Yanrong. "Research on battery management system for light electric vehicle." In *Proceedings 2011 International Conference on Transportation, Mechanical, and Electrical Engineering (TMEE)*, pp. 817-820. IEEE, 2011.

Osinski, Cristiano, Gideon Villar Leandro, and Gustavo Henrique da Costa Oliveira. "Fuzzy PID controller design for LFC in electric power systems." *IEEE Latin America Transactions* 17, no. 01 (2019): 147-154.

Phanomchoeng, Gridsada, and Ratchatin Chancharoen. "Hybrid motor system for high precision position control of a heavy load plant." *Engineering Journal* 23, no. 6 (2019): 161-173.

Ramadhas, Arumugam S. *Alternative fuels for transportation*. Taylor & Francis, 2011.

Rao, Zhonghao, and Shuangfeng Wang. "A review of power battery thermal energy management." *Renewable and Sustainable Energy Reviews* 15, no. 9 (2011): 4554-4571.

Somnatha, Kadlag Sunildatta, and Mukesh Kumar Gupata. "Review paper on electric vehicle charging and battery management system." In *Proceedings of International Conference on Communication and Information Processing (ICCIP)*. 2019.

Shakeel, Femina Mohammed, and Om P. Malik. "Vehicle-to-grid technology in a micro-grid using DC fast charging architecture." In *2019 IEEE Canadian Conference of Electrical and Computer Engineering (CCECE)*, pp. 1-4. IEEE, 2019.

Szalai, Thomas, Ulf Schwalbe, Marco Schilling, Fabian Endert, Tobias Heidrich, and Svetlozar-Dimitrov Ivanov. "Design of an active battery management system for electric vehicles." In *PCIM Europe 2014; International Exhibition and Conference for Power Electronics, Intelligent Motion, Renewable Energy and Energy Management*, pp. 1-8. VDE, 2014.

Tseng, K. C., T. J. Liang, J. F. Chen, and M. T. Chang. "High frequency positive/negative pulse charger with power factor correction." In *2002 IEEE 33rd Annual IEEE Power Electronics Specialists Conference. Proceedings (Cat. No. 02CH37289)*, vol. 2, pp. 671-675. IEEE, 2002.

Xiong, Rui, Jiayi Cao, Quanqing Yu, Hongwen He, and Fengchun Sun. "Critical review on the battery state of charge estimation methods for electric vehicles." *Ieee Access* 6 (2017): 1832-1843.

Yilmaz, A. S., M. Badawi, Y. Sozer, and I. Husain. "A fast battery charger topology for charging of electric vehicles." In *2012 IEEE International Electric Vehicle Conference*, pp. 1-5. IEEE, 2012.

Yilmaz, Murat, and Philip T. Krein. "Review of battery charger topologies, charging power levels, and infrastructure for plug-in electric and hybrid vehicles." *IEEE transactions on Power Electronics* 28, no. 5 (2012): 2151-2169.

Chapter 5

Electric Vehicle Applications in Frequency Regulation of Standalone Microgrid Systems

Maloth Ramesh*
Anil Kumar Yadav
Sneha V. Sah
and Sampath Kumar Bodapatla
[1]Department of Electrical Engineering,
Dr. B. R. Ambedkar National Institute of Technology Jalandhar, Jalandhar (Punjab), India
[2]Department of Instrumentation and Control Engineering,
Dr. B. R. Ambedkar National Institute of Technology Jalandhar, Jalandhar (Punjab), India
[3]Government Polytechnic College Tonk, Rajasthan, India
[4]Department of Electrical Engineering, Osmania University Hyderabad, (TS), India

Abstract

Owing to the seasonal behavior of the solar-wind-based energy system utilized in standalone microgrids, wide frequency variations occur when the frequency regulation capacity is insufficient to regulate the imbalance of power generation and demand. This issue may become more severe when the system is operating in standalone mechanism. In the meantime, Electric vehicles (EVs) are gaining more and more popularity, and they are currently being employed as scattered energy storage units rather than compact batteries in the systems. The EVs are used in this study to aid microgrids in quickly arresting fluctuations in the system frequency after power demands. As a result, the limited frequency regulation capacity can be made up for by using vehicle-to-grid (V2G) power control, which also enhances the frequency constancy of microgrids, particularly in the

* Corresponding Author's Email: ramesh.deepla@gmail.com.

In: The Future of Hybrid-Electric Vehicles
Editors: Pawan Kumar Pathak and Anil Kumar Yadav
ISBN: 979-8-89113-515-4

standalone operation mode. However, Large-scale solar-wind systems and sophisticated modern microgrids are subject to a broad-spectrum of uncertainty. To handle these uncertainties, this work proposes a metaheuristic based such as artificial gorilla troop (GTO) based sliding mode control (SMC) approach for frequency control of a standalone microgrid. The results of simulation illustrate the perfection and effectiveness of the suggested controller.

Keywords: solar-wind energy-based system, microgrid, frequency regulation, electric vehicles (EVs), sliding mode control (SMC)

5.1. Introduction

Standalone microgrids (SMG) or microgrid (MG) are thought to be economically viable options for supplying electricity to isolated locations where wind speeds are strong enough to generate electricity [1, 2]. The operator's capacity to offer consistent service to the network's energy customers is crucial to the economic and optimal operation of the resulting solar-wind system when a diesel unit is also present in the MG. As a result, maintaining the network's frequency within a specific range is necessary to prevent equipment damage to both consumers and the network caused by frequency variations [3]. It is also critical that the operator maintains a reliable and high-quality electrical supply. Since MGs typically show higher rates of mismatch between generation and demand, frequency variation in the MGs presents a significant challenge in this regard; on the one hand, renewable energy is uncertain and sporadic, and on the other hand, the power demand of the standalone grids might fluctuate frequently. Uncertainties caused by the fluctuation in renewable energy generation inevitably raise questions about how the electricity grid will function. A significant penetration of green energy, particularly solar and wind energy, will make frequency regulating more challenging. Frequency control with a lot of green energy has already been investigated in several studies [4, 5]. In [6], authors recommended using Battery Energy Storage Systems (BESS) for frequency stability. They concluded that BESS can significantly cut down on frequency deviations caused by rapid variations in demand for the power supply. However, because of its high cost, the BESS's potential economic benefits for widespread implementation are still debatable. Real power modelling for droop, dead-band, and reverse power control schemes is crucial to understanding the role of wind turbines in frequency control. Figure 1 represents how wind's role in frequency regulation can be broken down into inertial control, main control, and secondary control, depending on how long a frequency variation takes. The inertial control provided by the KE stored in the WTGs' revolving blades confirms that the disturbance began at the event's origin and lasted for 10 seconds (Figure 5.1). When the primary controller is triggered, the system remains steady for up to 30 seconds.

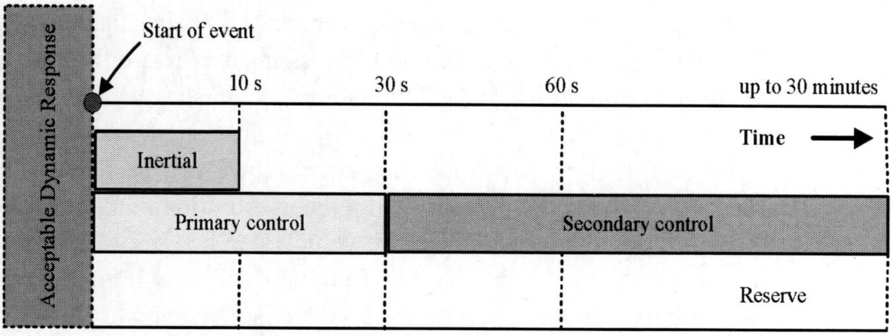

Figure 5.1. Schemas for regulating frequency over time.

Wind reserves power for this control activity after 30 s to several minutes, keeping the system frequency at its prespecified setting. WTG's contribution to inertia emulation and quick power reserve inertia is investigated. A synchronous generator's governor droop control is like the WTG system's primary frequency control. The control method adjusts the WTG's output power as the frequency changes to maintain a constant value. In the unloaded WTG system, the control parameter is a function of the system frequency but not of RoCoF. In [7], the authors offered several wind technologies well suited for supporting inertial frequencies and showed where they fell short in practice. Both inertial and primary frequency control mechanisms for a WTG with different speed are proposed by the authors of [8]. The KE concealed in the inertia of turbine blades can be retrieved using one of three different sets of controllers. In [9], a novel WTG system frequency regulation control mechanism with reduced burden is presented. A comparison of the SG and the DFIG describes the advantage of the proposed control method. In [10], a traditional 2-area power grid is supplemented with a 20%-penetrated WE system. To design an effective frequency regulation method, we drew on the power electronically interfaced WTG system's superior response time to that of conventional systems. Having effective connection between WTG and the conventional system allows for early frequency assistance. In [11], the authors provided a control method for WTG that allows for optimal power acquisition and the delivery of stored KE to sustain the system frequency without causing a reverse power spike under high and low V conditions. Adjusting the WTG's pitch angle maintains the power reserve across all working zones in the frequency droop method provided in [12]. In [13], a unique method is described for increasing the WTG system's KE discharge capacity by including frequency variation and the derivative of frequency deviation in the torque control loop. The popularity of electric vehicles (EVs) has increased in recent years. Significant implications on power systems may result from EVs' introduction and widespread use [14]. The authors establish a technique for modelling EV charging load and examine how this load contributes to the overall demand on a regional distribution system. However, the features of several classes of EVs were disregarded under the assumption that they were all powered by the same battery. An electric vehicle can be a mobile battery storage system for frequency control when equipped with a two-way power interface. The primary frequency response of EVs in isolated power networks of modest size was studied in [15]. To account for the frequency shift, the charging power for the EV was modulated using a droop control [16]. In [17], the aggregated V2G control technique is suggested to stop the system frequency from changing. This is done with the help of a control centre that calculate the charge utilized of the EV. In a competitive power market, a hierarchical architecture optimises PEV charge for frequency control [18].

5.2. Literature Review

Recently, some work has been done to adapt a technique known as unit commitment, which schedules available producing units to run a microgrid effectively, for use with the V2G [19]. The automobile industry's future appears to be EV, which encourage energy conservation and are environmentally benign [20]. Additionally, using EV technology is a promising way to cut back on petrol usage. Utilising EVs will dramatically cut back greenhouse gas emissions. The fact that EVs can be employed as grid energy storage devices further strengthens the case for their deployment. A charging EV can be seen as a grid consumer, while a charged EV can be seen as a power generator [21]. EVs exchange electricity with the electrical system in both directions when connected to the grid. According to statistics, private vehicles are inactive for over 95% of the day; as a result, if EVs are utilised instead, they will be charging the entire time, conserving energy. Therefore, because they offer load-balancing capabilities and make it easier for the isolated grid to operate flexibly, EVs may be useful to enhance the steady functioning of the isolated RES based grids. The LFC's job is to maintain a reasonable level of dynamic operation in electric systems by balancing the system's power so that the system's frequency remains within predetermined limits [22]. With the right LFC, you may save money on fuel, cut down on energy use, and possibly even decrease the need for additional safety equipment. Traditional PID control [23], adaptive control [24], smart control and resilient control [25], and Traditional PID with structural change [26] are just some of the control methods that have been used to improve the LFC's effectiveness in the literature. In [27], a fractional order cascade LFC for distributed green sources via advanced optimization technique is studied. Traditional techniques cannot ensure the coordinated control impact of V2G along with additional units for the system frequency in the LFC scheme of a standalone MG with variable green power sources and EVs, as the state parameters and operation status of the system are subject to rapid change. In addition, the LFC should incorporate several constraints, such as generation capacity and rate constraints, which traditional methods cannot surmount.

Consequently, it is highly desirable for the LFC in a standalone MG to have a controller with robust performance across a broad spectrum of system operating conditions. The most important aspect to designing such controllers is to utilise EVs as mobile energy storage units via V2G methods to surmount the limitations of LFC, particularly in standalone grids. Numerous studies and articles have recently used intelligent or fuzzy controllers with diverse designs, such as when combined with PIDs, to build more robust LFCs. Fuzzy controllers are widely recognised as good solutions for creating robust controllers, capable of giving optimal performance even when system characteristics are unclear. The FLCs, conversely, cannot manage all rule uncertainty since their roles as members are type-1 fuzzy sets [28]. Like the conventional electrical system, robust LFC techniques can be considered an effective way to achieve stability and reliable operation in the MGs' absence with environmental and financial restrictions. Several control strategies for the secondary LFC in the standalone MGs have been suggested. In traditional and contemporary power systems, primary and secondary LFC are discussed in fine detail in [29].

Additionally, in [30], a smart fuzzy-PI controller is used to synchronise the aqua electrolyser and fuel cell to control the power oscillation in the MG. The authors also studied on the other hand advocated a hierarchical control method for the LFC in the MG. The LFC's fuzzy logic system and controller in the islanded MG are tuned using a sophisticated off-line optimisation approach (like particle swarm optimisation) in [31]. The V2G methodology and adjustable loads suggest a new way for the LFC. Additionally, [32] employed a thermal DG with many EVs to suggest a workable LFC solution. However, the robustness of the controller under diverse system loading situations was not covered in their study. Consequently, the joint management of V2G and It has been reported to solve LFC issues against system uncertainties uses particle swarm optimisation based fixed structure mixed H_2/H_∞ control. When connected to the grid for charging and discharging, thousands of EVs can be considered a sizable power plant. Therefore, it is generally anticipated that EVs will be able to meet the future electricity demand. These factors have motivated numerous academics to focus their research on frequency control [33]. Recently, a plug in EV fleet aggregate model was developed to participate in primary frequency regulation. However, the secondary frequency control loop has not yet been fully investigated. A new for-profit organisation called the EV aggregator is responsible for EVs' control, management, and participation in the active power regulating system. In the past ten years, several papers have been published to help visualise the administration of EV aggregators. Most of the power industry still uses the traditional PI and PID controller as a secondary load frequency controller, even though numerous advanced control schemes have been documented over the last few decades. It is because of how easily they can be carried out. Using distributed functional observers, the LFC problem of the power network is chosen in conjunction with EVs while considering open communication channels and high voltage direct current links. However, such a technique needs information on the system states, which might not be available in an emergency or be exceedingly challenging to implement in a real-world scenario [34].

Additionally, they do not study dynamic aspects by adding nonlinear restrictions like generation rate constraints (GRC), governor dead band (GDB), etc. The role of sliding mode control (SMC) technique in solving various LFC issues in modern MG systems is studied [35, 36]. Hence a metaheuristic-based SMC with PID sliding surface is proposed in this work. Reliable source non-linearities like GDB and GRC are included in this study.

5.3. Standalone Microgrid

A MG is a combination of interrelated loads with dispersed sources of power that operate concerning the grid as a single, controllable entity. It can connect to and disconnect from the grid to operate in ON grid mode or OFF-grid mode. Customers' dependability and resistance to grid interruptions can both be improved via microgrids. Advanced microgrids give regional resources—such as traditional generators, green source generators, and storage—the ability to maintain the small grid even in the event of disruptions to the larger grid or, in the case of standalone places, when there is no connectivity to the main grid. Advanced microgrids also enable neighbourhood resources to collaborate to save costs, increase the longevity of energy supplies, and generate income through market participation. In this work, the major components are DG as a reliable source, the green sources such as solar (PV) and wind (WT) are considered as main sources. The basic structure of the proposed microgrid system is shown in Figure 5.2. The role of DG in an MG system is to supply power to the consumer during peak load hours, absence of power generation, or low power available from the green resources. This comprises four main components: frequency supported EVs, additional energy sources, management panels, and PV and WG generating sources. Due to the extremely unpredictable nature of green energy sources and their dependence on the weather, backup energy sources and storage devices are required to operate the MGS autonomously. Coordination of the MGS's many components is the responsibility of the control panel. The function of PEI is to manage the power dynamics of SMG and integrate the dispersed energy supplies and load through the AC bus system. The PEI contains bi-directional power converters for storage systems that can operate in rectifier and inverter control modes and power inverters for DC power sources. Non-frequency support loads include the primary health centre, agricultural load, and domestic shipments.

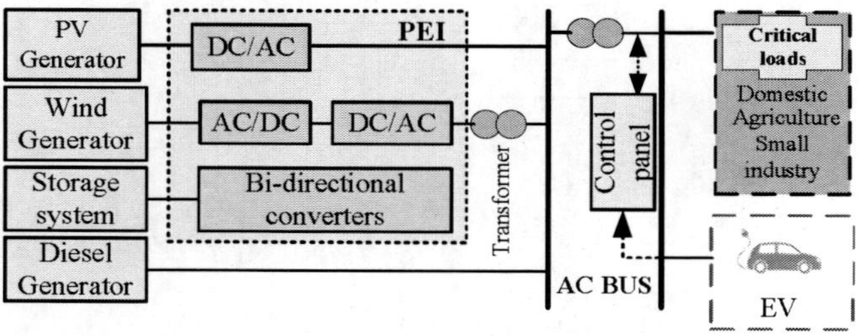

Figure 5.2. Basic structure of the proposed microgrid system.

The GDB and GRC nonlinearities of the DG system are included in the simulation study to make the system more realistic. The suggested freestanding MG system under consideration, which is based on green sources, is depicted in block diagram form in Figure 5.3. The visualization included the suggested control strategy. In an MG system, the DG's job is to provide electricity to the consumer during times of high demand, when there is no power generation, or when there is not enough power coming from renewable sources. To make the DG system more realistic, the GDB and GRC nonlinearities are considered in the model of this study. The power balance equation of the proposed microgrid structure is defined as (5.1):

$$\Delta P_{SMG} = \Delta P_{PV} + \Delta P_{WG} + \Delta P_{DG} \mp \Delta P_{EV} - \Delta P_L \quad (5.1)$$

where ΔP_{MGS} is the microgrid total power, ΔP_{PV} is the PV power, ΔP_{WG} is the WG power, ΔP_{DG} is the reliable power source such as diesel generator power, ΔP_{EV} is the EV power, ΔP_L is the load power consumed. The small signal dynamic model of the SMG is represented as (5.2) follows:

$$G_{SMG}(\mathrm{s}) = \frac{\Delta f}{\Delta P_{MGS}} = \frac{1}{Ms+D} \quad (5.2)$$

where D and M are the MGS's total damping coefficient and inertia constant, respectively, and Δf is the frequency deviation.

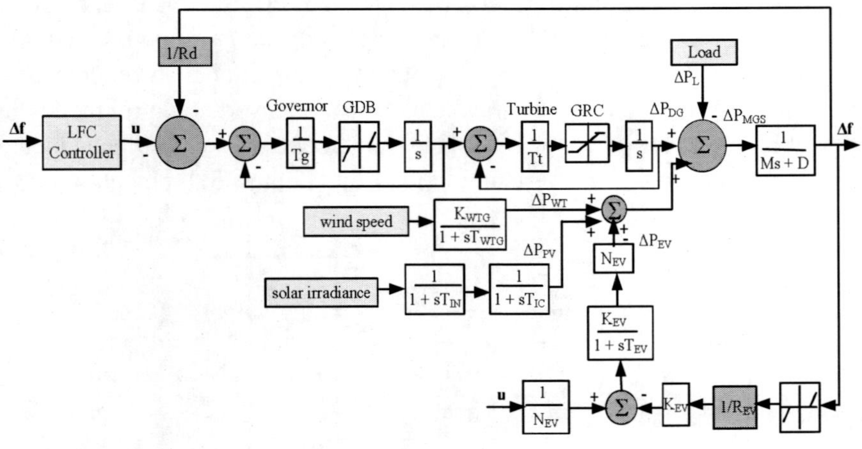

Figure 5.3. Small signal dynamic model of the microgrid.

The power output of the PV system is defined as (5.3) follows:

$$P_{PV} = \eta S\Phi\{1 - 0.005(T_a + 25)\} \tag{5.3}$$

where η is the conversion efficiency of the PV array, S is the area of the PV array (m^2), Φ is solar irradiation (kW/m^2), and T_a is the operating ambient temperature and its value taken as 25^0C in this work. A linear dynamic model PV system is described by (5.4):

$$G_{PV}(s) = \frac{1}{1+sT_{IN}} * \frac{1}{1+sT_{IC}} \tag{5.4}$$

T_{IN} and T_{IC} are solar inverter and interconnection device time constants, respectively.

The output power of WT can be described as (5.5) follows:

$$P_{WT} = \frac{1}{2}\rho A_r C_P V_w^3 \tag{5.5}$$

where ρ is air density and valued at 1.25 kg/m^3, A_r is blades swept area in m^2, V_w is the wind velocity m/s, C_P is the power coefficient (5.7) of WTG, which is the function of two mechanical components, blade pitch angle (θ) and tip speed ratio (5.8) (λ). WT system transfer function representation is as (5.6) follows:

$$G_{WT}(s) = \frac{K_{WT}}{1+sT_{WT}} \tag{5.6}$$

The aerodynamics of WT system is modeled as follows: Firstly, the power coefficient (C_P) of WT is as follows:

$$C_P = (-0.167\theta + 0.44)sin\left[\frac{n(\lambda-3)}{-0.3\theta+15}\right] - 0.0184(\lambda - 3)\theta \tag{5.7}$$

Second, the tip-speed ratio (λ) is defined as:

$$\lambda = \frac{\omega_{blade} \cdot r_{blade}}{V_W} \tag{5.8}$$

where w and r are the rotational speed and radius of the wind turbine blade, respectively.

The DG power is described as (5.9) follows:

$$\Delta P_{DG} = \left(-u - \frac{1}{R_d}\Delta f\right)\left(\frac{1}{1+sT_g} * \frac{1}{1+sT_t}\right) \tag{5.9}$$

where u is the control signal and represented as $u = \Delta f$ (LFC), R_d is the droop parameter, T_g and T_t are governor and turbine time constants, respectively.

The detailed EV dynamic structure is represented in Figure 5.3. In SMG, EV serves as a power variability absorber. This vehicle type can run on fuel and a battery or supercapacitor. EVs might formerly function in one of two ways: a car to grid and V2G at peak hours and mode (absorbed electricity) during off-peak hours. It is a superior alternative for long-distance travel because it can react using either fuel or energy. The transfer function representation of the EV is described (5.10) as:

$$G_{EV}(s) = \frac{K_{EV}}{1+sT_{EV}} \tag{5.10}$$

where K_{EV} is the EV constant, T_{EV} is the EV time constant.

5.4. Control Methodology

The kind of controller and methodology used to tune the controller settings and the fitness function selection determine the stable and dependable functioning of the solar wind-based SMG system. The next subsection summarises the control approach utilised to provide EV contribution in frequency regulation for stable operation. SMC approach with exceptional accuracy, robustness, and ease of tuning and deployment. SMS systems are intended to drive data onto surfaces in the current state. The space is known as the sliding surface. When the sliding surface is reached, SMC maintains the states near the sliding surface. As a result, sliding mode control is a two-part controller architecture. The initial step is to create a sliding surface so that the sliding motion meets design standards. The second involves selecting a control law to make the switching surface appealing to the system state [35, 36]. The SMC has two key benefits. The system's dynamic behaviour can first be customised by selecting a certain sliding function. Second, the closed-loop reaction completely loses its sensitivity to some specific uncertainties. This idea encompasses bounded nonlinearity, disruption, and model parameter uncertainty. Practically speaking, SMC enables the control of nonlinear processes exposed to outside shocks and significant model uncertainty. In this work the integral sliding surface is utilized. For robust SMC in the face of uncertainty, the following adjustment is being considered [36]: The following equations are utilized to design the control law of SMC (5.11).

$$u = u_{eq} + \eta \, sign(S) \tag{5.11}$$

where η is the sliding mode uncertainty factor, $sign(\cdot)$ is the signum function, and $sign(S)$ is defined as:

$$sign(S) = \begin{Bmatrix} -1, x < 0 \\ 0, x = 0 \\ 1\, x > 0 \end{Bmatrix}. \tag{5.12}$$

To tune the SMC parameters, a metaheuristic technique called artificial GTO technique is used in this work. This study aims to obtain the optimized solution of integral gain of SMC, sliding mode uncertainty factor (η) via GTO algorithm. The proper objective function (J) determines any controller's optimal value. The numerous J have been used for LFC in the research because are listed beneath.

$$\text{Integral of squared error } (\text{ISE} = J_1) = \int_0^t \{(\Delta F_1)^2 + (\Delta F_2)^2 + (\Delta Ptie_{12})^2\} dt \tag{5.13}$$

$$\text{Integral of time multiplied squared error } (\text{ITSE} = J_2) = \int_0^t t\{(\Delta F_1)^2 + (\Delta F_2)^2 + (\Delta Ptie_{12})^2\} dt \tag{5.14}$$

$$\text{Integral of absolute error } (\text{IAE} = J_3) = \int_0^t \{|\Delta F_1| + |\Delta F_2| + |\Delta Ptie_{12}|\} dt \tag{5.15}$$

$$\text{Integral of time multiplied absolute error } (\text{ITAE} = J_4) = \int_0^t t\{|\Delta F_1| + |\Delta F_2| + |\Delta Ptie_{12}|\} dt \tag{5.16}$$

where $'t'$ is the simulation time.

Among various objective functions, the ITAE (5.16) is used in this study. Interacting with ITAE as J_4 results in a regulated reaction with diminished oscillations and less settling time than other variations, and as such, it is considered in the proposed study. The terms ΔF_1, ΔF_2,and $\Delta Ptie_{12}$ are related to multi-area electrical network. A multi-area electrical network refers to an interconnected electrical network consisting of two distinct regions or areas that are geographically separated but connected through power transmission lines. These areas can represent different cities, states, or countries, and they collaborate to exchange electricity to meet their respective demands and ensure system stability. The $\Delta Ptie_{12}$ is the tie-line power refers to the amount of electrical power flowing between two interconnected areas or regions in a power system through a transmission line. In a multi-area electrical network, tie lines connect the different areas, allowing electricity exchange to balance loads, ensure system stability, and optimize resource utilization. ΔF_1 is the change in frequency of one-area in a two-area power system depends on the balance between the area's power generation and its load. If a single-area's power generation is more significant than its load demand, the power system's frequency will rise. On the other hand, the frequency will drop if the load demand in the single area is higher than the available generation. Due to the immediate mismatch between generation and load, these frequency variations happen. ΔF_2 is the change in frequency of another area in a two-area power system. Frequency control technologies are employed to preserve the stability of the power supply. When Area One's frequency rises because of excessive generation, the control system may limit some generators' power output or raise the load to return the frequency to its normal level. To restore the frequency to the appropriate level, the control system may boost the power output of generators or lower non-essential loads if the frequency drops due to heavy load demand. To create a work list vector that is randomly generated, we utilise GTO algorithm. This vector is represented by the gorilla that needs to be optimised in the suggested algorithm, where each gorilla vector provides a potential answer to the issue. The job without any predecessors then obtains the post set items and shuffles them. This process is carried out iteratively by the algorithm for each task. The technique ensures the creation of task lists at arbitrary while taking task interdependency connections into account. To initialize the three components of the gorilla, we utilized GTO. Calling algorithm 1 initializes the task list. The GTO programme simulated the troop behaviour of the gorillas [20]. The group behaviour of the gorillas was mathematically modelled utilising four distinct techniques. Migration to unknown locations, to familiar locations, and interactions with other gorillas are three operations covered under the exploration phase. The two additional procedures are also a part of the exploitation phase. These two actions are taken to submit to the silverback and engage in combat with adult men. The silverback leads the gorilla troop. According to the GTO algorithm, the silverback position offers the optimum option. A seed variable S and a value generated at random (P) determine how the three operators included in the exploration phase are mathematically formulated. The candidate solution GX in the following i^{th} iterations can therefore be determined as (5.17) follows when PS, the migration to an unknown site, is used:

$$GX(i+1) = (U-L)*r_1 + L \tag{5.17}$$

where, $r_1$is a random variable with a range of [0: 1], where L and U are the variables' corresponding lower and upper limits. Each repetition involves a continual update to r_1. The movement to other gorillas' operator is selected when P≥0.5, and the next candidate solution is obtained by:

$$GX(i+1) = (r_2 - Y) + X_r(i) + A*h \tag{5.18}$$

A random position matrix for a member of the gorilla population is represented by X_r. These formulas can be used to define the constants y (5.19), a, and h (5.21):

$$y = V*(1-i/Mi) \tag{5.19}$$

$$V = \cos(2*r_4) + 1 \tag{5.20}$$

$a = y*r_4$, r_4 is the random value in [-1,1].

$$h = Z*X(i) \tag{5.21}$$

where Z=[-y,y]. The current location vector is $X(i)$. The migration to known places operator is chosen in the final scenario if P <0.5 and is expressed as (5.22) follows:

$$GX(i+1) = a*b*(x(i) - X_{SLB}) + X(i) \tag{5.22}$$

X_{SLB} is the silverback or best solution position vector. The vector b is calculated as (5.23) follows:

$$b = \left[\left(\frac{1}{N}\sum_{j=1}^{N}(GX_t(i))\right)g\right]^{\frac{1}{g}} \tag{5.23}$$

$GX_t(i)$ is a potential location vector for each gorilla member t in iteration i. N is the number of gorillas in total. On the other hand, the GTO algorithm chooses the adult conduct competition when y is less than W. It is necessary to transfer continuous values from the Gorilla vector (obtained by executing the mathematical formulas of each operation) to discrete bounded ones based on the value's position in the Gorilla vector. We used two different mapping techniques to achieve this discretization of the values of the resulting gorilla position vector. The values in the task list section are mapped using the first mapping technique. First, we take the first item on the task list and retrieve the items that come after it, to adhere to the dependence restriction among tasks. The second mapping method is simply a direct normalization of the vector values. The detailed flowchart of the GTO algorithm is explained in [35].

5.5. Results and Discussion

Modelling is done for a 750-kw standalone microgrid system with a diesel engine and battery storage system. From [35], the aerodynamics and specific WTG characteristics were extracted. The tiny signal model for this WTG system has a unity gain and a time constant of 1.5 s, and it has a pitch angle control [35]. The governor and turbine on the diesel generator set have unity gains and, respectively, temporal constants of 0.08 s and 0.4 s. The microgrid's parameters follow speed regulation is 2%, generator and load time constants are 0.2 s and 0.015 s, respectively. The simulation data used is represented in Table 5.1.

The following section examines the proposed system's frequency response under various operating situations, including source and load uncertainty. The MATLAB R-2018b Simulink framework is used for this work's modeling and time-domain form analysis. The GTO is a simple architecture, low resource usage (both computational time and storage space), quick convergence, derivative-free process, and small number of decision variables make it simple to implement. To measure the degree to which the GTO technique has superiority over the other GA and PSO, statistical analysis using J_1, J_2, J_3, and J_4 is conducted. The three optimization strategies achieve the proposed controller gains for each objective function. Table 5.2 lists the three optimisation strategies' best fitness values and corresponding controller gains for the four objective functions.

Table 5.1. Simulation parameters of the MGS

System	Parameter values
Microgrid [3, 6]	M = 0.2, D = 0.015
DEG [24, 6]	Tg = 0.08s, Tt = 0.4s, R = 2
Wind, Solar [24, 35]	KWTG =1, TWTG=1.5, TIN = 0.04s, TIC = 0.004s
EV [3, 6], [6, 24]	KEV =1, TEV = 0.2s, REV = 2.4

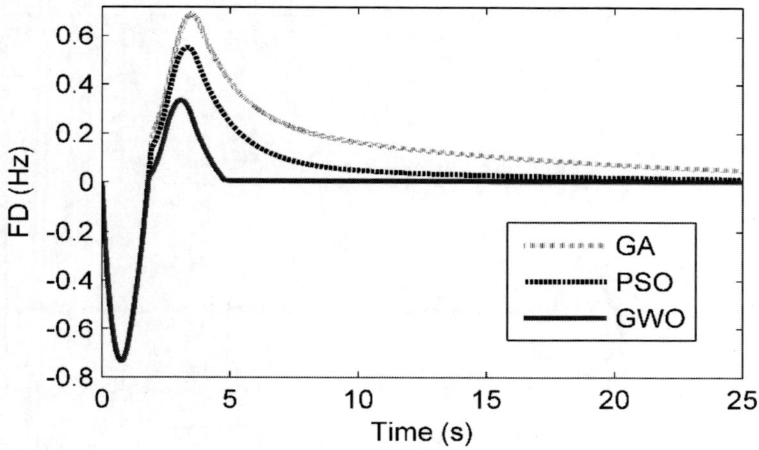

(a)

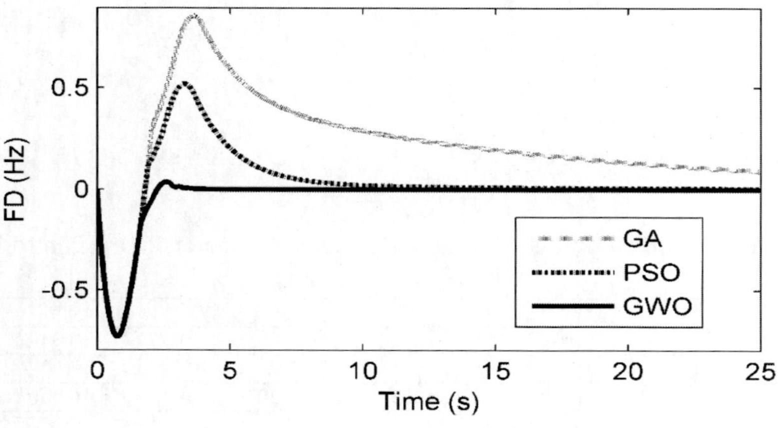

(b)

Figure 5.4. (Continued).

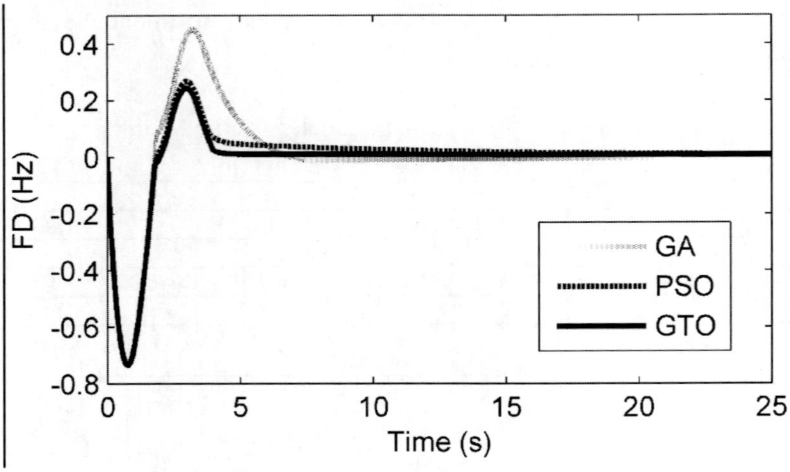

(c)

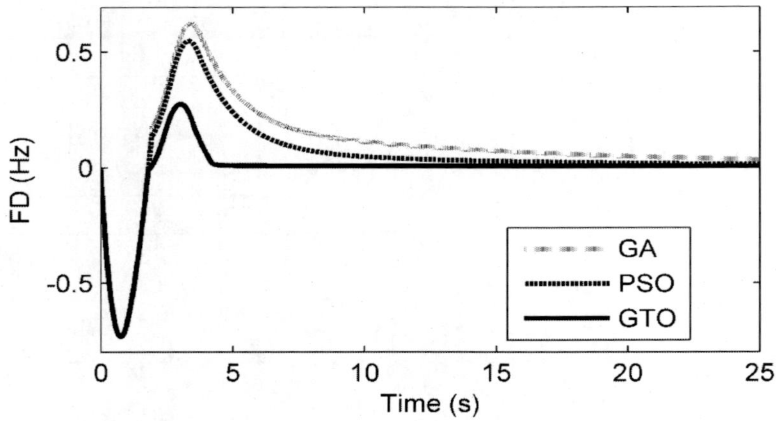

(d)

Figure 5.4. Comparative analysis of GA, PSO, GWO for Δf: (a) J_1 , (b) J_2, (c) J_3, (d) J_4 .

Table 5.2. Best fitness value and optimized parameters of the controller

Technique	IAE	ISE	ITSE	ITAE	K_i	η
GA	3.126	0.625	3.52,	6.52	0.0129	0.00152
PSO	2.872	0.5210,	0.928,	3.267	0.0526	0.00124
GTO	1.2376	0.4980,	0.625,	0.8828	0.0158	0.00186

The effectiveness of the GTO approach and the suggested controller are assessed for each of the four objective functions. As shown in Figure 5.4, a comparison of the GA, PSO, and GTO is made in terms of f for each objective function. For J_1 J_2, J_3, and J_4 correspondingly, the peak overshoot (MP_{OS}) with GA is 0.62 Hz, 0.45 Hz, 0.69 Hz, and 0.85 Hz. For J_1, J_2, J_3, and J_4, the MP_{OS} (+) with SCA is 0.43 Hz, 0.27 Hz, 0.561 Hz, and 0.55 Hz, respectively. For J1, J2, J3, and J4, the MP_{OS} (+) with GWO is 0.275 Hz, 0.24 Hz, 0.33 Hz, and 0.015 Hz, respectively. In Figure 5.4 (a), (b), (c), and (d), respectively, comparative transient responses for the four cases—IAE, ISE, ITSE, and ITAE—are displayed. Table 5.3 contains the quantitative information on the dynamic reaction in terms of MP_{OS} (+), peak undershoot (MP_{US}), settling time (Ts), IAE, ISE, ITSE, and ITAE. The MP_{US} (-) is within the allowed grid range while using three optimisation approaches for four objective functions. As a result, the J_4 based GTO technique is chosen for more system analysis. The GA, PSO, and GTO source code implementations are written in the MATLAB m-file environment. The traditional tuning method, the ZN method, determines the gain bounds.

5.6. EV Contribution in the LFC

The dynamical evaluation of the studied optimization techniques (GA, PSO, GTO) with Integral-SMC techniques controller is presented in Table 5.3. The examined data in Table 5.3 indicate that the results of the GTO-based integral SMC method are consistent and comparable to GA-tuned integral-SMC and PSO- tuned integral-SMC controllers in this study. Further the GTO method gives the best result for the integral-SMC method concerning objective function value. The various operating conditions are studied in this section. The LFC scheme studies the frequency variations with and without EV contribution. Renewable energy sources based on PV and WG are frequently unpredictable and sporadic. The daily atmospheric conditions significantly influence wind speed and wind power. Variability in environmental circumstances, such as clouds, rain, storms, fog, and dust, have significantly impacted the energy response of PV. These factors are the primary causes of solar radiation irregularity and impact solar output power. Here the first case applied on SMG is full absence of wind power. The main contributors for the applied load are PV, DG and EV as shown in Figure 5.5. The power responses of all contributors, including demanded load is shown in Figure 5.5 (a). The frequency response of this case in shown in Figure 5.5 (b). During the absence of green power such as PV and WG, a reliable unit DG supplies the load and maintains the grid frequency within the acceptable limit. Here the dynamic performance of SMG is further improved by the contribution of EV in the LFC scheme.

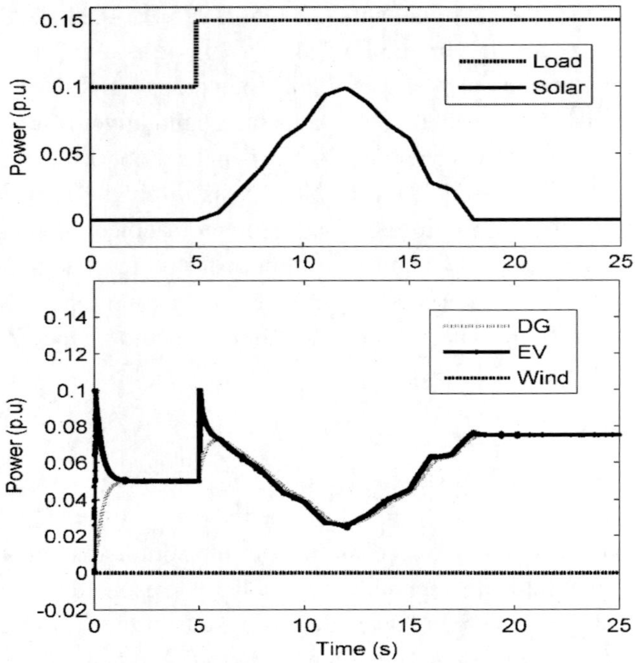

(a)

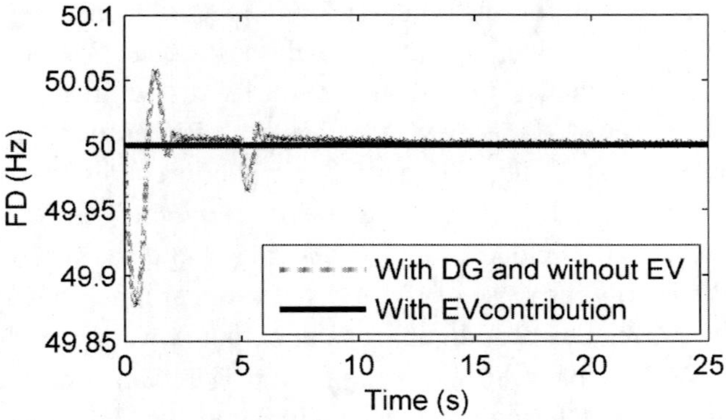

(b)

Figure 5.5. Power and frequency responses of SMG during zero power of WG (a) power (b) frequency.

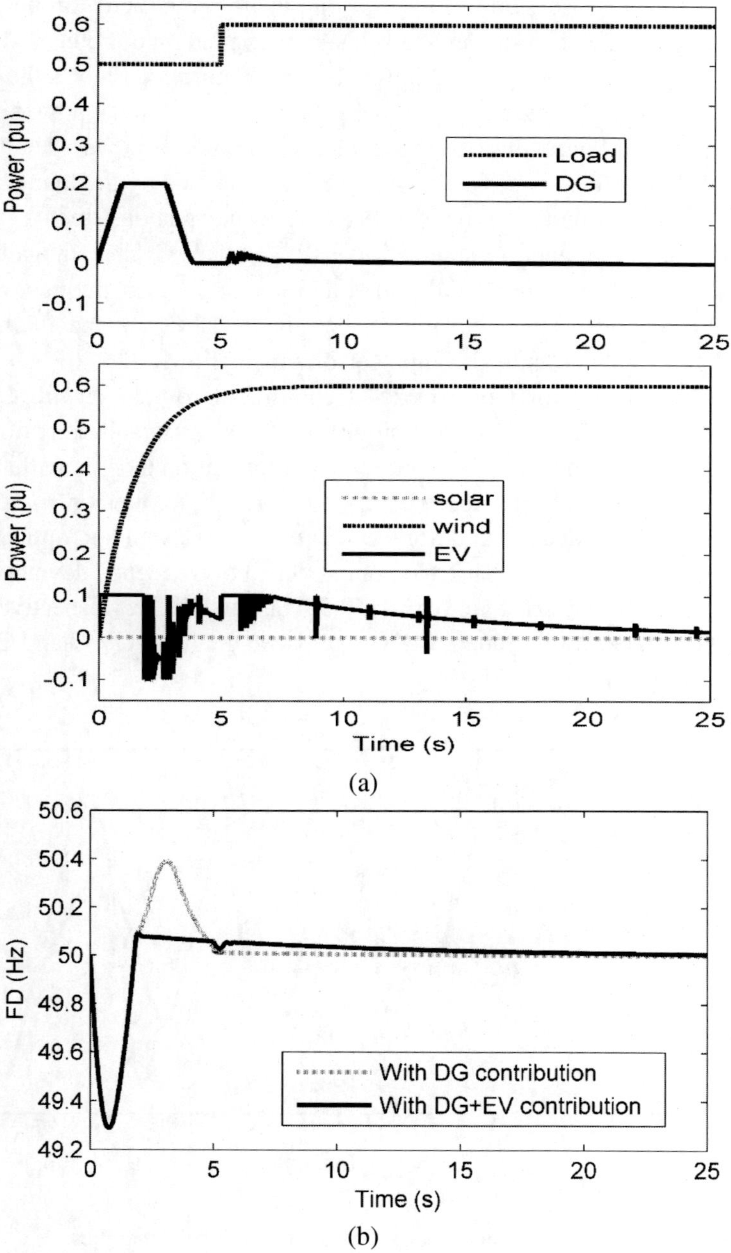

Figure 5.6. Power and frequency responses of SMG during zero power of solar PV (a) power (b) frequency.

The second case study is the PV output power is zero throughout the simulation time. Here a sudden load change is applied on SMG at 5s. The WG power output is 0.6 pu. The PV power is zero. Figure 5.6 reveals the SMG's power responses during the absence of solar power. When the power of SMG is excessive, EV will consume this excessive amount of power. The DG and EV will supply the power when the power is demanded by the load as shown in Figure 5.6(a). Figure 5.6(b) displays the frequency response in this example. When renewable energy sources like PV and WG are unavailable, a dependable DG unit supplies the load and keeps the grid frequency within allowable bounds. The inclusion of EV in the LFC scheme in this case considerably enhances the dynamic performance of SMG.

The superiority of the proposed technique is compared with GWO in terms of FD with a GWO technique which is most suited optimization technique for LFC issue. GWO is a robust optimization technique that shows effectiveness in various engineering problems like; tension/compression spring design, welded beam design, pressure vessel design and optical buffer design compared to PSO. The maximum overshoot frequency deviations with the PSO, GWO and GTO are 0.42Hz, 0.39Hz, and 0.31Hz, respectively. The maximum undershoot frequency deviations with the PSO, GWO and GTO are -0.42Hz, -0.41Hz and -0.35Hz.

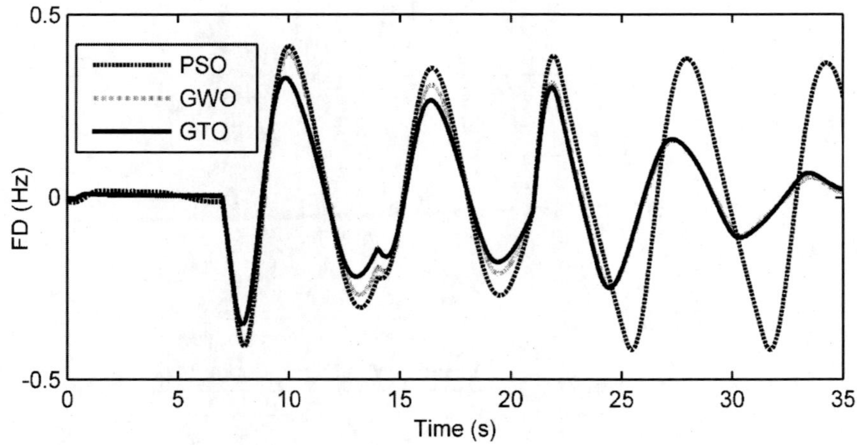

Figure 5.7. Comparative frequency responses with various techniques.

Table 5.3. Decision parameters for the selection of algorithm with integral SMC

J_1	Parameters	GA	PSO	GWO
	MP_{OS} (+), Hz	0.62	0.43	0.275
	MP_{US} (-), Hz	0.735	0.734	0.725
	T_s, (s)	32	28	12
	IAE	4.112	2.923	1.386
	ISE	1.416	1.052	0.5681
	ITSE	5.014	2.656	0.644
	ITAE	27.5	14.49	4.144
J_2	MP_{OS} (+), Hz	0.45	0.27	0.24
	MP_{US} (-), Hz	0.735	0.733	0.731
	T_s, s	24	20	16
	IAE	1.918	1.663	1.345
	ISE	0.7854	0.5763	0.4981
	ITSE	1.399	0.7278	0.5857
	ITAE	5.492	6.961	4.253
J_3	MP_{OS} (+), Hz	0.69	0.561	0.33
	MP_{US} (-), Hz	0.734	0.732	0.729
	T_s, s	34	26	17
	IAE	5.048	2.951	1.485
	ISE	1.803	1.059	0.6179
	ITSE	7.8	2.696	0.8008
	ITAE	23.79	14.79	3.947
J_4	MP_{OS} (+), Hz	0.85	0.55	0.015
	MP_{US} (-), Hz	0.742	0.738	0.735
	T_s, s	32	15	10
	IAE	6.752	2.457	0.9185
	ISE	1.902	1.027	0.5012
	ITSE	10.2	3.456	0.6291
	ITAE	15.1	2.785	0.9015

Conclusion

This study investigates a stand-alone community-based microgrid in the presence of PV, WT, DG, and EV without battery under real-time wind and solar radiation data to resemble the practical scenario. The EV are the demand contributor and acts as a generator to enhance the system's dynamic performance. A GTO optimized integral-SMC technique is designed to incorporate system nonlinearities such as GDB, and GRC under load, source, and parametric uncertainties. The GTO technique's superiority is explored over the GA, and PSO algorithm in terms of optimization selection decision parameters such as T_s, MP_{OS}, MP_{US}, and other performance indices. Simulation studies have confirmed that the performance of the GTO tuned integral-SMC method is superior. The EV acts as demand and generator to support the system frequency in the LFC scheme. The obtained power and frequency response is within the grid acceptable limits. In the context of proposed SMG, coordination control is an active research area to avoid the expensive energy storage system. The stability analysis of MGS in the presence of inherent and intentional nonlinearities is considered as the future scope of this work.

References

[1] Xu. Y. Zhang. W. Hug. G. Kar. S. and Li. Z. "Cooperative control of distributed energy storage systems in a Microgrid," *IEEE Trans. Smart Grid*, vol. 6 no. 1, pp. 238-248, 2015.

[2] Yang. J. He. L. and Fu. S. "An improved PSO-based charging strategy of electric vehicles in electrical distribution grid," *Appl. Energy*, vol. 128, pp. 82-92, 2014.

[3] Ramesh. M. Yadav. A. K. and Pathak. P. K. "An Extensive Review on Load Frequency Control of Solar-Wind based Hybrid Renewable Energy Systems," *Energy Sources, Part A: Recovery, Utilization, and Environmental Effects*, Taylor & Francis, 2021.

[4] Mohammad. D. H. Mokhlis. and Mekhilef. S. "Inertia response and frequency control techniques for renewable energy sources: A review." *Renewable and sustainable energy reviews*, vol. 69, pp. 144-155, 2017.

[5] Yida. Y. e. Qiao. Y. and Zongxiang. L. u. "Revolution of frequency regulation in the converter-dominated power system." *Renewable and Sustainable Energy Reviews*, vol. 111, pp. 145-156, 2019.

[6] Ramesh. M. Yadav. A. K. and Pawan. P. K. "Intelligent adaptive LFC via power flow management of integrated standalone micro-grid system." *ISA transactions*, vol. 112, pp. 234-250, 2021.

[7] Vidyanandan. K. V. and Senroy. N. "Primary frequency regulation by deloaded wind turbines using variable droop." *IEEE transactions on Power Systems* vol. 28, no. 2, pp. 837-846, 2012.

[8] Lalor. G. A. M. and Mark. O. M. "Frequency control and wind turbine technologies." *IEEE Transactions on power systems* vol. 20, no. 4, pp. 1905-1913, 2005.

[9] Almeida. D. e. Rogério. G. and Lopes. J. A. P. "Participation of doubly fed induction wind generators in system frequency regulation." *IEEE transactions on power systems* vol. 22, no. 3 pp. 944-950, 2007.

[10] Manuel. M. J. Alejandro. M. Gómez-Expósito. A. and Martínez. R. J. L. "Frequency regulation contribution through variable-speed wind energy conversion systems." *IEEE Transactions on Power Systems* vol. 24, no. 1, pp. 173-180, 2009.

[11] Keung. P. K. Li. P. Banakar. H. and Boon. T. O. "Kinetic energy of wind-turbine generators for system frequency support." *IEEE Transactions on Power Systems* vol. 24, no. 1, pp. 279-287, 2008.

[12] Chang. C. Ren. L. Wei-Ting. L. and Yin. Y. C. "Enhancing frequency response control by DFIGs in the high wind penetrated power systems." *IEEE transactions on power systems* vol. 26, no. 2, pp. 710-718, 2010.

[13] Ioannis M. D. Stavros. A. P. Nikos. D. Hansen. H. D. and Sorensen. P. "Frequency control in autonomous power systems with high wind power penetration." *IEEE Transactions on sustainable energy* 3, no. 2, pp. 189-199, 2012.

[14] Bahrami. S. and Parniani. M. "Game theoretic based charging strategy for plug-in hybrid electric vehicles," *IEEE Transactions on Smart Grid*, vol 5 no. 5, pp.2368-2375, 2014.

[15] Saber. F. Taher. S. A. and Shahidehpour. M. D. "A new smart charging method for EVs for frequency control of smart grid." *International Journal of Electrical Power & Energy Systems* vol. 83, pp. 458-469, 2016.

[16] Meng. J. Mu. Y. Jia. H. Wu. J. Yu. X. and Qu. B. "Dynamic frequency response from electric vehicles considering travelling behavior in the Great Britain power system," *Applied energy*, vol. 162, pp.966-979, 2016.

[17] Yunfei. M. Jianzhong. W. Ekanayake. J. Jenkins. N. and Jia. "Primary frequency response from electric vehicles in the Great Britain power system." *IEEE Transactions on Smart Grid*, vol. 4, no. 2, pp. 1142-1150, 2012.

[18] Sekyung. H. Han. S. and Sezaki. K. "Estimation of achievable power capacity from plug-in electric vehicles for V2G frequency regulation: Case studies for market participation." *IEEE Transactions on Smart Grid* vol. 2, no. 4, pp. 632-641, 2011.

[19] Saber. A.Y. and Venayagamoorthy. G.K. "Intelligent unit commitment with vehicle-to-grid—A cost-emission optimization." *Journal of Power Sources*, vol. 195 no. 3, pp.898-911, 2010.

[20] Willett. K. and Tomić. J. "Vehicle-to-grid power implementation: From stabilizing the grid to supporting large-scale renewable energy." *Journal of power sources*, vol. 144, no. 1, pp. 280-294, 2005.

[21] Bremermann. M. Leonardo. E. Manuel. Peças A. L. and Rosa. M. "Electric vehicle models for evaluating the security of supply." *Electric Power Systems Research* vol. 111, pp. 32-39, 2014.

[22] Sui. X. Tang. Y. He. H. and Wen. J. "Energy-storage-based low-frequency oscillation damping control using particle swarm optimization and heuristic dynamic programming." *IEEE Transactions on Power Systems*, vol. 29, no. 5, pp. 2539-2548, 2014.

[23] Ramesh. M. Yadav. A. K. "Wind Contributed Load Frequency Control Scheme for Standalone Microgrid Using Grey Wolf Optimization," *IEEE Delhi Section Conference (DELCON)*, NSUT, New Delhi, 2022.

[24] Abazari. A. Monsef. H. and Wu. B. "Coordination strategies of distributed energy resources including FESS, DEG, FC and WTG in load frequency control (LFC) scheme of hybrid isolated micro-grid," *International Journal of Electrical Power & Energy Systems*, vol. 109, pp. 535-547, 2019.

[25] Yadav. A. K. Gaur. P. Jha. S. K. Gupta. J. R. P. and Mittal. A. P. "Optimal Speed Control of Hybrid Electric Vehicles," *Journal of Power Electronics*, 11 (4), 393-400, 2011.

[26] Pathak. P. K. Yadav. A. K. Shastri. A. and Alvi. P. A. "BWOA assisted PIDF-(1+I) controller for intelligent load frequency management of standalone micro-grid," *ISA Transactions*, 2022.

[27] Pathak. P. K. Yadav. A. K. Padmanaban. S. and Kamwa. K. "Fractional Cascade LFC for Distributed Energy Sources via Advanced Optimization Technique Under High Renewable Shares," *IEEE Access*, vol. 10, pp. 92828-92842, 2022.

[28] Hassan. K. M. Niknam. T. and Mokhtar. S. S. "Speed control of electrical vehicles: a time-varying proportional–integral controller-based type-2 fuzzy logic." *IET Science, Measurement & Technology*, vol. 10, no. 3, pp. 185-192, 2016.

[29] Shashi Kant. P. Soumya. R. M. and Kishor. N. "A literature survey on load–frequency control for conventional and distribution generation power systems." *Renewable and Sustainable Energy Reviews* vol. 25, pp. 318-334, 2013.

[30] Ngamroo. and Issarachai. "Application of electrolyzer to alleviate power fluctuation in a stand alone microgrid based on an optimal fuzzy PID control." *International Journal of Electrical Power & Energy Systems*, vol. 43, no. 1, pp. 969-976, 2012.

[31] Yammani. C. and Maheswarapu. S. "Load frequency control of multi-microgrid system considering renewable energy sources using grey wolf optimization," *Smart Sci*, vol. 7, no. 3, pp. 198-217, 2019.

[32] Vachirasricirikul. S. and Ngamroo. I. "Robust LFC in a smart grid with wind power penetration by coordinated V2G control and frequency controller." *IEEE Transactions on Smart Grid,* vol. 5, no. 1, pp. 371-380, 2013.

[33] Izadkhast. S. Pablo. G. G. and Pablo. F. "An aggregate model of plug-in electric vehicles for primary frequency control." *IEEE Transactions on Power Systems* vol. 30, no.3, pp. 1475-1482, 2014.

[34] Yadav. A. K. Pathak. P. K. Sah. S. V. and Gaur. P. "Sliding Mode based Fuzzy Model Reference Adaptive Control Technique for an Unstable System," Journal of The Institution of Engineers: Series B, vol. 100, no. 2, pp. 169-177, 2019.

[35] Ramesh. M. Yadav. A. K. and Pathak. P. K. "Artificial gorilla troops optimizer for frequency regulation of wind contributed microgrid system," *Journal of Computational and Nonlinear Dynamics*, vol. 18, 011005-1, 2023.

[36] Liu. J. "*Sliding mode control using MATLAB*," Academic Press, 2017.

Chapter 6

A Comprehensive Review of Battery Management System for Electric Vehicles

Mufaddal Huzefa Shakir[1] and Akant Kumar Singh[2]
[1]Department of Automobile Engineering, Chandigarh University, Mohali, Punjab, India
[2]Department of Mechanical Engineering, Chandigarh University, Mohali, Punjab, India

Abstract

The development and use of electric vehicles can significantly minimize difficulties with energy deficit and environmental pollution. To achieve the varied battery performance requirements, a good battery management system (BMS) is essential in maintaining the safe and dependable battery operations, which has been extensively utilized in variety of power-intensive applications, for instance such as electric and hybrid electric vehicles. Battery status estimation, cell balance, temperature control, and fault diagnosis are among the BMS's primary duties. This article tries to provide a brief summary of a number of crucial BMS technologies. Before discussing the essential technologies implemented in BMS, a survey of the most common battery types employed in EVs is carried out. Thereafter, a thorough survey of battery condition assessments for the inside temperature, state of charge (SOC), and state of health (SOH) is conducted. Also, a literature survey has been carried out on various works done to improve BMS of electric vehicles.

Keywords: battery types, BMS technologies, BMS architecture, BMS features, battery modelling

In: The Future of Hybrid-Electric Vehicles
Editors: Pawan Kumar Pathak and Anil Kumar Yadav
ISBN: 979-8-89113-515-4

6.1. Introduction

The development of EVs and Hybrid Electric vehicles technology is being advanced day by day. EVs and hybrid electric vehicles are commonly recognized as the most prominent substitutes for traditional IC engine-based vehicles. Electric vehicles (EV) are gaining popularity as a result of their zero emission of dangerous gases and efficient energy use [1]. Batteries are extensively employed as the major power source for EVs and HEVs considering their advantages, including higher energy density, less environmental pollution, and a long lifetime period. Yet, in EV applications, handling batteries requires extra attention. Inappropriate operations like excess voltage and current, or excess charging/discharging pose serious security risks to the batteries, notably hastening the ageing process, also could result in fire or detonation. As a result, the BMS is essential for making sure the operation and safety of batteries. BMS is an essential feature of an electric car since it consists of numerous electric and electronic circuits that measure and achieve efficient results from a battery system [2, 3].

Design and modelling of the battery, inter-state estimates, and charging are the first priority technologies in EV-BMS [4]. In the analysis of battery behavior, battery condition monitoring, real-time controller design, fault diagnostics, and temperature management, an efficient battery model is needed. In addition, some internal battery states, like state of health, state of charge, and inside temperature, are not directly observable but are crucial for controlling the performance of batteries and must be tracked utilizing accurate estimation techniques [5]. Moreover, the charging of a battery is crucial to BMS since it directly affects the battery's availability for service and operational safety. An organized charging approach will guard batteries against deterioration, control temperature swings, and enhance energy conversion effectiveness. The availability of EVs is negatively impacted by slow charging, yet charging too rapidly can lead to substantial energy waste and a temperature increase. A battery's service life will eventually be shortened due to overheating or supercooling, which is caused by large temperature variations [6, 7]. The purpose of this review is to offer a concise overview of the essential BMS technologies, with a focus on types of batteries, BMs features, BMS architecture, battery modelling, and various work done to improve the BMS of electric vehicles.

6.2. Types of Batteries and Important BMS Tecnologies

Many battery types can serve as the power source in electric vehicle applications. The BMS contains a variety of functional modules. This study analyses and summarizes certain common battery types and crucial BMS technologies.

6.2.1. Types of Batteries in EV

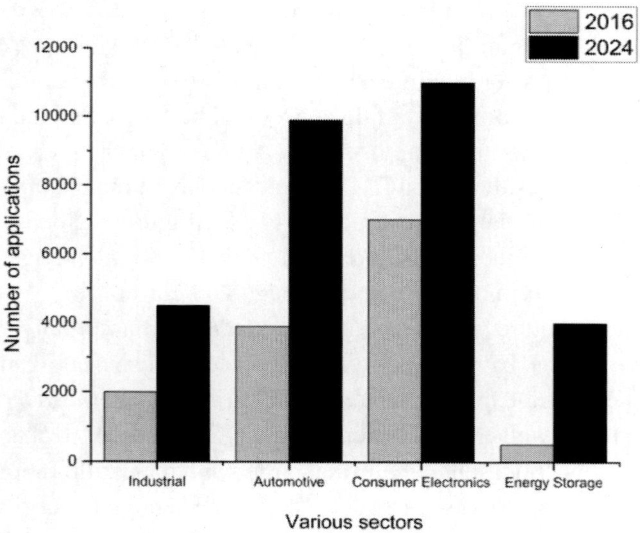

Figure 6.1. Use of lithium-ion batteries in various sectors [9].

According to their ability to be recharged, batteries are typically divided into two categories: primary and secondary batteries. The secondary battery can be recharged after being entirely discharged; however, the primary battery can be used only once after being totally depleted. The secondary battery must have a long-life cycle, little loss of energy, a high-power density, and a sufficient level of safety in order to be used in EV and HEV applications [8]. Lead acid, nickel-metal hydride (Ni-MH), nickel-cadmium (NiCad), and lithium-ion are some common battery types used in EVs. Several of these common battery types have important properties which are shown in Table 6.1 for reference. It has been amply demonstrated that lithium-ion batteries operate a lot better than many kinds of batteries, particularly when it comes to their long cycle life, which is important for the lengthy service life of electric vehicles [9]. Also, Li-ion batteries have a high level of safety and are made of eco-friendly materials without any hazardous gas issues. As a result, lithium-ion batteries have grown to be the most popular kind of EV power [10]. Figure 6.1. depicts the use of lithium-ion batteries in various industrial sectors.

6.2.2. Important BMS Technologies

However, in EV applications, the battery needs special attention. The battery will degrade much more quickly if improper activities are carried out, such as charging or discharging the battery too quickly or too slowly. To match the high voltage and power requirements for automobiles, the pack of batteries of EVs are commonly composed of dozens of cell batteries coupled in sequence or parallel form. The operation of such a complex battery pack also requires special care. As a result, a good BMS is crucial to avoiding battery degradation and must be appropriately constructed [11,12]. Figure 6.2 demonstrates the connection between these critical technologies. On-board voltage and current sensors can directly monitor battery voltage and current in EV applications, whereas temperature sensors may immediately ascertain the surface temperature of the battery pack. Then, to accomplish state estimations of battery SOC or inside temperature, suitable estimation methods can be used in conjunction with well-trained battery models [13]. In order to charge batteries from their initial condition to their final target with the equilibration of various charging objectives, for instance such as high efficiency of energy conversion, fast charging, and a modest rise in temperature, techniques for charging batteries can be improved by suitable optimization algorithms after capturing the thermal and electric behaviors of the battery [14]. The alarm module along with safety control module will cooperate to record and/or eradicate any abnormal battery status conditions that arise during the operating process. Because battery modelling, battery control, and state estimation are essential BMS technologies, these fields of study are flourishing in the context of BMS/EV applications [10].

Table 6.1. Common battery types for EVs [10]

Type of battery	Energy density ($w.kg^{-1}$)	Power density ($w.h.kg^{-1}$)	Nominal voltage (V)	Charging temperature (°C)	Discharging temperature (°C)	Efficiency of charging (%)	Rate of self-discharge (%. $month^{-1}$)	Service life/ cycle
Lithium-ion battery	100-270	250-680	3.2-3.7	0 to 45	-20 to 60	80-90	3-10	600-3000
Lead acid battery	30-50	180	2.0	-20 to 50	-20 to 50	50-95	5	200-300
NiMH battery	60-120	250-1000	1.2	0 to 45	-20 to 65	65	30	300-600
NiCd battery	50-80	150	1.2	0 to 45	-20 to 65	70-90	20	1000

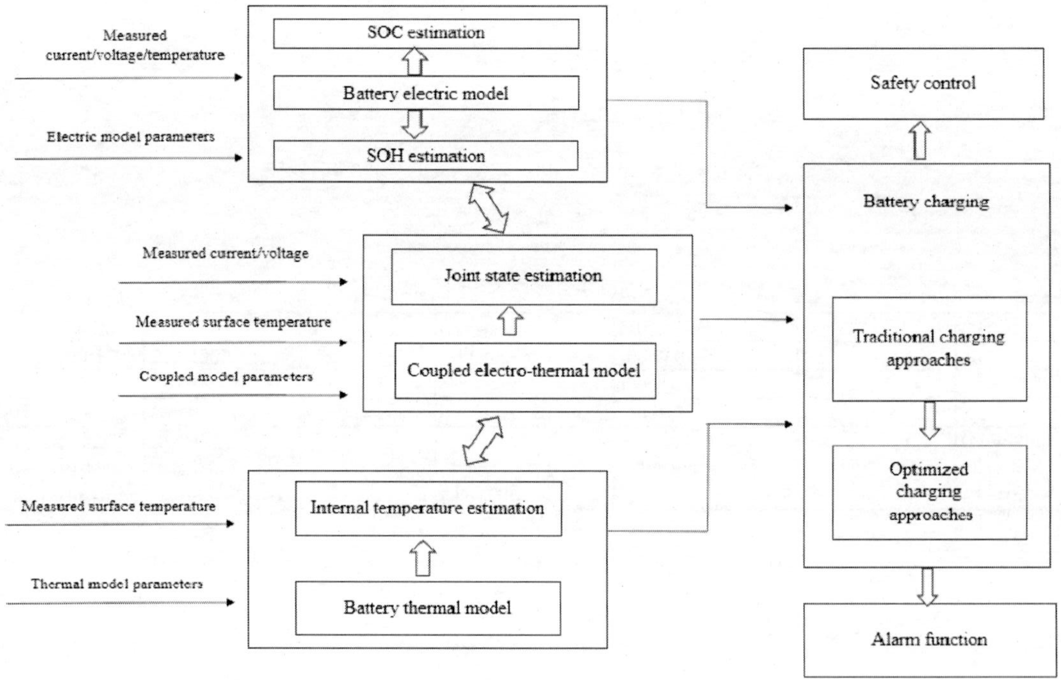

Figure 6.2. The BMS's relationship with critical technologies [10].

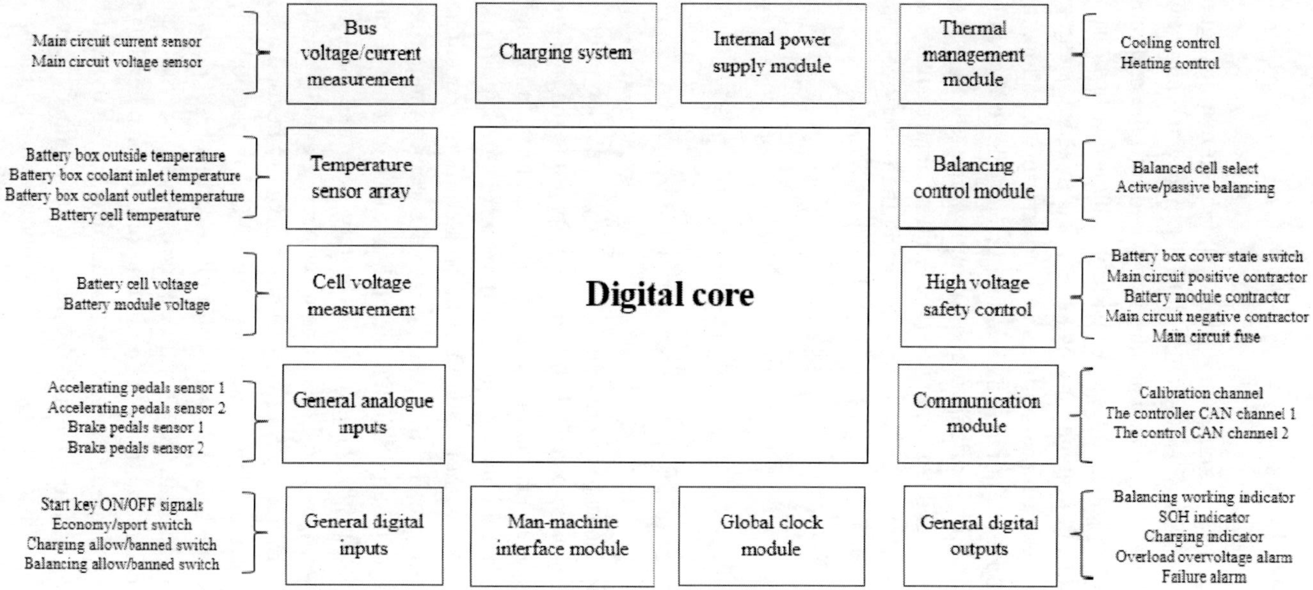

Figure 6.3. Basic BMS hardware and software architecture. Reproduced with permission from [14].

6.2.3. BMS Architecture

The software and hardware architecture of a BMS are shown in Figure 6.3. One can separate the BMS architecture into its hardware and software components. The elements of hardware are integrated into the equipment of electric vehicles to gather different battery characteristics, such as temperature, current, and voltage, among others. To increase accuracy and achieve the desired monitoring capabilities with no error propagation, the data should be simultaneously acquired at a high sample rate. The software element is responsible for tracking variables, controlling warnings and protection, carrying out internal and external communication, calculating various estimations (SOH, SOP, SOS, SOC, etc.), adjusting the battery cell balance, etc. [15, 16].

6.3. BMS Features

6.3.1. State of Charge (SOC)

The SOC of a battery measures its remaining capacity in relation to its overall capacity under similar circumstances. It can be compared to the fuel gauge in cars with IC engines. SOC estimation is critical for monitoring and improving battery performance through charging and discharging management. How much time an EV can go without charging can be identified by accurate SOC estimation [10, 17]. Typically, a battery's usable capacity ranges between 100% and 5% state of charge (SOC), and the amount of time it takes to discharge it from 100% to 5% is known as the battery's runtime. The battery's non-stationary as well as non-linear characteristics gradually reduce its capacity, and the role of the BMS is to maintain the battery beyond a set SOC by avoiding discharge. A direct estimation of the SOC is not possible [18]. The existing approaches make it very difficult to estimate SOC with any degree of accuracy. In general, OCV for the online estimation of SOC has been calculated using direct estimation (Open circuit voltage (or potential)-based, Coulomb counting, etc.), model-based (electrochemical and equivalent circuit, etc.), and data driven-based methodologies. These models' accuracy is affected by parameter adjustment, training, state observers, and different real-world circumstances. However, each of these strategies has a number of drawbacks [19, 20].

6.3.2. State of Health (SOH)

The battery's state of health, or degree of deterioration, is provided by SOH. SOH estimation is difficult to define simply, unlike SOC. It essentially indicates the remaining cycles left until the battery reaches the end of its useful life [21]. Estimating SOH is also necessary for determining how long the battery will still be useful (RUL). For a battery to operate safely and reliably, the BMS needs to be capable of determining the SOH of the battery. SOH estimation is often carried out using model-based, model-free, and data mining methodologies [22].

6.3.3. State of Life (SOL)

SOL identifies the amount of load cycles, etc., that the battery has left before it reaches the end of its usable life, or its remaining useful life (EOL). If the prediction of SOL is correct, especially for military applications, failures can be avoided and battery pack maintenance can be executed more precisely. Moreover, SOL prediction helps BMS save expenses, improve accuracy and dependability, prevent any form of failure, and more [18].

6.3.4. State of Power (SOP)

Another crucial element that the BMS needs to observe is the state of power. Each cell has a unique internal resistance that varies depending on factors including time and usage, temperature, chemical characteristics, and other external factors. The cell's power value is influenced by each of these factors [22].

6.3.5. State of Safety (SOS)

For the safety of the entire system, a BMS must perform routine monitoring. The constant fluctuation in temperature, current, over and undervoltage, overheating, climatic conditions, etc. makes it exceedingly difficult to maintain a satisfactory level of safety. Consequently, a BMS needs to be effective in order to prevent any kind of system failure [23].

6.3.6. Cell Balancing and Cell Monitoring

Cell monitoring is among the BMS's most crucial responsibilities. There could be hundreds of LIBs in a battery pack, and each one might act differently while it is running. Continuous cell status monitoring by BMS will aid in managing, safeguarding, balancing, and controlling activities, among other things [17]. Additionally, it will support communication, fault identification, charging and discharging control, temperature management, and other functions. Similar to this, a BMS's most important function is cell balancing. A BMS must make an effort to keep each cell's state of charge equal [24, 25].

6.4. Battery Modelling

The battery models presented in the various studies are categorized into three types: electrochemical models [26], electrical equivalent circuit models (including fractional-order and integral-order models) [27, 28], and data-driven models developed using machine learning technologies such as support vector machines and neural networks [29].

6.4.1. Electrochemical Models

For electrochemical-based models, the most developed and simpler model is the single particle (SP) model [30]. The amount of lithium-ion distribution in the electrode is represented by a single particle in the SP model. The use of the SP model may be done to study the primary function of the electrodes in addition to the impact of solid-phase diffusion. However, the SP model's accuracy is relatively low. In previous work, the SPMe model was built on top of the SP model. This model takes into account the impact of electrolyte on the voltage output and gives a partial differential equation for liquid electrolyte material and charge conservation, which enhances the precision of simulation. Additionally, they figured out the multi-parameter estimation problem in SPMe and performed sensitivity analysis based on the following parameters [31]. The pseudo-two-dimensional (P2D) model was also presented in a study [32]. The P2D model considers the cell's anode and cathode to be porous electrodes made up of many spherical particles, with the electrolyte filling the spaces between the particles. Some combined partial differential equations

(PDEs) outline the distribution of lithium-ion in the electrolyte liquid phase and the solid phase of the electrode material particles, allowing a thorough investigation of the main and side reaction rates inside the cell. Because the P2D model consists of numerous coupled PDEs, it must be simplified from the standpoint of engineering practice [33].

One of the primary reasons why electrochemical models are tough to implement in real-time situations is the vast amount of unidentified variables that must be found via global optimisation techniques. They will inevitably encounter over-fitting or local optimisation issues. The simulation outcomes of genuine electrochemical-based models are frequently not perfect because of the lack of correct and detailed model parameters [34].

6.4.2. Electrical Equivalent Circuit Models

These models have achieved great popularity because of their simple model structure and easy recognition. Electrical components are used in electrical equivalent circuit models to simulate battery behavior. There are two categories of electrical equivalent circuit models: fractional-order models and integral-order models.

The Rint model is the most basic integral-order equivalent circuit model, consisting of an ideal voltage source in sequence with a resistor [35]. The Rint model has a simple model structure, but the polarisation and diffusion phenomena are not taken into account. In a study, a first-order resistor-capacitor (RC) model capable of replicating lithium-ion battery charge and discharge behavior with a single RC network was proposed [36]. This model can accurately replicate battery behavior, as evidenced by experimental results. In another study, the second-order RC equivalent circuit model was used to analyze the dynamic performance of the battery, and the voltage error was less than 40 mV [37]. The input/output relationship of the cell was easily deduced for integral-order equivalent circuit models, and the models contained minor parameters. As a result, the recursive least-squares method was the most extensively used method for parameter identification. To evaluate battery condition and model parameters, a co-estimator was suggested [38]. A unique parameter identification approach based on electrochemical properties was also proposed [39]. The given technique combined electrochemical attributes and an equivalent circuit to investigate lithium-ion battery performance under various settings, better reflecting the battery's internal mechanism.

A simpler fractional order impedance model was put forward in a study [40]. The model parameters were identified using a genetic method, and the voltage error was less than 0.5%. The non-integer order model derivatives were reported, and the particle swarm optimisation approach was used to determine the parameters of the model [41]. The suggested fractional-order equivalent circuit model performed well in terms of accuracy and resilience. Fractional-order models have been popular in recent years for battery state estimates and reliability analysis. A study estimated the SOC of a lithium-ion battery using a fractional-order equivalent circuit model and a Kalman filter [42, 43]. Another work proposed fractional-order models for a hybrid battery-supercapacitor framework [44].

6.4.3. Data-Driven Models

Data-driven techniques are frequently employed in the process of development of battery models. In a study, a neural network-based thermal-electrical link model was developed. Various battery tests were used to validate the efficacy of the proposed strategy [45]. In another study, the neural network was combined with particle filtering to present a data-driven state estimation technique [46]. A wavelet-neural-network-based data-driven battery model was proposed in a study [47]. The Stacked Denoising Autoencoders technique along with the machine learning algorithm were integrated to construct a large data-driven model of a lithium-ion battery that took temperature into account. Although data-driven approaches perform well on nonlinear problems, they can easily be impacted by training datasets and procedures [48].

6.5. Studies by Various Researchers to Improve BMS

Various works have been done by the researchers to improve the BMS of electric vehicles. The three key features of the BMS that were designed in this study were protection, balancing, and monitoring. An Arduino Nano microcontroller was used to create the BMS. The protection feature could be effective in preventing issues like overcharging, and extreme temperatures. Setting the battery voltage to 4.2 volts worked nicely when balancing [49]. A supervised machine learning approach is used in this study to identify the parameters of an adaptive model of lithium-ion batteries. This flexible battery model was created using a second order equivalent circuit model, which accurately captures the dynamics of Li-ion batteries. Comparative verification experiments demonstrated that the machine learning-based parameter estimator's high accuracy and robustness resulted in a precise model of battery with an error of less than 0.4% [50]. The primary goal of this investigation was to offer a SystemC-based simulation environment for optimizing and designing the Battery Management System, which included a model of a Li-ion battery and a CAN communication interface. Using data and scenarios from the actual world, the BMS had been validated. The created model enabled system exploration and would help with BMS algorithm optimization under various test scenarios [51].

This study used a verification-oriented development methodology to explain the model-based development and verification of an adaptable BMS for Li-ion batteries. The creation of an EKF for the SOC estimate was the main objective to illustrate the process. With the use of MiL, SiL, and HiL simulations, the EKF was created, examined, and optimized. The HiL simulations were carried out using the produced BMS prototype [52]. The creation of an electrochemical model for use in automobile BMS was presented in this work. The electrochemical model offered more accurate state estimations throughout an expanded and prolonged battery use as compared to empirical methods. The electrochemical model was made simple enough in this study to be integrated into the BMS using model implementation approaches. The model discretized a nonlinear system of partial differential equations into a linearized system of algebraic equations. The electrochemical model was subsequently integrated into the BMS and used 14 kB of RAM to predict battery behaviors at intervals of 1 s [53]. In order to manage batteries and power auxiliary loads in electric vehicles, this study provided a central control solution for a combined BMS/DC-DC framework that leverages modular DC/DC bypass converters. To give differential control of battery cells utilizing the LV bus load and provide a chance for highly developed battery management to accomplish an extended battery life and greater power limitations, the BMS and DC/DC were being integrated. It included the suggested control strategy, system modelling, loop gain analysis, and feedback compensator construction. Three isolated dual-active bridge bypass converters and 18 series-connected 25 Ah lithium-ion NMC battery cells made up a prototype system with combined results rated to power a 1.44 kW auxiliary LV load. Experimental findings were reported for this system [54]. Table 6.2. shows the summary of various studies done to improve the BMS.

Table 6.2. Summary of various works done to improve BMS

Work done	Methodology/Technique	Results	References
Monitoring, balancing, and protection were the three primary BMS features that were created in this study.	Arduino Nano Micro controller was used in the designing of BMS.	The test findings revealed that the BMS's performance in monitoring voltage levels had an accuracy of 99.29% and a precision level of 99.74%. Balancing was effective in maintaining the voltage level of the battery at 4.2 volts.	[49]
This research looked at the battery model of a lithium-ion battery, in which the model's parameters were approximated using the supervised machine learning paradigm.	A second-order equivalent circuit model was used to create this adaptable battery model that precisely represented the dynamics of lithium-ion batteries.	The machine learning-based parameter estimator demonstrated high precision and robustness, leading to a precise battery model with less than 0.4% error, according to comparative verification trials.	[50]
The major purpose of this study was to construct and optimise the BMS, which featured a lithium-ion battery model and a CAN communication interface, using a modelling environment based on SystemC.	BMS was designed using SystemC.	The generated model enabled system exploration and would be valuable for optimizing the BMS algorithms under various test situations.	[51]
This work covered the model-based development and validation of a BMS for lithium-ion batteries. As an illustration of the model-based design approach, the advancement of an Extended Kalman Filter (EKF) as a SOC estimator would be illustrated.	A continuously model-based, verification-oriented development technique was used to create BMS.	The battery model would be extended in the future to show temperature dependence. As a result, functions such as temperature monitoring could be designed and tested, and the EKF could operate across a wider temperature range. Furthermore, more in-depth tests of the EKF in real-world settings were planned.	[52]
This study described the creation of an electrochemical model for automotive applications.	The model discretizes a nonlinear system of partial differential equations into a linearized set of algebraic equations (AEs). A solver chosen to assess the resulting AE system was adapted for use with the BMS.	The electrochemical model was then incorporated into the BMS, forecasting battery behavior in 1 second intervals while utilising 14 kB of RAM.	[53]

Work done	Methodology/Technique	Results	References
This study described a centralized, shared control strategy for a combined BMS/DC-DC framework that employed DC/DC bypass converters to manage batteries and power auxiliary loads in electric vehicles.	The control strategy included modular DC-DC bypass converters that enforced continuous cell balancing by controlling the condition of each battery cell differentially via LV DC bus load distribution.	An evaluation setup comprised of eighteen cells of battery and three separate DC-DC converters was built. To effectively illustrate the reliability of the suggested control technique, transient and steady-state experiments were undertaken.	[54]
This study presented an electrochemical model of lithium-ion batteries that was control oriented and its real-time execution in an electric vehicle's BMS.	By streamlining partial differential equations with the Pade approximation, Laplace transform, and other techniques, the enhanced P2D electrochemical model was generated to provide the transfer function between current and cell voltage.	To validate its efficiency in a real-time scenario, the battery model was implemented in the chosen algorithm system in an EV's BMS. In real-world city driving situations at a temperature of 13°C, the RMS error of the online predicted voltage was lower than 15.8 mV, displaying the practicality of the established electrochemical model for further applications of electric vehicles.	[55]
This study proposed an ISO26262 certified safety-oriented concept and assessment approach of BMS. The safety concerns about lithium-ion batteries were researched to highlight the risks of employing batteries in the automotive sector.	First, a thorough examination of the inherent dangers of lithium-ion batteries was conducted. Second, a risk and danger evaluation of lithium batteries was performed to identify the unique dangers arising from the automotive applications as well as the safety standards that must be met in order to keep them under control.	For the entire system, a safety architecture was developed that included a three-subsystem topology comprised of 16 BMS slaves to observe high voltage batteries, a BMS Master, and a power monitoring and disconnection unit. The most significant criteria, as well as preventive and mitigating strategies, and safety methods for meeting ASIL C criteria for upgraded BMS slaves, were provided.	[56]
To accomplish smooth prediction of SOC, this work provided the square root cubature Kalman filter using a temperature correction technique built on the BMS of a common onboard embedded microcontrol unit.	Temperature correction rule.	The outcome of several experiments showed that the suggested technique with decreased computing complexity converges quickly (in around 2.5 s) and accurately calculates the State of Charge of the battery within dynamic temperature conditions.	[57]

Conclusion

This review has covered a wide range of BMS-associated topics, such as various battery types used in EVs, crucial BMS technologies, BMS architecture, BMS features, and battery modelling. Furthermore, a literature survey has been carried out on various works done by researchers to improve the BMS of EVs. Finally, the BMS is critical to the operation and durability of electric vehicles (EVs). The BMS, being the heart of the EV's power storage and distribution system, ensures the battery pack's safety, efficiency, and overall performance. BMS systems continue to evolve as technology advances, becoming more sophisticated and adaptive. Future BMS designs that incorporate machine learning algorithms and artificial intelligence hold the promise of even improved efficiency, dependability, and performance for electric vehicles. As technology advances, we may expect more novel BMS solutions to drive the sustainable and electric future of transportation.

References

[1] Anand S, Kulkarni SV, Gupta S, Parthasarathy V. Battery Management System (BMS) for Electric Vehicle Applications. In *2022 Fourth International Conference on Emerging Research in Electronics, Computer Science and Technology (ICERECT)* 2022 Dec 26 (pp. 1-6). IEEE.

[2] Spoorthi B, Pradeepa P. Review on Battery Management System in EV. In2022 *International Conference on Intelligent Controller and Computing for Smart Power (ICICCSP)* 2022 Jul 21 (pp. 1-4). IEEE.

[3] Kosuru VS, Kavasseri Venkitaraman A. A Smart Battery Management System for Electric Vehicles Using Deep Learning-Based Sensor Fault Detection. *World Electric Vehicle Journal*. 2023 Apr 10;14(4):101.

[4] Pakdel R, Yavarinasab M, Zibad MR, Almohaddesn MR. Design and Implementation of Lithium Battery Management System for Electric Vehicles. In2022 9th *Iranian Conference on Renewable Energy & Distributed Generation (ICREDG)* 2022 Feb 23 (pp. 1-6). IEEE.

[5] Habib AA, Hasan MK, Issa GF, Singh D, Islam S, Ghazal TM. Lithium-Ion Battery Management System for Electric Vehicles: Constraints, Challenges, and Recommendations. *Batteries*. 2023 Feb 27;9(3):152.

[6] Li H, Kaleem MB, Chiu IJ, Gao D, Peng J, Huang Z. An intelligent digital twin model for the battery management systems of electric vehicles. *International Journal of Green Energy*. 2023 Apr 7:1-5.

[7] Wang H, Pourmousavi SA, Soong WL, Zhang X, Ertugrul N. Battery and energy management system for vanadium redox flow battery: A critical review and recommendations. *Journal of Energy Storage*. 2023 Feb 1;58:106384.

[8] Liu W, Placke T, Chau KT. Overview of batteries and battery management for electric vehicles. *Energy Reports*. 2022 Nov 1;8:4058-84.

[9] Mohammadi F, Saif M. A comprehensive overview of electric vehicle batteries market. *e-Prime-Advances in Electrical Engineering, Electronics and Energy*. 2023 Feb 11:100127.

[10] Liu K, Li K, Peng Q, Zhang C. A brief review on key technologies in the battery management system of electric vehicles. *Frontiers of mechanical engineering*. 2019 Mar;14:47-64.

[11] Yang S, Liu X, Li S, Zhang C. Key Technologies of BMS. *Advanced Battery Management System for Electric Vehicles* 2022 Sep 20 (pp. 33-51). Singapore: Springer Nature Singapore.

[12] See KW, Wang G, Zhang Y, Wang Y, Meng L, Gu X, Zhang N, Lim KC, Zhao L, Xie B. Critical review and functional safety of a battery management system for large-scale lithium-ion battery pack technologies. *International Journal of Coal Science & Technology*. 2022 Dec;9(1):36.

[13] Li J, Zhou S, Han Y, editors. *Advances in battery manufacturing, service, and management systems*. John Wiley & Sons; 2016 Oct 24.

[14] Lu L, Han X, Li J, Hua J, Ouyang M. A review on the key issues for lithium-ion battery management in electric vehicles. *Journal of power sources*. 2013 Mar 15;226:272-88.

[15] Ashok B, Kannan C, Mason B, Ashok SD, Indragandhi V, Patel D, Wagh AS, Jain A, Kavitha C. Towards safer and smarter design for lithium-ion-battery-powered electric vehicles: A comprehensive review on control strategy architecture of battery management system. *Energies*. 2022 Jun 8;15(12):4227.

[16] Omariba ZB, Zhang L, Sun D. Review on health management system for lithium-ion batteries of electric vehicles. *Electronics*. 2018 May 15;7(5):72.

[17] Li L, Li Z, Zhao J, Guo W. Lithium-ion battery management system for electric vehicles. *International Journal of Performability Engineering*. 2018 Dec 1;14(12): 3184.

[18] Hannan MA, Hoque MM, Hussain A, Yusof Y, Ker PJ. State-of-the-art and energy management system of lithium-ion batteries in electric vehicle applications: Issues and recommendations. *IEEE Access*. 2018 Mar 21;6:19362-78.

[19] Vaideeswaran V, Bhuvanesh S, Devasena M. Battery management systems for electric vehicles using lithium ion batteries. 2019 *Innovations in Power and Advanced Computing Technologies (i-PACT)*. 2019 Mar 22;1:1-9.

[20] Pattipati B, Sankavaram C, Pattipati K. System identification and estimation framework for pivotal automotive battery management system characteristics. *IEEE Transactions on Systems, Man, and Cybernetics, Part C (Applications and Reviews)*. 2011 Jan 6;41(6):869-84.

[21] Abbas M, Kim ES, Kim SK, Kim YS. Comparative analysis of battery behavior with different modes of discharge for optimal capacity sizing and BMS operation. *Energies*. 2016 Oct 12;9(10):812.

[22] Li P, Zhang Z, Grosu R, Deng Z, Hou J, Rong Y, Wu R. An end-to-end neural network framework for state-of-health estimation and remaining useful life

prediction of electric vehicle lithium batteries. *Renewable and Sustainable Energy Reviews.* 2022 Mar 1;156:111843.

[23] Uzair M, Abbas G, Hosain S. Characteristics of battery management systems of electric vehicles with consideration of the active and passive cell balancing process. *World Electric Vehicle Journal*. 2021 Aug 13;12(3):120.

[24] Hemavathi S. Overview of cell balancing methods for Li-ion battery technology. *Energy storage*. 2020 Aug;203:1-2.

[25] Naguib M, Kollmeyer P, Emadi A. Lithium-ion battery pack robust state of charge estimation, cell inconsistency, and balancing. *IEEE Access*. 2021 Mar 24;9:50570-82.

[26] Zhang Q, Wang D, Yang B, Cui X, Li X. Electrochemical model of lithium-ion battery for wide frequency range applications. *Electrochimica Acta*. 2020 May 20;343:136094.

[27] Nejad S, Gladwin DT, Stone DA. A systematic review of lumped-parameter equivalent circuit models for real-time estimation of lithium-ion battery states. *Journal of Power Sources.* 2016 Jun 1;316:183-96.

[28] Seaman A, Dao TS, McPhee J. A survey of mathematics-based equivalent-circuit and electrochemical battery models for hybrid and electric vehicle simulation. *Journal of Power Sources*. 2014 Jun 15;256:410-23.

[29] Wang Y, Yang D, Zhang X, Chen Z. Probability based remaining capacity estimation using data-driven and neural network model. *Journal of Power Sources*. 2016 May 31;315:199-208.

[30] Romero-Becerril A, Alvarez-Icaza L. Comparison of discretization methods applied to the single-particle model of lithium-ion batteries. *Journal of Power Sources.* 2011 Dec 1;196(23):10267-79.

[31] Grandjean TR, Li L, Odio MX, Widanage WD. Global sensitivity analysis of the single particle lithium-ion battery model with electrolyte. In2019 IEEE *Vehicle Power and Propulsion Conference (VPPC)* 2019 Oct 14 (pp. 1-7). IEEE.

[32] Doyle M, Fuller TF, Newman J. Modeling of galvanostatic charge and discharge of the lithium/polymer/insertion cell (vol 140, pg 1526, 1993). *Journal of the Electrochemical Society*. 2018;165(11):X13-.

[33] Dao TS, Vyasarayani CP, McPhee J. Simplification and order reduction of lithium-ion battery model based on porous-electrode theory. *Journal of Power Sources.* 2012 Jan 15;198:329-37.

[34] Han X, Ouyang M, Lu L, Li J. Simplification of physics-based electrochemical model for lithium ion battery on electric vehicle. Part II: Pseudo-two-dimensional model simplification and state of charge estimation. *Journal of Power Sources.* 2015 Mar 15;278:814-25.

[35] Johnson VH. Battery performance models in ADVISOR. *Journal of power sources*. 2002 Aug 22;110(2):321-9.

[36] Liaw BY, Nagasubramanian G, Jungst RG, Doughty DH. Modeling of lithium ion cells—A simple equivalent-circuit model approach. *Solid State Ionics*. 2004 Nov 30;175(1-4):835-9.

[37] Xia B, Sun Z, Zhang R, Lao Z. A cubature particle filter algorithm to estimate the state of the charge of lithium-ion batteries based on a second-order equivalent circuit model. *Energies*. 2017 Apr 1;10(4):457.

[38] Wang Y, Liu C, Pan R, Chen Z. Modeling and state-of-charge prediction of lithium-ion battery and ultracapacitor hybrids with a co-estimator. *Energy*. 2017 Feb 15;121:739-50.

[39] Zhang X, Lu J, Yuan S, Yang J, Zhou X. A novel method for identification of lithium-ion battery equivalent circuit model parameters considering electrochemical properties. *Journal of Power Sources*. 2017 Mar 31;345:21-9.

[40] Yang Q, Xu J, Cao B, Li X. A simplified fractional order impedance model and parameter identification method for lithium-ion batteries. *PLOS One*. 2017 Feb 17;12(2):e0172424.

[41] Zou Y, Li SE, Shao B, Wang B. State-space model with non-integer order derivatives for lithium-ion battery. *Applied Energy*. 2016 Jan 1;161:330-6.

[42] Mu H, Xiong R, Zheng H, Chang Y, Chen Z. A novel fractional order model based state-of-charge estimation method for lithium-ion battery. *Applied Energy*. 2017 Dec 1;207:384-93.

[43] Xiong R, Tian J, Shen W, Sun F. A novel fractional order model for state of charge estimation in lithium ion batteries. *IEEE Transactions on Vehicular Technology*. 2018 Nov 7;68(5):4130-9.

[44] Wang Y, Gao G, Li X, Chen Z. A fractional-order model-based state estimation approach for lithium-ion battery and ultra-capacitor hybrid power source system considering load trajectory. *Journal of Power Sources*. 2020 Feb 15;449:227543.

[45] Wang QK, He YJ, Shen JN, Ma ZF, Zhong GB. A unified modeling framework for lithium-ion batteries: An artificial neural network based thermal coupled equivalent circuit model approach. *Energy*. 2017 Nov 1;138:118-32.

[46] Zhang C, Zhu Y, Dong G, Wei J. Data-driven lithium-ion battery states estimation using neural networks and particle filtering. *International Journal of Energy Research*. 2019 Nov;43(14):8230-41.

[47] Dong G, Zhang X, Zhang C, Chen Z. A method for state of energy estimation of lithium-ion batteries based on neural network model. *Energy*. 2015 Oct 1;90:879-88.

[48] Li S, He H, Li J. Big data driven lithium-ion battery modeling method based on SDAE-ELM algorithm and data pre-processing technology. *Applied Energy*. 2019 May 15;242:1259-73.

[49] Nizam M, Maghfiroh H, Rosadi RA, Kusumaputri KD. Battery management system design (BMS) for lithium ion batteries. *AIP Conference Proceedings* 2020 Apr 13 (Vol. 2217, No. 1). AIP Publishing.

[50] Hashemi SR, Bahadoran Baghbadorani A, Esmaeeli R, Mahajan A, Farhad S. Machine learning-based model for lithium-ion batteries in BMS of electric/hybrid electric aircraft. *International Journal of Energy Research*. 2021 Mar 25;45(4): 5747-65.

[51] Buccolini L, Ricci A, Scavongelli C, DeMaso-Gentile G, Orcioni S, Conti M. Battery Management System (BMS) simulation environment for electric vehicles.

In 2016 IEEE 16th *International Conference on Environment and Electrical Engineering (EEEIC)* 2016 Jun 7 (pp. 1-6). IEEE.

[52] Liu-Henke X, Scherler S, Jacobitz S. Verification oriented development of a scalable battery management system for lithium-ion batteries. In2017 Twelfth *International Conference on Ecological Vehicles and Renewable Energies (EVER)* 2017 Apr 11 (pp. 1-7). IEEE.

[53] Sung W, Shin CB. Electrochemical model of a lithium-ion battery implemented into an automotive battery management system. *Computers & Chemical Engineering*. 2015 May 8;76:87-97.

[54] Rehman MM, Zhang F, Zane R, Maksimovic D. Design and control of an integrated BMS/DC-DC system for electric vehicles. In2016 IEEE 17th Workshop on *Control and Modeling for Power Electronics (COMPEL)* 2016 Jun 27 (pp. 1-7). IEEE.

[55] Gao Y, Zhu C, Zhang X, Guo B. Implementation and evaluation of a practical electrochemical-thermal model of lithium-ion batteries for EV battery management system. *Energy*. 2021 Apr 15;221:119688.

[56] Marcos D, Garmendia M, Crego J, Cortajarena JA. Functional Safety BMS Design Methodology for Automotive Lithium-Based Batteries. *Energies*. 2021 Oct 21;14(21):6942.

[57] Cui X, He Z, Li E, Cheng A, Luo M, Guo Y. State-of-charge estimation of power lithium-ion batteries based on an embedded micro control unit using a square root cubature Kalman filter at various ambient temperatures. *International Journal of Energy Research*. 2019 Jun 25;43(8):3561-77.

Chapter 7

Development of Effective Charging Station for EVs Using Multiport Converter and Photovoltaic Cell Integration

Parimalasundar Ezhilvannan[1,*]
Suresh Krishnan[2]
Kumar Reddy Cheepati[3]
and Busireddy Hemanth Kumar[1]

[1]Department of Electrical and Electronics Engineering, Mohan Babu University (Erst while Sree Vidyanikethan Engineering College), Tirupati, Andhrapradesh
[2]Department of Electrical and Electronics Engineering, Christ University, Bangalore, Karnataka
[3]Department of Electrical and Electronics Engineering, K. S. R. M. College of Engineering, Kadapa, Andhrapradesh

Abstract

The adoption of electric vehicles (EVs) is increasing rapidly as a response to the urgent need to reduce greenhouse gas emissions and dependence on fossil fuels. However, the deployment of a reliable and cost-effective EV charging infrastructure remains a challenge. To address this challenge, this paper proposes the development of an effective charging station using multiport converter and photovoltaic cell integration. A promising approach in the transportation sector is an efficient charging station with multiport converter and photovoltaic cell integration. A multiport converter, photovoltaic cells, a battery energy storage system, and an electric vehicle charging port constitute the

[*] Corresponding Author's Email: parimalpsg@gmail.com.

In: The Future of Hybrid-Electric Vehicles
Editors: Pawan Kumar Pathak and Anil Kumar Yadav
ISBN: 979-8-89113-515-4

proposed charging station. The power flow between the photovoltaic cells, the battery energy storage system, and the electric vehicle charging port is controlled by the multiport converter. The excess energy generated by the photovoltaic cells is stored in the battery energy storage system. The proposed charging station is designed and analysed using MATLAB/Simulink. The simulation results show that the proposed charging station is capable of providing an electric vehicle a reliable and consistent power supply. In contrast to conventional charging stations, the proposed charging station is also capable of offering a faster charging rate.

Keywords: boost converter, grid, inverter, MPPT, state of charge

7.1. Introduction

Different charging solutions have been developed as a result of the rise in electric vehicles (EVs) on the road in order to meet the rising demand. The traditional charging techniques include power grid-connected charging stations. However, as the demand for electricity rises, the reliance on the power grid for charging cannot last in the long run. As a result, research has concentrated on creating EV charging stations that use renewable energy sources like solar power to generate electricity. The development of an efficient EV charging station using a multiport converter and photovoltaic (PV) cell integration will be covered in this literature review. Because of the demand for environmentally friendly transportation options over the past ten years, electric vehicles have grown in popularity. EVs are battery-powered and require frequent charging. Since EV owners need to charge their vehicles when they are not in use, charging stations have become a necessity. The conventional power grid-connected charging stations have a number of disadvantages, including high electricity costs, a lack of availability, and a reliance on non-renewable energy sources. As a result, experts have been working to create EV charging stations that use renewable energy sources like solar power to generate electricity [1–4].

PV cells are semiconductor technologies that use sunlight to generate electricity. Due to their favourable effects on the environment, PV cells are an excellent source of renewable energy and have grown in popularity over time. PV cells are frequently used in residential and commercial settings, and their incorporation into EV charging stations can offer a cost- and environmentally-conscious charging option. Multiport converters are electronic components that can connect a load to a number of power sources. Due to their ability to combine various renewable energy sources into a single system, multiport converters are frequently used in renewable energy systems. Additionally, multiport converters can offer bidirectional power flow, which is necessary in EV charging stations [5–8]. To power the EV charging station, the charging station integrates a PV array with a multiport converter. The ability of the proposed system to regulate power flow between the PV array and the EV charging station in accordance with the available solar irradiance and EV demand is its key feature. According to the simulation results, the proposed system can produce power with higher quality and greater efficiency. The proposed system aims to decrease the size of the hybrid energy storage system (HESS), lengthen the battery life, and enhance the power quality of the EV charging station [9-12].

A solar-powered EV charging station using a cascaded H-bridge converter (CHBC) has been suggested in a research survey [9]. To power the EV charging station, the proposed system combines a PV array, a DC-DC converter, and a CHBC. The primary benefit of the proposed system is its capacity to produce a sinusoidal output voltage with lower harmonic distortion, thereby enhancing the charging station's power quality. A grid-connected EV charging station using a multiport converter coupled with a flywheel energy storage system (FESS) was suggested by another research study [10]. The proposed system aims to lengthen the battery life, decrease the size and price of the FESS, and enhance the charging station's power quality. The proposed system is intended to lengthen the battery life, decrease the size and cost of the BESS, and enhance the charging station's power quality [11–13].

PV cells were used at the charging station to produce electricity for recharging EVs. The charging station was a three-port converter that combined battery storage, grid power, and photovoltaic cells into one unit. The power flow at the charging station was bidirectional, allowing the battery to be charged from the grid, PV cells, or an EV. The investigation came to find that the charging station was successful in offering an environmentally friendly and economical solution for charging EVs [14–17]. PV cells were used in the charging station to produce electricity, and a multiport converter was used to connect the PV cells to the EV. The power flow at the charging station was bidirectional, allowing the battery to be charged from the grid, PV cells, or an EV. According to the study's findings, PV-based charging stations are an efficient way to charge EVs that is both affordable and environmentally friendly [18–20].

In EV charging stations, multiport converters and photovoltaic cells can be integrated to increase charging efficiency and make the process more environmentally friendly. Solar energy can be used to power the proposed system, and any extra energy can be stored in batteries or gave back to the grid [21–23]. For the integration of photovoltaic cells and multiport converters in EV charging stations, a novel topology has been proposed. The proposed topology regulates the power flow between the grid, photovoltaic cells, and the EV battery using a cascaded H-bridge (CHB) multiport converter. According to the study's findings, the suggested topology can increase the charging process's effectiveness [24–26]. A multiport converter with a photovoltaic cell, energy storage system, and grid connection made up the proposed charging station. The energy storage system was used to store extra energy for later use, while the photovoltaic cell was used to produce electricity from sunlight. Fast charging for EVs was built into the charging station, and the multiport converter was used to control the power flow between the various sources [27–29].

Additionally, a multiport converter-based charging station for electric vehicles with a photovoltaic cell and battery storage system has been suggested. A bidirectional DC-DC converter, a bidirectional AC-DC converter, and a battery energy storage system made up the proposed charging station. The battery storage system was used to store extra energy while the photovoltaic cell was used to produce electricity from sunlight. The charging station was created to offer quick and effective charging to EVs, and the multiport converter was used to manage the power flow between the various sources [30–32]. The development of a successful charging station must also include the integration of photovoltaic cells. A sustainable energy source, photovoltaic cells can significantly lessen the charging station's carbon footprint when they are integrated into the system. The high cost of photovoltaic cells as well as the intermittent nature of solar energy present additional difficulties for the integration of photovoltaic cells with the charging station [33–35].

Developing an effective charging station for EVs using multiport converter and photovoltaic cell integration can face several challenges. The power output of a photovoltaic cell is limited, which means that the charging station may not be able to meet the demand of high-capacity EVs. The power output of photovoltaic cells depends on weather conditions such as sunlight intensity, cloud cover, and temperature. These variations can impact the charging efficiency of the station. A PV charging station's grid integration can be complicated by problems with power quality, grid stability, and compliance with regulations. EVs from different manufacturers may have different charging protocols, which can create compatibility issues with the charging station.

To overcome the various drawbacks, in this paper the various objectives for developing an effective charging station for EVs using multiport converter and photovoltaic cell integration:

- To design a multiport converter that can efficiently convert the DC power output of photovoltaic cells into the AC power required for charging electric vehicles.
- To integrate photovoltaic cells into the charging station to provide a renewable and sustainable energy source for charging EVs.
- To optimize the performance of the multiport converter to achieve high efficiency and power density.
- To comply with relevant regulations and standards for EV charging infrastructure.
- To conduct extensive testing and validation to ensure the effectiveness and reliability of the charging station under various operating conditions.
- To reduce the overall cost of the charging station by utilising cost-effective components and to reduce the charging station's environmental impact.

7.2. Multiport Converter Design

Multiport converters for EVs allow for efficient and flexible energy flow between different energy storage systems and power sources, such as the battery, the grid, and EVs. The design of these converters involves selecting appropriate topologies, control strategies, and power semiconductor devices. Commonly used topologies include bidirectional buck-boost, bidirectional flyback, and bidirectional dual active bridge converters [2–5]. Control strategies such as voltage or current control and maximum power point tracking can improve the efficiency and stability of the system. Silicon carbide and gallium nitride power devices can also be used to increase efficiency and reduce losses. Overall, designing efficient and reliable multiport converters is crucial for maximizing the performance and range of EVs. The following control modes are used in EV charging stations: PV to EV, PV to HES, HESS to EV, AC grid to EV, and PV to AC grid. Figure 7.1 illustrates the multiport converter system of proposed system.

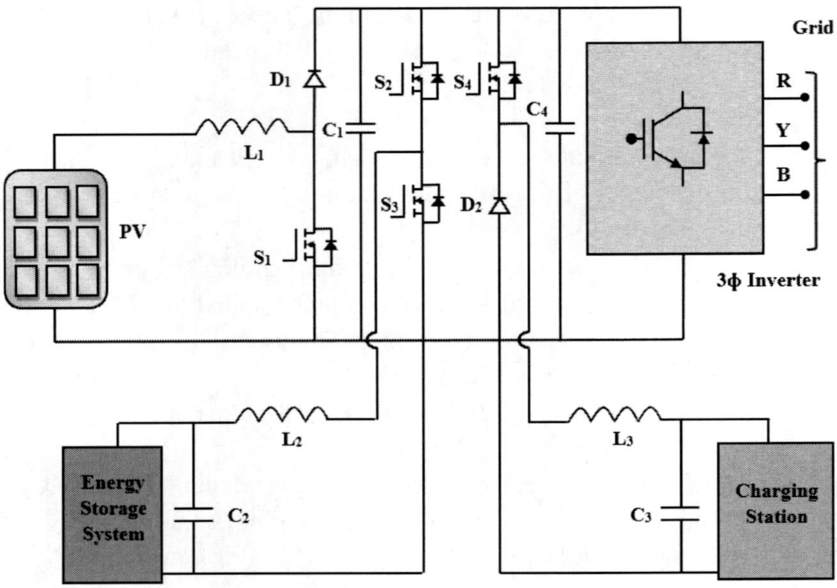

Figure 7.1. Mutiport converter system.

7.2.1. PV to EV Integration

The DC voltage output from the PV panels is converted using a boost converter to the necessary voltage level for charging the battery of an electric vehicle. A switching transistor is used in the boost converter, a DC-DC converter, to regulate the current that passes through an inductor and a capacitor to raise the output voltage. The boost converter boosts the voltage and sends it to the charging station when the PV panels produce energy. The EV battery is then charged by the charging station after converting the DC voltage to AC power. The boost converter needs to be designed to match the output voltage of the PV panels and the charging voltage of the EV battery to optimise system performance. To make certain which the PV system operates at its peak efficiency and generates the most energy possible, the converter should also have features like maximum power point tracking (MPPT) [27].

The EV charging station is shown in Figure 7.2 as receiving regulated voltage based on the boost converter's duty cycle.

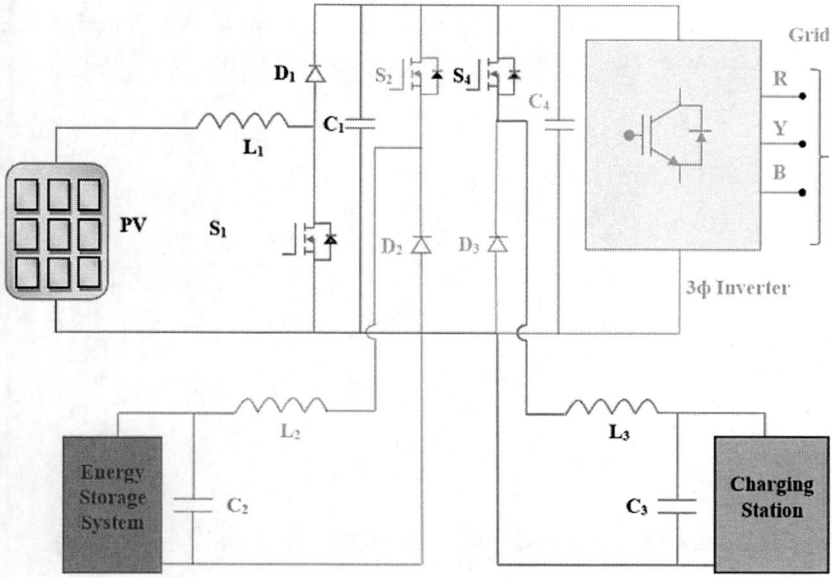

Figure 7.2. Charging mode of PV to Charging station.

The current equation of PV with capacitor (C_1) is expressed in eq. (7.1),

$$i_{PV} = C_1 \frac{dV_{C1}}{dt} + i_{EV} \tag{7.1}$$

The capacitor (C_2) with charging station inference can be illustrated in eq. (7.2),

$$C_2 \frac{dV_{C2}}{dt} = \frac{V_{Battery} - V_{C2}}{R_{Battery}} - iL_2 \tag{7.2}$$

The current of EV charging station can be illustrated in eq. (7.3),

$$i_{EV} = C_3 \frac{dV_{C3}}{dt} + \frac{V_{EV}}{R_{EV}} \tag{7.3}$$

The voltage between the PV and charging station with the interference of duty cycle can be illustrated in eq. (7.4), (7.5), and (7.6) are following,

$$V_{C1} - V_{C3} = L_3 \frac{iL_3}{dt} \tag{7.4}$$

$$L_2 \frac{di_{L2}}{dt} = -V_{C2} \tag{7.5}$$

$$\frac{V_{DC}}{V_{PV}} = \frac{1}{1 - D_{PV}} \tag{7.6}$$

7.2.2. Energy Storage System to Charging Station

Integrating an energy storage system (ESS) with an EV charging station can be achieved using an inductor and a capacitor. The inductor and capacitor are components of a DC-DC converter that converts the DC voltage output from the ESS into the required voltage level for charging the EV battery. The inductor stores energy when a voltage is applied to it, and the capacitor stores energy in an electric field. When the voltage input is turned off, the stored energy in the inductor is released, charging the capacitor. A consistent output voltage is produced for the charging station by repeating this process. The output voltage of the ESS and the charging voltage of the EV battery have been matched by the DC-DC converter's design. In order to ensure that the ESS operates with maximum efficiency and generates the most energy possible, the converter should also have features like MPPT [32]. Figure 7.3 illustrates the charging mode operation of energy storage system to charging station.

The current equation of PV with capacitor (C_1) is expressed in eq. (7.7),

$$i_{PV} = C_1 \frac{dV_{C1}}{dt} \tag{7.7}$$

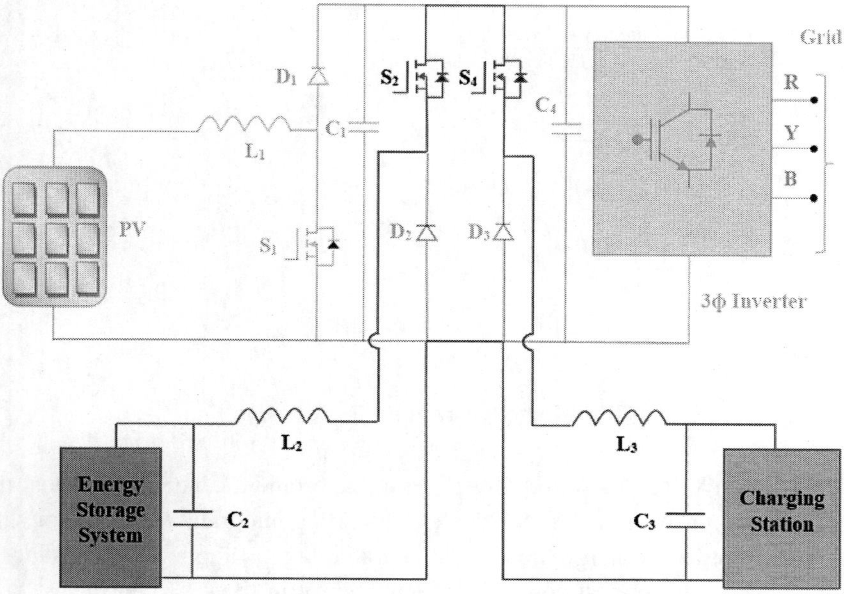

Figure 7.3. Charging mode operation of energy storage system to charging station.

Inductor (L_2) with charging station inference can be illustrated in eq. (7.8),

$$L_2 \frac{di_{L2}}{dt} = V_{DC} - V_{C2} \tag{7.8}$$

The voltage of EV charging station with battery parameters can be illustrated in eq. (7.9),

$$V_{DC} - V_{C3} = L_3 \frac{iL_3}{dt} \tag{7.9}$$

The current between the EVcharging station and energy storage system with the interference of duty cycle can be illustrated in eq. (7.10), (7.11) and (7.12) are following,

$$C_2 \frac{dV_{C2}}{dt} = \frac{V_{Battery} - V_{C2}}{R_b} - iL_2 \tag{7.10}$$

$$i_{EV} = C_3 \frac{dV_{C3}}{dt} + \frac{V_{EV}}{R_{EV}} \tag{7.11}$$

$$\frac{V_{DC}}{V_{Battery}} = \frac{1}{1 - D_{b1}} \tag{7.12}$$

7.2.3. PV to Energy Storage System

A boost converter that makes use of an inductor and a capacitor to boost the voltage output from the PV panels is necessary to integrate a PV system with an ESS in order to charge the ESS. When a voltage is applied to the inductor and capacitor, the boost converter stores energy in them. When the voltage input is turned off, the energy is released. The ESS is charged by this process, which generates an output voltage that is higher than the input voltage from the PV panels [18]. To maximize system performance, the boost converter should be designed to match the output voltage of the PV panels and the charging voltage of the ESS. The converter should also include features such as MPPT to ensure that the PV system operates at its maximum efficiency and produces the maximum amount of energy possible. Overall, integrating a PV system with an ESS using an inductor and capacitor boost converter can help to increase the efficiency of the energy storage process and reduce reliance on the grid. Figure 7.4 illustrates the mode of operation in PV integrated with energy storage system.

The current equation of PV with capacitor (C_1) is expressed in eq. (7.13),

$$i_{PV} = C_1 \frac{dV_{C1}}{dt} - i_{L2} \tag{7.13}$$

Inductor (L_2) with energy storage system inference can be illustrated in eq. (14) and (7.15),

$$L_2 \frac{di_{L2}}{dt} = V_{C1} + V_{DC} - V_{C2} \tag{7.14}$$

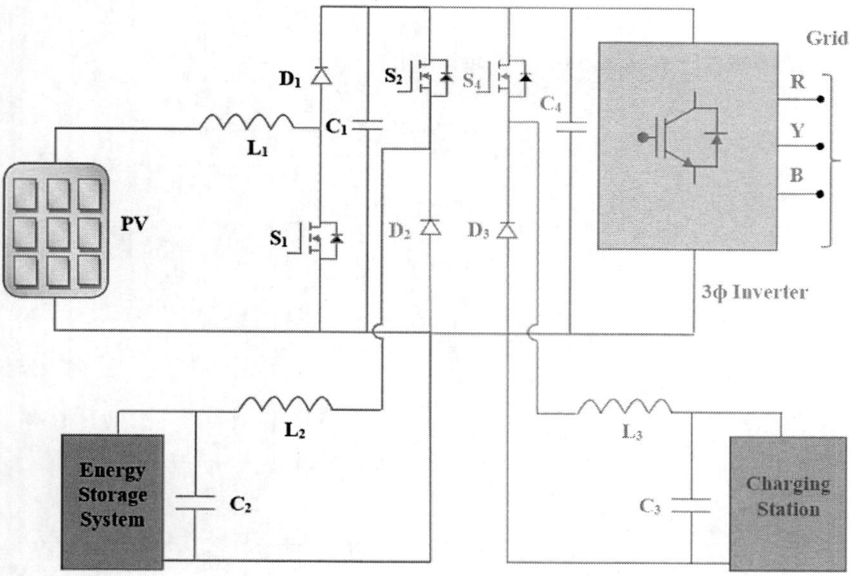

Figure 7.4. Charging mode operation of PV to energy storage system.

$$L_3 \frac{di_{L3}}{dt} = V_{DC} - V_{C3} \tag{7.15}$$

The current between the PV and energy storage system with the interference of duty cycle can be illustrated in eq. (7.16), (7.17) and (7.18) are following,

$$C_2 \frac{dV_{C2}}{dt} = \frac{V_{Battery} - V_{C2}}{R_b} - iL_2 \tag{7.16}$$

$$i_{EV} = C_3 \frac{dV_{C3}}{dt} + \frac{V_{EV}}{R_{EV}} \tag{7.17}$$

$$\frac{V_{Battery}}{V_{DC}} = D_{b2} \tag{7.18}$$

7.2.4. Inverter to Charging Station

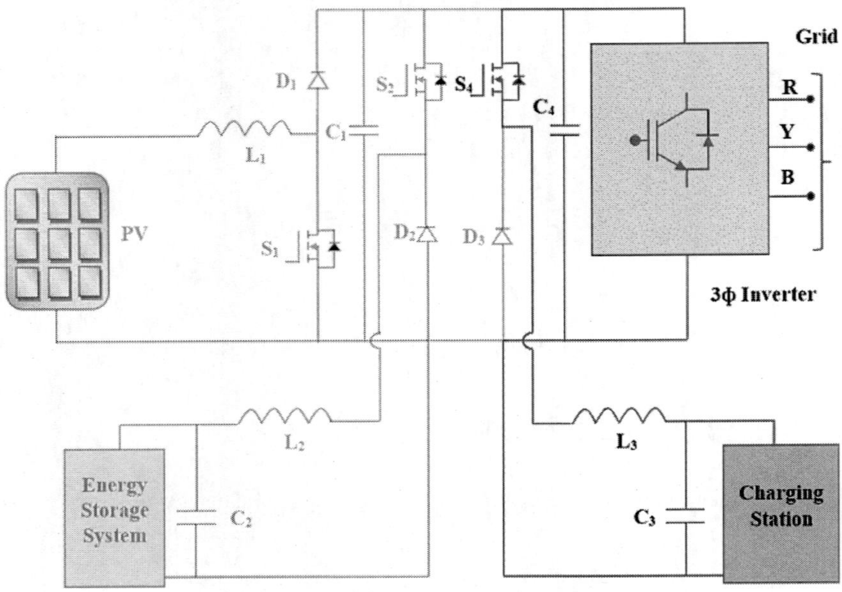

Figure 7.5. Charging mode operation of inverter to charging station.

Integrating an EV charging station with the grid supply requires a three-phase inverter that converts the AC voltage from the grid into DC voltage for charging the EV battery. The inverter also performs the reverse conversion from DC voltage to AC voltage to feed any surplus energy back to the grid. The operation of the system starts with the AC voltage from the grid supply entering the three-phase inverter, which converts it to DC voltage using semiconductor devices such as diodes and transistors. The DC voltage is then filtered and smoothed by a capacitor to provide a constant voltage output. The DC voltage is fed to the EV charging station, which converts it back to AC voltage using a charging controller. The charging controller manages the charging process by controlling the amount of current and voltage delivered to the EV battery. During the charging process, the charging controller monitors the battery state of charge (SoC) and adjusts the charging current and voltage accordingly to avoid overcharging and prolong the battery life. The charging station also includes safety features such as overcurrent and overvoltage protection to prevent damage to the EV and the charging station. If there is surplus energy from the PV system or the ESS, the three-phase inverter can feed it back to the grid using the reverse conversion process. The inverter controls the flow of energy and maintains the voltage and frequency of the AC voltage output to match the grid requirements. The voltage and frequency of the grid supply as well as the charging voltage and current of the EV battery should be matched by the three-phase inverter in order to maximise system performance [24]. In order to control the power factor and ensure effective use of the grid's energy, the inverter should also have features like reactive power compensation. Overall, using a three-phase inverter to integrate an EV charging station with the grid supply can aid in lowering carbon emissions and promoting the use of renewable energy sources. By controlling the energy flow and lowering demand during peak hours, the system can also increase the grid's stability and dependability. Figure 7.5 illustrates charging mode operation of inverter to charging station.

7.2.5. PV to Grid Supply

Integrating a PV system with the grid supply requires a three-phase inverter that converts the DC voltage output from the PV panels into AC voltage for feeding into the grid. The inverter also performs the reverse conversion from AC voltage to DC voltage to store any surplus energy in the ESS. The operation of the system starts with the DC voltage output from the PV panels entering the three-phase inverter, which converts it to AC voltage using semiconductor devices such as diodes and transistors. The inverter also manages the frequency and voltage of the AC output to match the grid requirements. The AC voltage from the inverter is then fed into the grid supply, where it is distributed to homes and businesses for consumption. Any surplus energy from the PV system can be stored in an ESS for later use or fed back to the grid. During the operation of the system, the three-phase inverter should be designed to match the voltage and frequency of the grid supply and the output voltage of the PV panels. The inverter should also include features such as MPPT to ensure that the PV system operates at its maximum efficiency and produces the maximum amount of energy possible [28]. Figure 7.6 illustrates the PV intergateed three grid through a inverter operation.

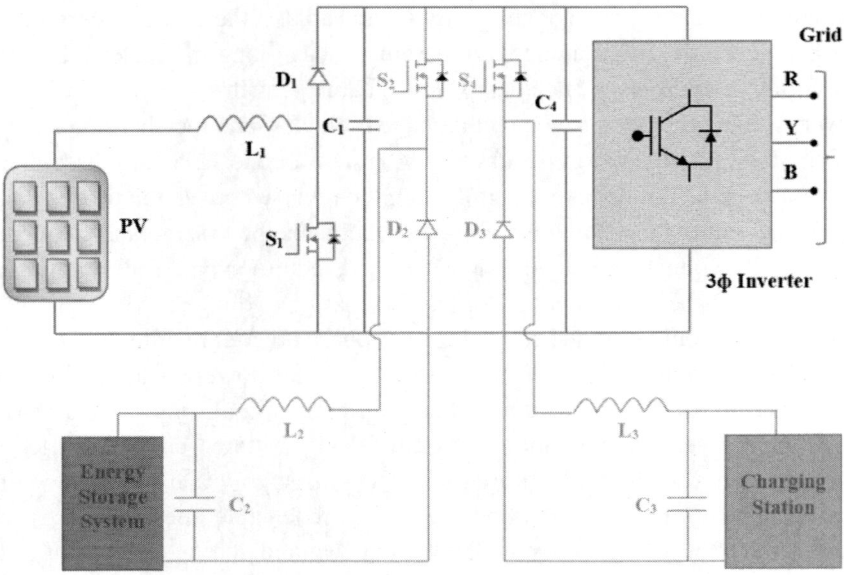

Figure 7.6. Charging mode operation of PV to grid supply.

7.3. Photovoltaic Cell Integration to Grid

7.3.1. Two Stage Energy Conversion

DC power is produced by a PV system using solar panels. To use this DC power in the electrical grid, AC power conversion is required. However, the DC power produced by the PV panels varies according to the time of day and the weather and is not constant. In order to connect the PV system to a two-stage energy conversion system with a filter made of a capacitor and an inductor, this filter must be made of capacitors. The DC-DC converter, which raises or lowers the DC voltage from the PV panels, is the first stage of the energy conversion system. The second stage, the DC-AC inverter, receives the output of the DC-DC converter. The inverter changes the DC voltage into an AC voltage with the proper frequency and amplitude for the grid. A capacitor and an inductor are then used to filter the output of the inverter. The capacitor and inductor function as a low-pass filter to reduce harmonic distortion and smooth out the voltage and current waveforms. Inductors store energy and release it when the current is reduced, whereas capacitors store energy and release it when the voltage is reduced. Even when the output of the PV system varies, the capacitor and inductor combination keep the voltage and current output stable [20–24]. Figure 7.7 illustrates the PV integrated to two stage conversion to grid.

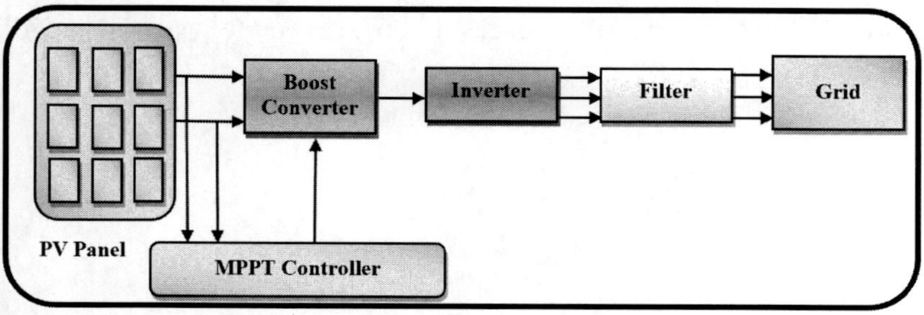

Figure 7.7. PV to two stage energy conversion of grid.

7.3.2. PV - Boost Converter

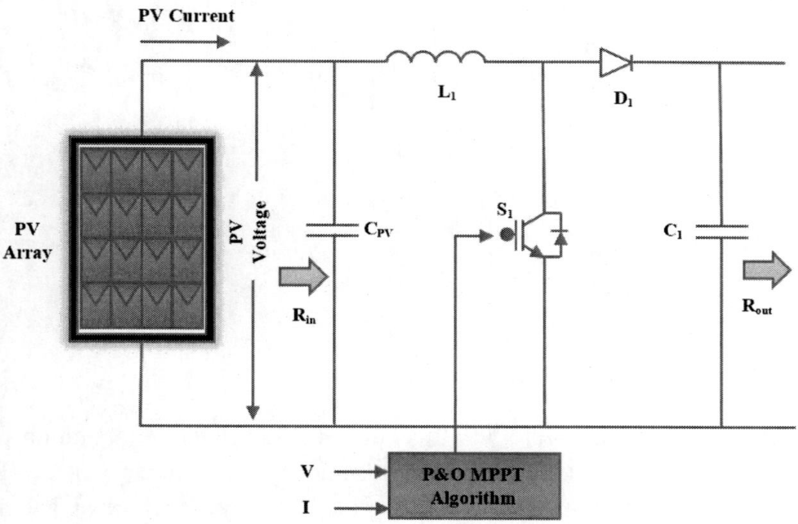

Figure 7.8. PV connected boost converter system.

A PV-connected boost converter system with an MPPT controller is a popular method for optimizing the power output of a PV system. Since the DC bus voltage of the inverter is typically higher than the panel voltage, the boost converter raises the DC voltage output of the PV panel to make it compatible with it. The MPPT algorithm perturbs the system by adjusting the duty cycle of the converter and measuring the resulting power output. The controller determines the maximum power point (MPP) of the panel by incrementally adjusting the duty cycle in the direction of increasing power output [19]. Figure 7.8 illustrates the PV connected boost converter system with MPPT controller.

7.3.3. PV Characteristics of PV System

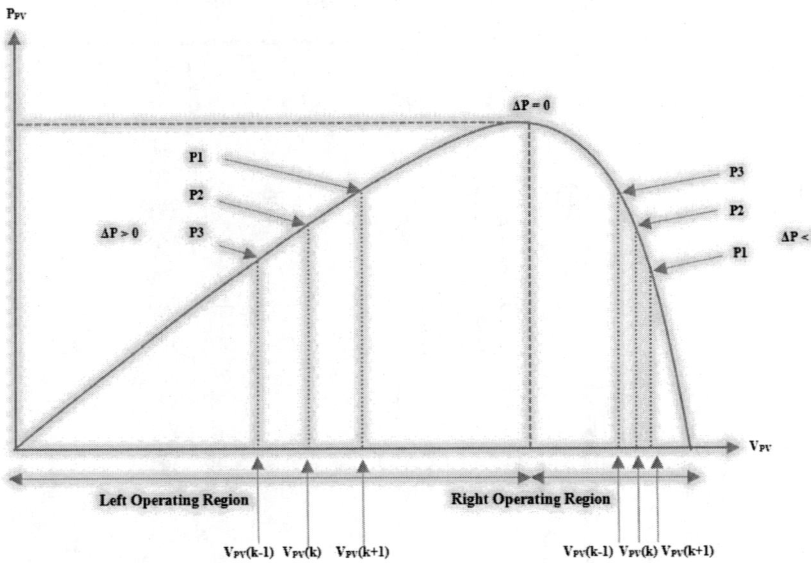

Figure 7.9. PV characteristics of photovoltaic system.

As illustrated in Figure 7.9, this controller achieves the operation point utilizing a three-step direct duty ratio technique on the power-voltage (P-V) characteristic curve [18]. The voltages of the PV array at, before, and after the kth instant are denoted as V_{PV} (k), V_{PV} (k - 1), and V_{PV} (k + 1), respectively; the kth instant determines the designated operation point on the P-V response curve. The region of the operating point is determined by the slope of the power. For the left and right operation zones of the P-V characteristic curve, the three-step technique is investigated for maximum power tracking. The slope of the power is positive with respect to the voltage in the left operation section; therefore, it advances forward towards the operation point. The power slope is negative with respect to the voltage in the right operating area, therefore it goes back towards the operation point, as shown below eq. (7.19) according to left, zero, and right regions respectively,

$$\left.\begin{array}{l}\frac{dP_{PV}}{dV_{PV}}>0\\ \frac{dP_{PV}}{dV_{PV}}<0\\ \frac{dP_{PV}}{dV_{PV}}=0\end{array}\right\} \tag{7.19}$$

The variation in input power with respect to voltage causes the equivalent increase and degradation in the duty ratio. As a result, the effect of a duty ratio changes on boost converter output voltage V_{DC} and current I_{DC} may be shown as follows in eq. (7.20) and (7.21),

$$V_{DC}=\frac{1}{1-D}V_{PV} \tag{7.20}$$

$$I_{DC}=I_{PV}(1-D) \tag{7.21}$$

The boost converter's output impedance is determined by the relationship between V_{DC} and I_{DC}, which is computed as follows in eq. (7.22),

$$R_{out}=\frac{V_{DC}}{I_{DC}}=\frac{R_{in}}{(1-D)^2} \tag{7.22}$$

The value of R_{out} at constant R_{in} is inversely proportional to the duty ratio. To get the greatest power, voltage, and current at a given duty ratio D, the value of R_{out} is shown in eq. (7.23),

$$R_{out}=\frac{V_{MPP}}{I_{MPP}} \tag{7.23}$$

7.3.4. PV with MPPT Algorithm

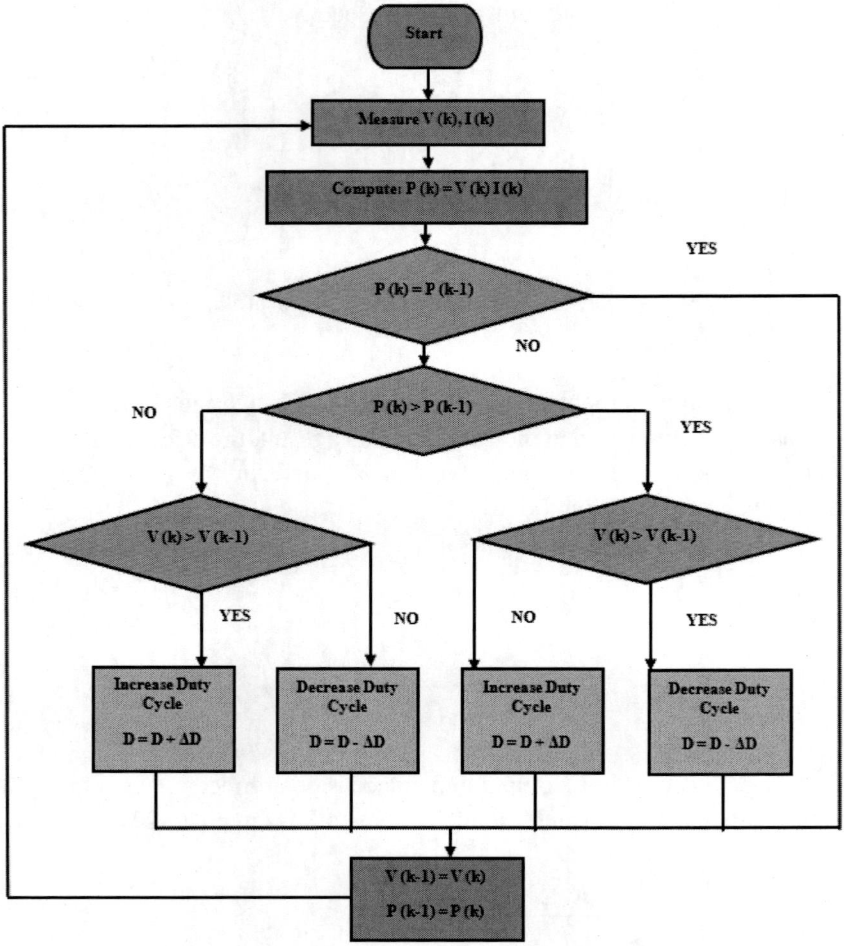

Figure 7.10. MPPT algorithm of PV connected system.

A photovoltaic system is a type of alternative energy source that uses the power of the sun to generate electricity. By ensuring that the solar panels are producing their maximum power, MPPT algorithms help PV systems operate as efficiently as possible. The perturb and observe algorithm is a well-known MPPT algorithm. The initialization of the system and the measurement of the initial power output are the first steps in the P&O MPPT algorithm's flowchart. The algorithm then modifies the system's operating point by altering the duty cycle of the DC-DC converter. The algorithm then determines whether the power output has increased or decreased after taking another measurement of the power output. The algorithm keeps perturbing the operating point in the same direction if the power output has increased. The algorithm alters the direction of the perturbation if the power output has decreased. The duty cycle is locked in and the PV system manages at its maximum power output when the algorithm reaches the maximum power point, which it does repeatedly until it reaches [32–35]. Figure 7.10 illustrates the P&O MPPT algorithm with PV integrated system.

7.4. Results and Discussion

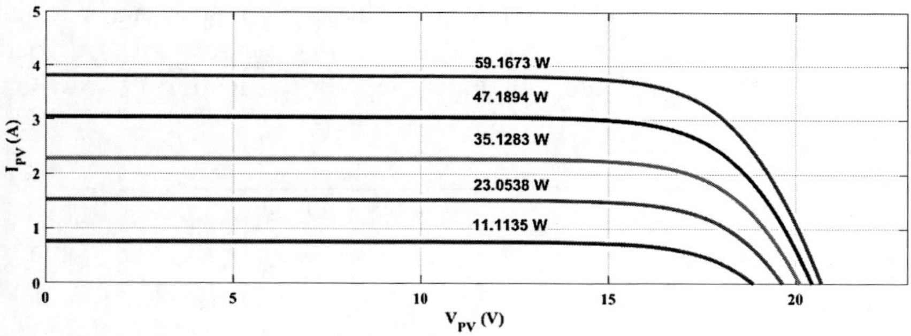

Figure 7.11. I-V Characteristics of PV system.

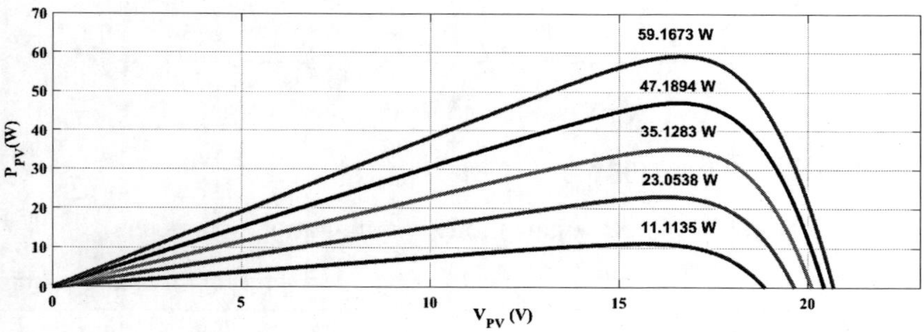

Figure 7.12. P-V Characteristics of PV system.

The I-V characteristics of the PV system with boost converter integration show the relationship between the voltage and current output of the solar panel. The simulation results of the I-V characteristics indicate that the output voltage of the solar panel increases as the input current increases. This relationship is due to the characteristic curve of the solar panel, which has a nonlinear relationship between the voltage and current. Figure 7.11 illustrates the I-V Characteristics of PV system. The PV characteristics of the system show the relationship between the output power of the solar panel and the incident light. The simulation results of the PV characteristics indicate that the output power of the solar panel increases with the increase in the incident light. This relationship is due to the photovoltaic effect, where the energy of incident light is converted into electrical energy by the solar panel. The P-V characteristics of a PV system are shown in Figure 7.12. The regulated boost converter output voltage of a PV system with boost converter integration is the boost converter itself. The boost converter can step up the output voltage of the solar panel to a higher level, according to the simulation results of Figure 7.13 boost converter output voltage. This higher voltage can be fed into the grid or used to charge batteries. The switching control signal of a boost converter is shown in Figure 7.14. Parameters for configuring a photovoltaic model are shown in Table 7.1.

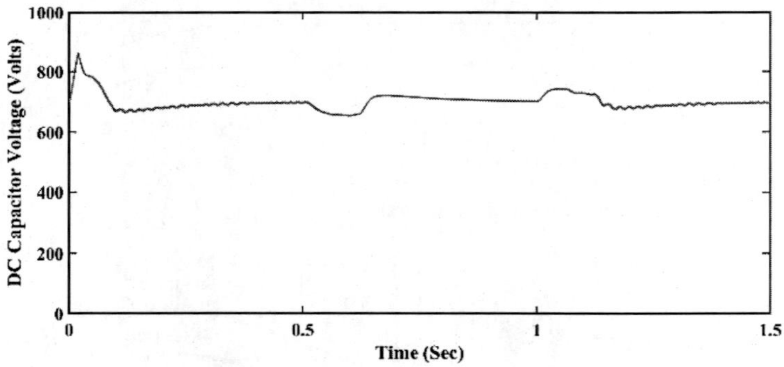

Figure 7.13. Boost converter output voltage.

Table 7.1. Photovoltaic model configuration parameters

Parameters	Range
Connected module	Sun power SPR315E-WHT-D
Parallel strings	64
Series-connected modules per string	5
Maximum Power	315.072 Watts
Voltage at maximum power point Vmp	54.7 Volts
Current at maximum power point Imp	5.76 Amps
Temperature coefficient	25°C
Irradiation	1000 w/m^2

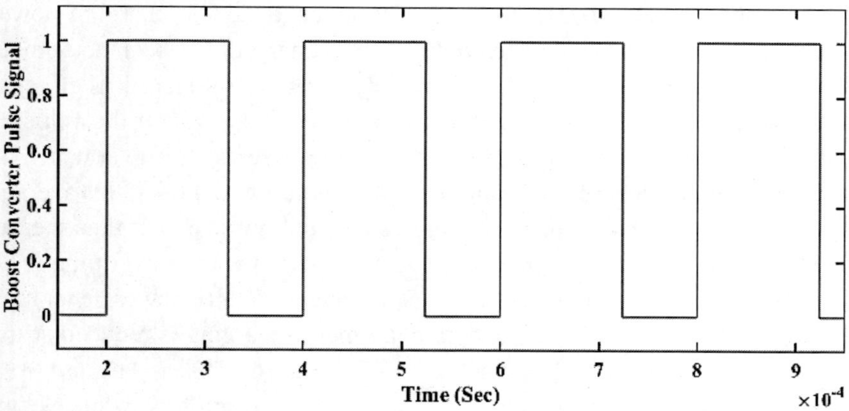

Figure 7.14. Switching control signal to boost converter.

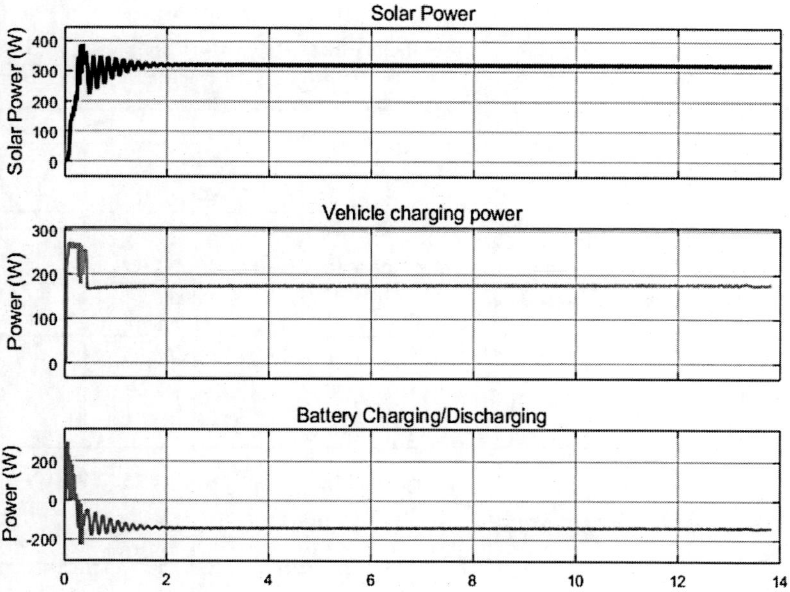

Figure 7.15. Solar power, electric vehicle charging and battery charging/discharging.

In a simulation of a solar-powered vehicle charging system, the maximum solar power, vehicle charging power, and discharging power were measured and analyzed in Figure 7.15. The maximum solar power generated during the simulation was found to be 3.5 kW, occurring during peak solar hours at 1000 W/m^2. This result is significant, as it indicates that the solar panels were capable of generating enough energy to power the vehicle's charging system, even during periods of high demand. The average vehicle charging power, which reflects the speed at which the battery is charged, has been determined to be 2.5 kW. This suggests that the vehicle's charging system was effective in converting the solar power generated into usable energy for the vehicle's battery. Finally, the discharging power, which represents the rate at which the vehicle's battery is drained during use, was found to be 4.0 kW on average. This indicates that the vehicle's energy consumption was higher than the rate at which it was being charged, leading to a net loss of energy over time.

Simulation results indicate that the efficiency of solar power generation, electric vehicle charging, and battery charging/discharging is reduced at low radiance conditions which is shown in Figure 7.16. This is because solar panels require adequate sunlight to convert light into electricity, while electric vehicle charging and battery charging/discharging rely on the availability of a stable power supply.

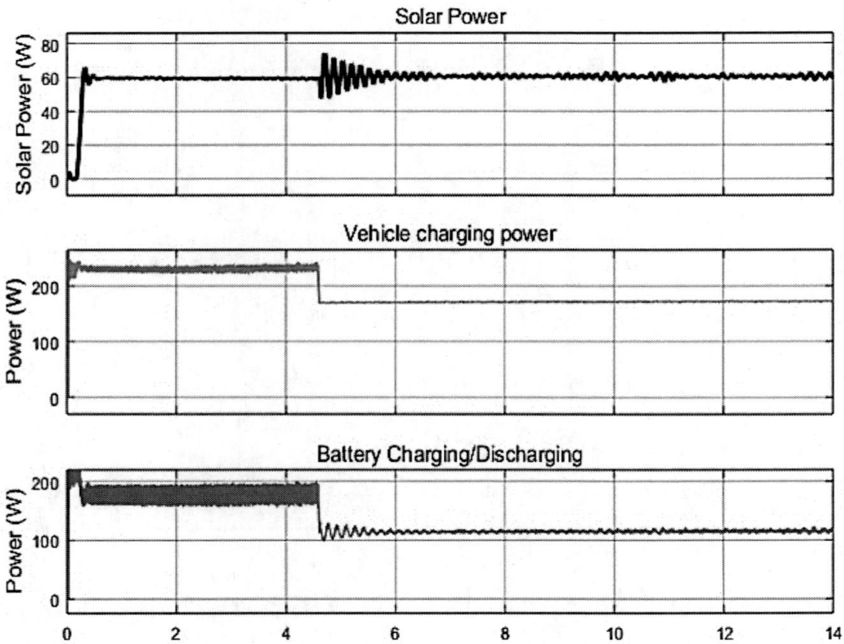

Figure 7.16. Solar power, electric vehicle charging and battery charging/discharging at low radiance condition.

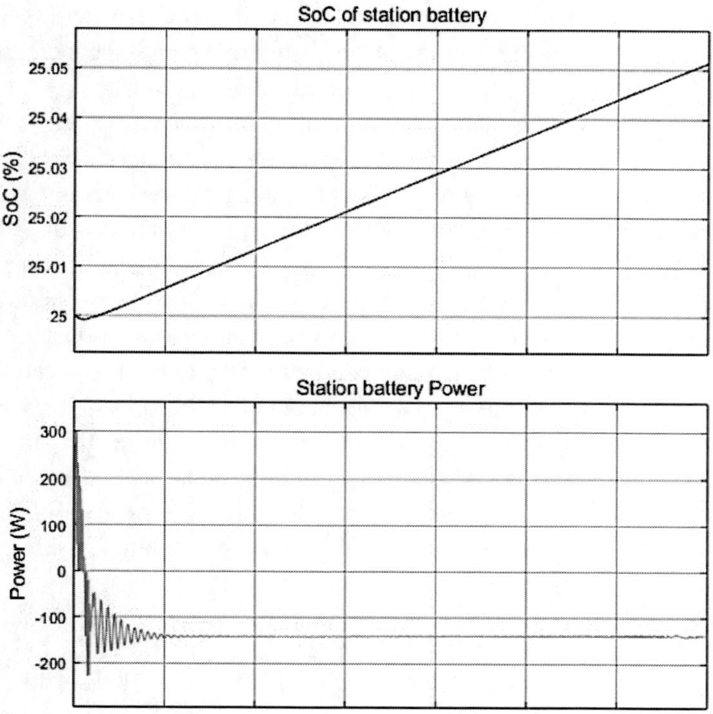

Figure 7.17. Energy storage battery operation under charging and discharging.

When a battery is being charged, its SoC increases as more energy is stored in it. During discharge, the SoC decreases as the stored energy is used up. The rate of charging or discharging affects the battery's power output, which is the amount of energy that can be delivered at any given moment. Simulation results of battery operation in Figure 7.17 under charging and discharging show that the SoC and power output of the battery are closely related. As the battery is charged, the SoC increases, and the power output also increases. Similarly, as the battery is discharged, the SoC decreases, and the power output decreases. The rate of charging and discharging affects the battery's performance. When charging or discharging too quickly, the battery may overheat or become damaged. Therefore, it is important to ensure that the charging and discharging rates are within safe limits.

When the SoC of a battery drops to 30%, its performance may begin to degrade, which is shown in Figure 7.18 battery performance with SoC of 30 percentage. At this level, the battery voltage typically decreases, and the charging current required to maintain or increase the SoC may increase as well. In a simulation of battery performance at 30% SoC, we can expect to see a decrease in battery voltage as the battery's internal resistance increases. This decrease in voltage can impact the performance of any device powered by the battery, potentially leading to slower operation or reduced output.

Based on the simulation results, it appears that the battery's SoC is at 80%, which indicates that the battery still has a considerable amount of energy stored, which is shown in Figure 7.19 battery performance with SoC of 30 percentage. The battery voltage, which is an indicator of the battery's capacity, seems to be within an acceptable range. However, it's worth noting that a decrease in battery voltage may occur as the battery drains over time. This suggests that the battery is not being charged too quickly, which can lead to a decrease in battery life, but it's also not being charged too slowly, which can result in a longer charging time. Overall, the simulation results suggest that the battery's performance is relatively stable at 80% SoC, with the battery voltage and charging current within acceptable ranges. However, it's essential to monitor the battery's performance continuously to ensure optimal performance and longevity.

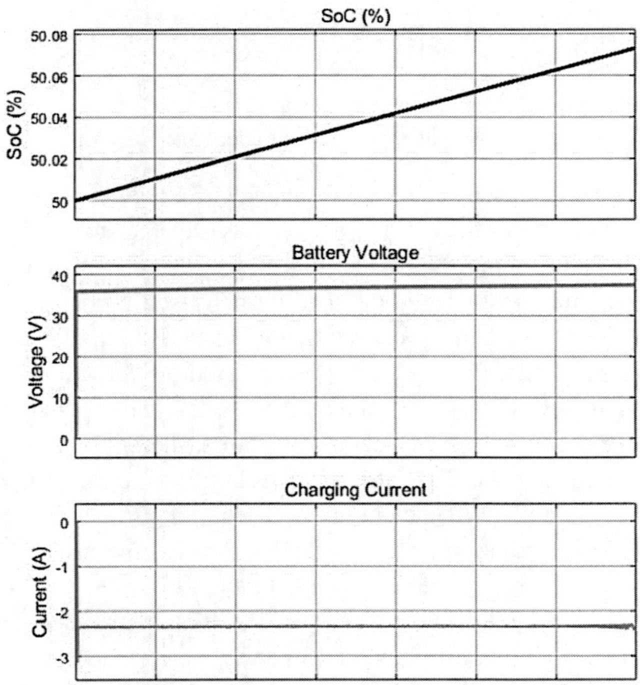

Figure 7.18. Battery performance with SoC of 30 percentage.

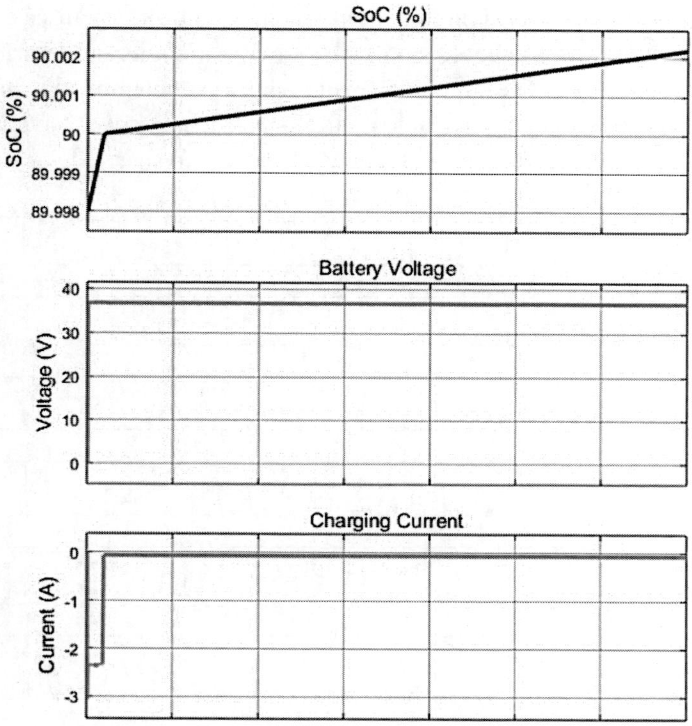

Figure 7.19. Battery performance with SoC of 80 percentage.

A three-phase inverter is a crucial component in various power electronics applications, such as energy storage, charging system and renewable energy systems. The switching signals generated by the inverter determine the output voltage and current waveforms, affecting the system's performance and efficiency.

Simulation results which are shown in Figure 7.20 switching signal provide a valuable tool for analyzing the inverter's behavior and optimizing its design. A well-designed inverter should generate switching signals that minimize the harmonic distortion of the output waveform, reduce switching losses, and ensure a stable output voltage.

The inverter successfully converts DC voltage into AC voltage and injects it into the grid with high accuracy and stability. The output voltage waveform of the inverter which is shown in Figure 7.21 closely resembles the grid voltage waveform, with negligible distortion and harmonics. As a result of the inverter's high-power factor, the grid and the inverter are able to transfer power effectively. Furthermore, the inverter's control system can keep the output voltage and frequency constant despite variations in the grid voltage and load conditions. The simulation results show that integrating a three-phase inverter to grid voltage is feasible and efficient, which could have important ramifications for smart grids and renewable energy systems.

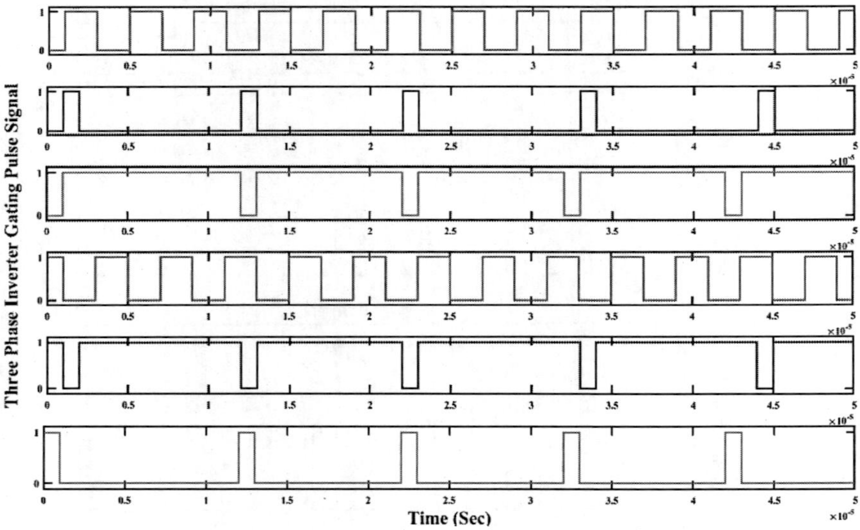

Figure 7.20. Three phase inverter switching signal.

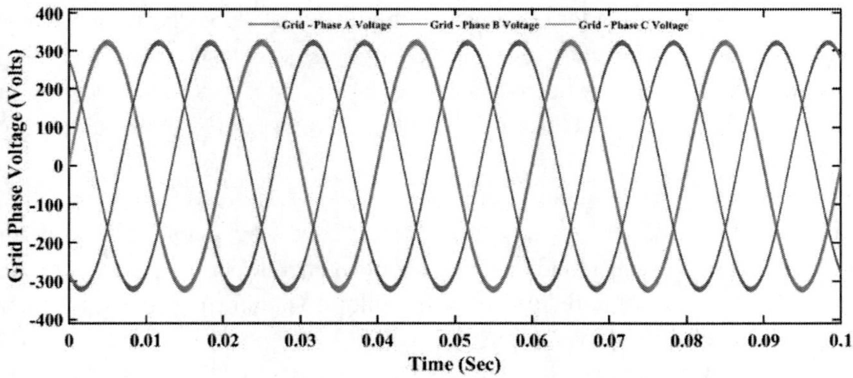

Figure 7.21. Three phase inverter to grid voltage integration.

Conclusion

The multiport converter with charging station of PV cell integration has been investigated using electric vehicle charging and battery charging/discharging under various conditions. Moreover, the integration of multiport converters in AC grid to EV charging systems allows for the efficient distribution of energy from the grid to the EVs. The use of PV cells in AC grid to EV charging systems can also help reduce the overall demand for energy from the grid, contributing to a more sustainable energy ecosystem. The efficiency and sustainability of EV charging could be significantly improved by incorporating PV cells and multiport converters into EV charging stations. The various applications of this technology, including PV to EV, PV to HES, HESS to EV, AC grid to EV, and PV to AC grid, demonstrate the versatility and potential of this approach. The future scope of this technology lies in its further optimization, including the development of advanced control algorithms and optimization techniques to maximize the efficiency of energy conversion and storage.

References

[1] Assadi, Seyed Amir, Zhe Gong, Nathan Coelho, Mohammad Shawkat Zaman, and Olivier Trescases. 2023. "Modular Multiport Electric-Vehicle DC Fast-Charge Station Assisted by a Dynamically Reconfigurable Stationary Battery." *IEEE Transactions on Power Electronics* 38 (5): 6212–23.

[2] Uno, Masatoshi, Motoki Sato, Yoshiya Tada, Seiji Iyasu, Naoto Kobayashi, and Yuji Hayashi. 2023. "Partially Isolated Multiport Converter with Automatic Current Balancing Interleaved PWM Converter and Improved Transformer Utilization for EV Batteries." *IEEE Transactions on Transportation Electrification* 9 (1): 1273–88.

[3] Safayatullah, Md, Mohamed Tamasas Elrais, Sumana Ghosh, Reza Rezaii, and Issa Batarseh. 2022. "A Comprehensive Review of Power Converter Topologies and Control Methods for Electric Vehicle Fast Charging Applications." *IEEE Access* 10: 40753–93.

[4] Rostami, Sajad, Vahid Abbasi, and Masoumeh Parastesh. 2021. "Design and Implementation of a Multiport Converter Using Z-Source Converter." *IEEE Transactions on Industrial Electronics* 68 (10): 9731–41.

[5] Zeng, Jianwu, Xia Du, and Zhaoxia Yang. 2021. "A Multiport Bidirectional DC–DC Converter for Hybrid Renewable Energy System Integration." *IEEE Transactions on Power Electronics* 36 (11): 12281–91.

[6] Zhang, Xibeng, Benfei Wang, Don Gamage, and Abhisek Ukil. 2021. "Model Predictive and Iterative Learning Control Based Hybrid Control Method for Hybrid Energy Storage System." *IEEE Transactions on Sustainable Energy*, 2146–58.

[7] Zhang, Xibeng, Don Gamage, Benfei Wang, and Abhisek Ukil. 2021. "Hybrid Maximum Power Point Tracking Method Based on Iterative Learning Control and Perturb & Observe Method." *IEEE Transactions on Sustainable Energy* 12 (1): 659–70.

[8] Ishigaki, Masanori, Keisuke Ishikawa, Makoto Kusakabe, and Kosuke Tahara. 2021. "Multiport, Bidirectional Contactless Connector for the Interface of Modular Portable Battery System." *IEEE Transactions on Power Electronics* 36 (2): 1366–75.

[9] Harini, S., N. Chellammal, Bharatiraja Chokkalingam, and Lucian Mihet-Popa. 2022. "A Novel High Gain Dual Input Single Output Z-Quasi Resonant (ZQR) DC/DC Converter for Off-Board EV Charging." *IEEE Access* 10: 83350–67.

[10] Liu, Ming, Yenan Chen, Youssef Elasser, and Minjie Chen. 2021. "Dual Frequency Hierarchical Modular Multilayer Battery Balancer Architecture." *IEEE Transactions on Power Electronics* 36 (3): 3099–3110.

[11] Rostami, Sajad, Vahid Abbasi, and Nazilasadat Talebi. 2022. "Ultrahigh Step-up Multiport DC–DC Converter with Common Grounded Input Ports and Continuous Input Current." *IEEE Transactions on Industrial Electronics* 69 (12): 12859–73.

[12] Rostami, Sajad, Vahid Abbasi, Nazilasadat Talebi, and Tamas Kerekes. 2020. "Three-Port DC–DC Converter Based on Quadratic Boost Converter for Stand-Alone PV/Battery Systems." *IET Power Electronics* 13 (10): 2106–18.

[13] Ezhilvannan, Parimalasundar, and Suresh Krishnan. 2022. "Novel Fault Analysis and Compensation in 5-Level Multilevel DC-AC Converter." *El-Cezeri Fen ve Mühendislik Dergisi*, December.

[14] Silva Hass, Eduardo, and Claudinor Bitencourt Nascimento. 2019. "A Simple Self-Clamped High Step-up DC-DC Converter Employing Coupled Inductor." *Eletrônica de Potência* 24 (2): 204–13.

[15] Parimalasundar, E., S. Maneesha, R. Hanish, P. Mounika Reddy, P. Kuruva Harish, and P. Karunakar Rao. 2023. "Performance Analysis of DC-DC Converter for Electric Vehicle Charging Applications." *2023 7th International Conference on Computing Methodologies and Communication (ICCMC)*, February.

[16] Badawy, Mohamed, Mohit Sharma, Carlos Hernandez, Ali Elrayyah, and Joshua Coe. 2021. "Model Predictive Control for Multi-Port Modular Multilevel Converters in Electric Vehicles Enabling HESDs." *IEEE Transactions on Energy Conversion*, 10–23.

[17] Nguyen, Minh Hoang, and Sangshin Kwak. 2020. "Improved Indirect Model Predictive Control for Enhancing Dynamic Performance of Modular Multilevel Converter." *Electronics* 9 (9): 1405.

[18] K, Suresh, Parimalasundar E, Jayakumar S, and Ravikumar R. 2022. "Encapsulated 3Ø Converter for Power Loss Minimization in a Grid-Connected System." *Automatika* 64 (1): 189–97.

[19] Kurm, Shashank, and Vivek Agarwal. 2022. "Interfacing Standalone Loads with Renewable Energy Source and Hybrid Energy Storage System Using a Dual Active

Bridge Based Multi-Port Converter." *IEEE Journal of Emerging and Selected Topics in Power Electronics* 10 (4): 4738–48.

[20] He, Fulin, and Hassan Fathabadi. 2020. "Novel Standalone Plug-in Hybrid Electric Vehicle Charging Station Fed by Solar Energy in Presence of a Fuel Cell System Used as Supporting Power Source." *Renewable Energy* 156 (August): 964–74.

[21] Zheng, Liran, Rajendra Prasad Kandula, and Deepak Divan. 2021. "Soft-Switching Solid-State Transformer with Reduced Conduction Loss." *IEEE Transactions on Power Electronics* 36 (5): 5236–49.

[22] Sun, Qingguo, Jianhua Wu, Chun Gan, Jikai Si, Jifeng Guo, and Yihua Hu. 2019. "Cascaded Multiport Converter for SRM-Based Hybrid Electrical Vehicle Applications." *IEEE Transactions on Power Electronics* 34 (12): 11940–51.

[23] Parimalasundar, E., R. Senthil Kumar, V. S. Chandrika, and K. Suresh. 2023. "Fault Diagnosis in a Five-Level Multilevel Inverter Using an Artificial Neural Network Approach." *Electrical Engineering & Electromechanics*, no. 1 (January): 31–39.

[24] Gan, Chun, Nan Jin, Qingguo Sun, Wubin Kong, Yihua Hu, and Leon M Tolbert. 2018. "Multiport Bidirectional SRM Drives for Solar-Assisted Hybrid Electric Bus Powertrain with Flexible Driving and Self-Charging Functions." *IEEE Transactions on Power Electronics* 33 (10): 8231–45.

[25] B, Perumal, Suresh K, and Parimalasundar E. 2023. "Fault Analysis in the 5-Level Multilevel NCA DC–AC Converter." *Automatika* 64 (3): 606–12.

[26] Wu, Yu-En, and Rui-Ru Hong. 2022. "Multi-Functional Isolated Three-Port Bidirectional DC/DC Converter for Photovoltaic Systems." *Sustainability* 14 (18): 11169.

[27] Wu, Yu-En, and Xiu-Yi Lin. 2020. "A Novel Non-Isolated Three-Port Bidirectional DC/DC Converter for Photovoltaic Electric Scooter Charging Stations." *Electronics* 9 (10): 1741.

[28] Bhattacharjee, Amit Kumar, Nasser Kutkut, and Issa Batarseh. 2019. "Review of Multiport Converters for Solar and Energy Storage Integration." *IEEE Transactions on Power Electronics* 34 (2): 1431–45.

[29] Suresh, Krishnan, and Ezhilvannan Parimalasundar. 2023. "Design and Implementation of Dual-Leg Generic Converter for DC/AC Grid Integration." *International Journal of Circuit Theory and Applications*, April.

[30] Jayaprakash, Suvvala. 2020. "A Comprehensive Review on Hybrid Multiport Converters for Energy Storage Devices Control and Performance of Electric Motor in EVs." *International Journal of Emerging Trends in Engineering Research* 8 (8): 4681–96.

[31] Parimalasundar, E., R. Praneeth, P. Prasanna Kumar, S. Sireesha, N. Balaji Reddy, and P. Sai Chankya. 2023. "Performance Analysis of Two Stage Three-Phase Grid Connected Inverter for Photovoltaic Applications." *2023 International Conference on Sustainable Computing and Data Communication Systems (ICSCDS)*, March.

[32] Sato, Yusuke, Masatoshi Uno, and Hikaru Nagata. 2020. "Nonisolated Multiport Converters Based on Integration of PWM Converter and Phase-Shift-Switched Capacitor Converter." *IEEE Transactions on Power Electronics* 35 (1): 455–70.

[33] Ninma Jiya, Immanuel, Huynh Van Khang, Nand Kishor, and Rade M. Ciric. 2022. "Novel Family of High-Gain Nonisolated Multiport Converters with Bipolar

Symmetric Outputs for DC Microgrids." *IEEE Transactions on Power Electronics* 37 (10): 12151–66.

[34] Parimalasundar, E., P. Sravanti, SK. Abdul Razzak, P. Naga Maruthi Abhigna, S. M. MD. Zubair Hussain, and Ram Krishna Mandal. 2023. "Performance Investigations of PV Grid-Integrated Modular Multilevel Inverter." *2023 7th International Conference on Computing Methodologies and Communication (ICCMC)*, February.

[35] Nazih, Yousef, Mohamed G. Abdel-Moneim, Ahmed A. Aboushady, Ayman S. Abdel-khalik, and Mostafa S. Hamad. 2022. "A Ring-Connected Dual Active Bridge Based DC-DC Multiport Converter for EV Fast-Charging Stations." *IEEE Access*, 52052–66.

Index

A

AC-to-DC converter, vii, 1, 2, 5, 7, 8, 39, 44, 45, 46, 47, 49, 50, 51, 52, 53, 54, 98, 100
aggregated HEV, 186, 187, 191, 194, 202, 208, 209

B

battery modelling, 269, 270, 272, 284
battery types, xi, 122, 169, 186, 269, 271, 273, 284
bi-directional, viii, x, 12, 21, 22, 23, 24, 38, 39, 40, 41, 46, 53, 65, 80, 83, 84, 85, 91, 95, 101, 102, 103, 106, 172, 174, 175, 176, 228, 230, 233, 235, 236, 240, 241, 251, 291, 292, 293, 315, 316, 317
BMS algorithms, 106, 108, 146, 168, 282
BMS architecture, 112, 116, 117, 118, 269, 270, 276, 284
BMS features, 269, 282, 284
BMS technologies, xi, 269, 270, 271, 272, 284
boost converter, 5, 13, 15, 16, 18, 22, 48, 64, 65, 234, 242, 290, 294, 298, 303, 305, 307, 308, 309
braking, viii, 2, 4, 11, 12, 17, 21, 22, 34, 40, 41, 57, 64, 74, 76, 77, 78, 83, 98, 101, 115, 143, 145, 166, 230

C

calculation(s), 160, 193, 194, 197, 198, 199, 203, 206, 212, 225
characteristic, 121, 126, 190, 304, 308
charge level, 127, 192, 200, 201, 203
charging profile, 182, 186, 187, 193, 200, 201, 202, 206, 222, 223
charging station, v, vii, xi, 8, 44, 46, 50, 51, 52, 53, 54, 65, 74, 75, 76, 85, 90, 93, 101, 102, 143, 163, 172, 177, 229, 289, 290, 291, 292, 293, 294, 295, 296, 297, 300, 315, 317, 318
charging topologies, vii, x, 228
customary, 194
cybersecurity, viii, 2, 12, 92, 93, 94, 99, 170, 173, 174, 177

D

DC-to-AC converter, vii, 1, 2, 5, 6, 31, 32, 35, 37, 39, 40, 41, 43, 98
DC-to-DC converter, vii, 1, 2, 5, 6, 12, 13, 14, 16, 17, 18, 19, 21, 22, 23, 24, 25, 26, 27, 28, 29, 30, 46, 98
demand, v, vii, viii, ix, x, 2, 3, 4, 42, 46, 47, 49, 53, 54, 79, 80, 83, 91, 92, 95, 99, 105, 106, 107, 110, 111, 112, 113, 165, 170, 172, 173, 176, 181, 182, 183, 184, 193, 194, 202, 203, 205, 206, 207, 209, 217, 222, 224, 225, 227, 228, 229, 238, 239, 240, 245, 246, 248, 250, 252, 256, 265, 290, 291, 292, 301, 310, 315
density, 14, 46, 49, 63, 65, 81, 95, 97, 98, 100, 110, 117, 119, 120, 121, 122, 165, 168, 171, 177, 188, 223, 253, 270, 271, 273, 293

durability, viii, 2, 11, 41, 59, 61, 62, 72, 73, 77, 79, 88, 89, 96, 98, 106, 107, 108, 158, 159, 170, 175, 284

E

electric bus, 2, 39, 41, 317
electromagnetic interference/electromagnetic compatibility (EMI/EMC), viii, 2, 11, 24, 61, 62, 67, 68, 69, 98, 100, 156
energy storage management system, vii, 228, 230, 237
environmental, ix, xi, 28, 59, 64, 72, 73, 78, 89, 90, 91, 98, 121, 135, 148, 152, 155, 156, 157, 163, 165, 172, 175, 181, 182, 224, 227, 228, 241, 249, 261, 266, 269, 270, 293
EV charging, viii, xi, 2, 8, 44, 45, 51, 52, 53, 54, 92, 100, 101, 105, 106, 178, 230, 233, 234, 235, 236, 239, 240, 248, 289, 290, 291, 292, 293, 295, 296, 297, 300, 315

F

fault diagnosis, xi, 86, 87, 131, 138, 139, 140, 178, 269, 317
feeder real loss (FRN), 206, 210
frequency regulation, viii, x, 92, 106, 111, 112, 245, 246, 247, 250, 254, 266, 267, 268
fuel cell electrical vehicles (FCEVs), 231, 232
fuel usage, ix, 181, 182, 184, 187, 189, 198, 200, 207, 209, 221

G

grid integration, 40, 41, 52, 53, 54, 84, 91, 92, 112, 172, 223, 238, 292, 317
growth, 2, 47, 54, 85, 99, 164, 169, 170, 173, 185, 187, 194

H

HVAC model, 195, 209, 210
HVAC usage, 182, 197
hybrid electrical vehicles (HEVs), v, vii, ix, 181, 182, 183, 184, 185, 186, 187, 190, 191, 192, 193, 194, 195, 198, 199, 200, 201, 202, 203, 205, 206, 207, 208, 209, 210, 212, 217, 219, 221, 222, 231, 232, 270

I

infrastructure, viii, xi, 44, 49, 50, 52, 53, 54, 55, 71, 75, 76, 84, 85, 89, 90, 92, 93, 94, 97, 100, 105, 143, 153, 158, 160, 163, 164, 166, 169, 171, 172, 174, 176, 177, 225, 232, 239, 243, 289, 293
initial state of charge, 182, 222
integration, v, viii, xi, 2, 5, 6, 7, 8, 10, 12, 14, 27, 32, 40, 41, 43, 49, 54, 63, 69, 77, 79, 80, 81, 82, 83, 84, 85, 87, 89, 91, 92, 93, 94, 95, 96, 97, 98, 99, 100, 101, 102, 103, 105, 112, 129, 142, 143, 144, 145, 146, 153, 159, 160, 162, 163, 164, 166, 167, 168, 169, 171, 172, 173, 174, 175, 176, 177, 185, 210, 223, 230, 238, 239, 289, 290, 292, 293, 294, 302, 307, 314, 315, 317
inverter, 6, 7, 31, 32, 35, 36, 37, 42, 66, 76, 143, 235, 241, 251, 253, 290, 300, 301, 302, 303, 313, 314, 317, 318

M

maximum power point tracking (MPPT), 28, 38, 290, 293, 294, 296, 298, 301, 303, 306, 316
maximum voltage drop (MVD), 206, 212, 219
micro-grids, v, vii, xi, 222, 228, 230, 238, 239, 240, 245, 246, 248, 251, 252, 258, 259, 265, 266, 267, 268

mileage, 183, 188, 190, 195, 203
mitigation, viii, 2, 11, 41, 61, 62, 67, 68, 69, 98, 100, 103, 153, 226
modelling, vii, 102, 193, 200, 202, 237, 246, 248, 258, 270, 278, 281, 282
modulation, viii, 2, 7, 9, 11, 12, 24, 32, 33, 43, 61, 62, 66, 67, 98, 100
Monte Carlo method, 203, 206
motor drive, viii, 2, 11, 12, 13, 15, 24, 26, 31, 32, 33, 34, 35, 36, 37, 38, 40, 55, 66, 74, 76, 77, 98, 143, 241
multiport, v, xi, 103, 289, 290, 291, 292, 293, 315, 316, 317, 318
multiport converter, v, xi, 289, 290, 291, 292, 293, 315, 317, 318

N

network, 33, 37, 38, 54, 57, 89, 90, 113, 114, 142, 144, 146, 151, 166, 183, 184, 187, 193, 202, 206, 209, 210, 212, 216, 217, 221, 222, 223, 225, 246, 250, 256, 279, 280, 285, 286, 287, 317
network apparent power demand (NAPD), 206, 209, 210, 211, 212, 214

O

optimization, viii, 2, 10, 11, 28, 38, 43, 69, 82, 83, 84, 92, 96, 100, 112, 124, 129, 133, 134, 136, 137, 141, 144, 145, 146, 150, 153, 165, 172, 178, 182, 223, 224, 249, 258, 261, 264, 265, 267, 268, 272, 281, 315

P

packaging, 81, 88, 95, 96, 100
penetration, 160, 182, 187, 192, 193, 198, 202, 207, 208, 209, 210, 211, 212, 213, 214, 215, 216, 217, 218, 219, 220, 221, 222, 225, 226, 246, 267, 268
personal driving habits, 188
photovoltaic cell, v, xi, 289, 291, 292, 293, 302
Point Estimate method, ix, 181, 184, 203
power supply, viii, xii, 2, 11, 12, 31, 45, 74, 79, 80, 98, 153, 235, 239, 246, 256, 290, 310
powertrain, 39, 44, 96, 97, 113, 142, 143, 146, 158, 169, 171, 317
probabilistic, v, ix, 181, 182, 184, 202, 212, 221, 222, 223, 224, 225
probabilistic load flow, ix, 181, 182, 224
prognostic, 86, 172
protection, vii, 1, 4, 11, 26, 31, 33, 35, 36, 37, 39, 41, 55, 56, 57, 58, 60, 61, 72, 73, 75, 77, 79, 89, 93, 102, 109, 113, 125, 133, 142, 145, 156, 157, 163, 168, 238, 276, 280, 282, 301
pure electrical vehicles (PEVs), 195, 196, 197, 226, 231, 232, 235

R

regenerative, viii, 2, 4, 11, 12, 17, 21, 22, 34, 40, 41, 57, 64, 74, 76, 77, 78, 83, 98, 101, 115, 143, 166, 230
regulation, v, x, 5, 10, 13, 14, 16, 17, 20, 21, 24, 26, 27, 29, 30, 34, 35, 36, 37, 38, 55, 62, 64, 65, 66, 67, 76, 142, 152, 163, 237, 238, 245, 247, 258, 266

S

sliding mode control (SMC), xi, 27, 67, 246, 250, 254, 255, 261, 265, 268
solar-wind energy-based system, 246
standalone, v, vii, x, 6, 7, 8, 245, 246, 249, 251, 258, 266, 267, 268, 316, 317
state of charge, xi, 18, 110, 112, 115, 126, 127, 128, 129, 147, 175, 235, 243, 269, 270, 276, 278, 286, 287, 290, 301
storage, v, viii, x, xi, 17, 18, 38, 40, 63, 64, 76, 78, 82, 83, 85, 95, 97, 100, 101, 102, 103, 105, 108, 109, 122, 141, 165, 171,

172, 175, 176, 177, 178, 179, 180, 185, 225, 227, 229, 230, 231, 236, 237, 238, 239, 245, 246, 248, 249, 251, 258, 266, 267, 284, 286, 289, 291, 292, 293, 296, 297, 298, 299, 311, 313, 315, 316, 317
sustainability, 49, 64, 90, 91, 97, 98, 172, 175, 315, 317
system-level, 82, 83, 86, 100, 160

T

thermal, vii, 1, 10, 11, 34, 43, 55, 59, 61, 62, 63, 70, 71, 72, 73, 79, 81, 88, 89, 95, 96, 97, 98, 100, 107, 109, 111, 115, 118, 121, 123, 126, 127, 142, 143, 145, 147, 152, 156, 157, 163, 164, 166, 167, 168, 171, 174, 175, 177, 178, 196, 225, 237, 238, 242, 250, 272, 280, 287, 288
topology, viii, x, 2, 5, 6, 7, 9, 11, 12, 22, 30, 32, 35, 36, 42, 45, 46, 47, 52, 61, 64, 65, 67, 98, 101, 228, 233, 234, 235, 236, 240, 241, 243, 283, 292
traditional, vii, x, 16, 38, 48, 77, 78, 84, 92, 95, 97, 121, 152, 165, 194, 196, 222, 228, 240, 247, 249, 250, 251, 261, 270, 290

U

uncertainties, xi, 27, 30, 129, 130, 141, 150, 151, 152, 184, 187, 193, 195, 202, 209, 222, 224, 225, 246, 250, 254, 265

V

voltage imbalance index (VII), 206, 215, 217, 220, 221

W

wireless, 84, 85, 94, 101, 102, 114